ALCOHOLISM AND DRUG DEPENDENCE

A Multidisciplinary Approach

ALCOHOLISM AND DRUG DEPENDENCE

A Multidisciplinary Approach

Edited by

J. S. Madden
Mersey Regional Addiction Unit
Chester, England

Robin Walker
Walton Hospital and Fazakerley Hospital
Liverpool, England

and

W. H. Kenyon
Merseyside Lancashire & Cheshire Council on Alcoholism
Liverpool, England

PLENUM PRESS · NEW YORK AND LONDON

Library of Congress Cataloging in Publication Data

International Conference on Alcoholism and Drug Dependence, 3d, Liverpool, 1976.
Alcoholism and drug dependence.

"Proceedings of the third International Conference on Alcoholism and Drug Dependence held in Liverpool, England, April 4-9, 1976."
Includes index.
1. Alcoholism–Congresses. 2. Drug abuse–Congresses. I. Madden, John Spencer. II. Walker, Robin. III. Kenyon, W. H. IV. Title. [DNLM: 1. Alcoholism–Congresses. 2. Drug abuse–Congresses. 3. Drug dependence–Congresses. W3 IN166 1976a/ WM274 I6123 1976a]
HV5009.I58 1976 362.2'9 76-30574
ISBN 0-306-31019-8

Proceedings of the Third International Conference on Alcoholism and Drug Dependence held in Liverpool, England, April 4–9, 1976

A Division of Plenum Publishing Corporation
227 West 17th Street, New York, N.Y. 10011

Printed in the United States of America

FOREWORD

It is for me, as President of the Merseyside Lancashire & Cheshire Council on Alcoholism, a signal privilege to write a brief foreword to these Proceedings of the Third International Conference on Alcoholism and Drug Dependence.

During the week experts, from the world over, in all those disciplines which have a contribution to make to the solution of the problems of alcoholism and drug dependence, surveyed recent advances in our knowledge of the aetiology, epidemiology, early recognition, management, and the social and industrial implications of these twin scourges of the contemporary scene. The discussions were most helpful and enlightening even when a wide spectrum of conflicting views and experiences were revealed.

The justification for adding yet another volume to the growing literature in this field is two-fold; first, that those taking part in the conference can study at leisure and in depth the transactions of the conference; and secondly, that those who were unable to attend might benefit from the most recent views of international experts in the diverse facets of one of the major current problems in world health.

LORD COHEN OF BIRKENHEAD

CH, PRSH, MD, DSc, ScD,
LLD, DCL, FRCP, DL, JP.

CONTENTS

Introduction xiii
J.S. Madden, R. Walker, and W.H. Kenyon

Opening Address to the Conference xv
M. Meacher

SECTION I

GENERAL

Sexual Disorders in Male Alcoholics 3
M.J. Akhtar

Vitamin B_{12} and Folic Acid in Chronic Alcoholics 15
E. Heilmann

Porphyria Cutanea Tarda in Chronic Alcoholics 19
E. Heilmann

Physical Complications of Alcohol Excess-Metabolism of Alcohol 23
D. Owens

Is Illegal Drugtaking a Problem? 35
M.A. Plant

Alcoholism and Drug Dependence - A Multidisciplinary Problem: The Sociologist's Viewpoint 47
D. Robinson

Alcohol and the General Physician 57
R. Walker

Denial of the Hidden Alcoholic in General Practice 63
R.H. Wilkins

SECTION II

AETIOLOGY AND EPIDEMIOLOGY

The Validity of Per Capita Alcohol Consumption as an Indicator of the Prevalence of Alcohol Related Problems: An Evaluation Based on National Statistics and Survey Data 71
A.K.J. Cartwright, S.J. Shaw, and T.A. Spratley

On the Need to Reconcile the Aetiologies of Drug Abuse 85
C. Fazey

Drinking Patterns of Young People 95
A. Hawker

Alcoholism and Psychology - Some Recent Trends and Methods 105
G. Lowe

A Psychiatric View of Substance Dependency 115
J.S. Madden

How Important is Alcohol in "Alcoholism"? 123
R.J. McKechnie

Social Circumstances of Non-Convicted vs Convicted Drug Users 139
C.E. Reeves

The Pathological and the Subcultural Model of Drug Use - A Test of Two Contrasting Explanations . . . 151
K.H. Reuband

Unemployment and Sickness Absenteeism in Alcoholics 171
E.S.M. Saad

Alcohol Problems in Women 181
A.B. Sclare

The Aetiology of Dependency 189
F.A. Seixas

SECTION III

TREATMENT

A Pilot Controlled Drinking Out-Patient Group 199
D. Cameron

Programming Alcoholism Treatment: Historical Trends 209
R.M. Costello

A Young Problem Drinkers Programme as a Means of Establishing and Maintaining Treatment Contact 227
B. Coyle and J. Fischer

Family Focused Treatment and Management: A Multi-Discipline Training Approach 239
J.P. Flanzer and G.M.St.L.O'Brien

The Young Alcoholic - Approaches to Treatment 263
P.D.V. Gwinner

Detoxification - The First Step 271
J.R. Hamilton

The Treatment of Drug Dependence - A Taxonomy of Approaches 277
I. Hindmarch

Aims of Treatment 293
B.D. Hore

Controlled Drinking in the Alcoholic - A Search for Common Features 297
T. Levinson

A Programme of Group Counselling for Alcoholics 309
J.S. Madden

The Ontario Detoxication System: An Evaluation of its Effectiveness 321
R.G. Smart

SECTION IV

REHABILITATION

The Role of the Probation Service in the Treatment of Alcoholism 331
T. Crolley

An Analysis of Clients Using Alcoholic Agencies within One Community Service 335
S.D. Delahaye

Co-ordination and Co-operation 351
B.D. Hore

The Problem - Its Magnitude and a Suggested Community Based Answer to Alcoholism 355
W.H. Kenyon

Methodological Problems in Evaluating Drug Misuse Intervention Programmes 369
M.A. Lavenhar

Planning for the Future - Developing a Comprehensive Response to Alcohol Abuse in an English Health District 379
T.A. Spratley, A.K.J. Cartwright, and S.J. Shaw

The Featherstone Lodge Project - Phoenix House. One Method of Rehabilitation 387
D. Warren-Holland and S. Warren-Holland

The Co-ordination of Care in the Field of Alcoholism 395
E. Wilkes

SECTION V

PREVENTION AND EDUCATION

The Role of Legislation in Diminishing the Misuse of Alcohol 401
C. Clayson

The Need for and Some Results of Evaluation of English Drug Education 409
N. Dorn

Developing a Co-ordinated Approach to Interprofessional Education 417
M. Grant

Alcohol Control Policy as a Strategy of Prevention: A Critical Examination of the Evidence 425
J. de Lint

Parents, Children and Learning to Drink 451
R.J. McKechnie

Alcohol and Education 457
E.B. Ritson

Chairmen 465

Contributors 467

Index 471

INTRODUCTION

The allied conditions of alcoholism and drug misuse are attracting increasing attention in an expanding number of countries. In Liverpool a tradition has been established of holding conferences on these subjects at local, national and international level. The list of sponsoring organisations for the larger conferences includes the International Council on Alcohol and Addictions, Lausanne and Liverpool University Department of Psychiatry and was strengthened for the third international conference held in 1976, by the Alcohol Education Centre, London.

This conference drew participants from 18 countries. Its theme centred on the importance of a multidisciplinary approach, and so allowed a wide spread of subjects to be covered by lecturers from many professional backgrounds, who sometimes held challengingly diverse points of view. The speakers included a legislator, psychiatrists, specialists in internal medicine, general medical practitioners, psychologists, social workers, sociologists and an anthropologist. Their topics were ranged from causation and prevention, education, and prevalence patterns of alcohol and drug usage, to treatment and rehabilitation.

The reader will find much evidence to support the view that the per capita rate of alcohol consumption of a nation intimately affects its incidence of alcohol problems. The hopeful corollary of this concept is that the amount of beverage alcohol consumed within a state is amenable to government intervention by legal and fiscal measures. But the situation is not clear cut; contributors from Scotland point out that their country has a distinctly more extensive alcoholism problem than neighbouring England and Wales, yet Scotland has more restrictive licensing laws and only a marginally higher level of alcohol consumption. Different attitudes and culturally determined reactions to the taking of alcohol contribute to the differing incidences of alcoholism between communities, and must not be overlooked in the search for preventative devices.

Attitudes to drug misuse are also discussed. The 'snowballing' technique of gaining introduction to a wide range of drug misusers by winning the initial confidence of a few subjects has confirmed what was long suspected, that drug misusers show a continuum of social and medical problems from minimal to severe. Society therefore should not over react in the way that occurred when drug misuse first developed some prestige among young people; this is not to undervalue the necessity for legal restraints that are flexible and neither unrealistically lax nor inhumane. Less reassuring are data presented at the conference which reveal a rising trend in alcohol misuse, especially among youngsters.

Some authors give an overview of alcoholism or drug dependence from the point of view of their own profession. Others concentrate on an issue of current or growing interest. Alcohol detoxification centres, for instance, although sporadically distributed, have existed in some cities long enough to allow some assessment of their advantages and disadvantages. Several papers review treatment programmes and their co-ordination, while family therapy and the controversial goal of controlled drinking for some alcoholics receive specific attention. A welcome feature of the conference has been the emphasis by general physicians on the organic complications of chronically excessive alcohol intake.

The surge of drug misuse has lost impetus, but many countries are faced with a growing alcoholism problem that has so far aroused neither determined counter-measures in prevention nor innovatory techniques in treatment. Progress in counteracting alcoholism and drug dependency requires an understanding of emerging insights and areas of dispute. The editors hope that the present series of papers will in some measure promote understanding of the themes that are attracting international attention and may stimulate effective preventative and remedial programmes.

Acknowledgements are made to the Merseyside, Lancashire and Cheshire Council on Alcoholism as the organisation mainly responsible for the conference, and to the contributors for the speedy provision of papers. Especial thanks are due to Margot Smith and Hazel Moon for cheerful acceptance of typing and countless secretarial tasks, and to Eric Bell for proof reading and advice.

J.S. MADDEN, R. WALKER AND W.H. KENYON

OPENING ADDRESS TO THE CONFERENCE

M. MEACHER, M.P.

Parliamentary Under Secretary of State
Department of Health and Social Security

INTRODUCTION

1. I am very glad to have the opportunity to speak at the opening of this third conference organised by the Merseyside Lancashire & Cheshire Council on Alcoholism so that people from several countries can gather to learn and discuss matters of considerable and increasing public concern. I say this not only because this is a distinguished gathering and from previous conferences we have come to have high expectations of this one too, I say it also because alcoholism in particular is a matter to which more attention must be devoted. At present services for alcoholics are all too often ill-co-ordinated and patchy, and the all important area of prevention has been neglected or left to voluntary initiative where financing may be precarious. Follow-up support also may be inadequate and after-care facilities may be sparse. More evaluation is needed of different methods and settings for treatment. It is against this background we need to look at the magnitude of the problem.

INCREASE IN PROBLEMS ASSOCIATED WITH ALCOHOL

2. Consumption of alcohol has increased dramatically in recent years. In England and Wales between 1969 and 1974 the consumption of wines almost doubled and of spirits increased by 80 per cent. Many believe that such increases inevitably lead to a rise in the incidence of alcohol-related problems and there is some evidence to support this. Hospital admissions for alcoholism have been increasing at 10½% or more since 1969. If admissions continue increasing at the present rate, they would account for nearly a quarter of all mental illness admissions by 1985. Furthermore, the

the numbers of people convicted of drunkenness offences and of drink and driving offences are increasing steadily, with a relatively greater increase in the proportion of young people involved.

SERVICES FOR ALCOHOLISM

3. Facilities for the treatment of alcoholism have improved considerably in recent years. In England there are now 23 specialised hospital units and one or two are still in the pipeline. Some of these units are developing out-patient services and provide a focus of expertise, training and research for all services concerned with alcoholics. Hospitals continue treatment on an in-patient, or day or out-patients basis to many of the alcoholics in their catchment areas. Hostels for alcoholics have increased considerably since my Department introduced a scheme to fund them for five years from 1973. This scheme is now being amended so that grants are paid for new hostels for five years from the date they are opened; in addition, grants can be sought in certain circumstances to meet the cost of fire precautions and essential repairs. We hope that hard-pressed local authorities will be able to support more schemes from voluntary organisations than they would otherwise have done. At present there are 47 hostels in England, providing 515 places for alcoholics.

4. Voluntary organisations also provide information, counselling and advice services for the public, for alcoholics and their families. The Regional Councils on Alcoholism are particularly active in these roles, and a very good example is here in Merseyside. There are 21 Regional Councils on Alcoholism in Great Britain, and several more are planned. I would like to pay tribute to their dedicated staff and also to the work of many other national and local agencies, including Alcoholics Anonymous, Al-Anon and Alateen which help many alcoholics and their families.

5. Increasingly the need is to develop a network of services for the identification of alcoholics and more effective follow-up services. We have to think through what the early identification of problems means. It means helping those with drinking problems to seek help before they reach a stage of chronic illness, and it means recognising that the family doctor, health visitors, social workers, policemen and probation officers, employers, also the friends, relations and work-mates of those with problems all have an important part to play. As to the role of my own Department, we are glad to be able to make grants to the National Council on Alcoholism, the Alcohol Education Centre and Health Education Council so that they can help to inform and educate professionals and the public.

DETOXIFICATION CENTRES

6. My Department is also supporting financially two or possibly three experimental detoxification centres to provide an alternative to the penal system, and an entry point to treatment and rehabilitation services for habitual drunken offenders. The first centre, a community-based project, will open in Leeds in May, and a hospital-based centre with a new alcoholism unit will open in Manchester later in the year. Elsewhere we believe that "drying out" services could be provided either within existing hospital facilities or on a modest scale in hostels or other accommodation. But whatever the nature of the detoxification service, it is unlikely to succeed unless there is the full co-operation of police, health and social services and also adequate after-care for its clients in the community.

ALCOHOLISM ADVISORY COMMITTEE

7. Last year, Health Ministers in England and Wales set up an Alcoholism Advisory Committee to advise them on services for alcoholism and where necessary to promote their development. At their first meeting they set up a Sub-Group charged with promoting services for homeless alcoholics which started its activities here in Merseyside last October. Members of the group have since held meetings in London and Hampshire and will shortly be visiting Hull, Newcastle, Manchester and Birmingham. Other Sub-Groups have been set up to consider the overall pattern of services for alcoholism and strategies for prevention. A fourth is likely to start late in the year to consider the training given to those whose work brings them in contact with alcoholics. I turn now to the future.

THE WAY AHEAD

8. I believe we must move towards a better co-ordinated network of services with earlier identification of those in need of help and more effective follow-up services. I am sure this will be a keynote of this Conference. The need is not so much to develop more specialised hospital or community services as for a greater awareness of alcoholism and of the needs of alcoholics or of those with drinking problems among all those whose responsibilities bring them into contact with this group. This involves not only staff of the health and social services, but also of other fields such as probation, police and others in contact initially with them. This in turn means that a source of expertise should be available in each district, achieved by having a nucleus of staff including members of health, social, educational and penal services, with a sufficient understanding and experience of the problems of alcoholics, if early detection and prevention are to become realities.

9. The precise pattern of service could vary from district to district, depending upon local circumstances and need. The recent reorganisation of the N.H.S. and of local authorities provides the opportunity to experiment with ways in which services could be provided. The Joint Consultative Committees and the Community Health Councils could play crucial parts in ensuring the necessary collaboration between all the local services involved.

THE DRUG PROBLEM

10. The same urgent need for improved co-ordination applies equally to managing the problem of drug dependence. Many of those addicted to, or misusing drugs are young people with a variety of problems including homelessness and often a history of emotional deprivation, disturbance and separation in the family and perhaps institutionalisation. Both legally and illicitly obtained drugs are misused in a variety of ways - whether by young experimenters, middle-aged therapeutic addicts, or the multi-drug misuser including those who use narcotics amongst other drugs and regularly overdose. The current number of active narcotic addicts known to be receiving narcotic drugs is probably around 2,000, and the trend, though not steep, is upwards. Multiple drug misuse among young people appears to be increasing and the use of barbiturates is giving cause for concern.

PREVENTION

11. Drug misuse has to be seen I think as part of the wider problem of our reliance on drink, cigarettes, and the general misuse of drugs by society, especially sleeping pills and tranquillisers. Very few become addicted compared with the number who may experiment with drugs. For this reason all those who deal with young people need more skill in recognising the possible signs of regular drug misuse. It may be that the availability of advisory and therapeutic services to help young people with personality needs and inadequacies would be more helpful than some form of health education where there is a danger of stimulating interest in drugs. Health education on drug misuse should I believe aim to help young people to cope with their problems without resorting to drugs: it should be part of the wider problem of health about which young people should be informed.

SERVICES FOR DRUG DEPENDENTS

LOCAL SERVICE

12. Hitherto health services have been almost wholly concerned with treating the established narcotic addict, whether by withdrawl or

maintenance. In only one or two areas has there been any attempt to see the treatment clinic as part of a range of services locally available, and as a resource for local social workers, teachers, youth workers and others in contact with young people in the area. With N.H.S. reorganisation bringing together the preventive and treatment services and providing new machinery for co-ordination with the local authority services and social services and education in particular, there is a strong case I believe for encouraging clinics to widen their approach to drug problems. A wider approach is necessarily a more locally-based approach.

13. It is perhaps inconsistent with these objectives and with the reorganisation of the N.H.S. for my Department to continue to assume a co-ordinating responsibility for the strategic planning and operation of local services. For this reason Regional Health Authorities are being encouraged to assume these responsibilities. Gradually those in the treatment services should become part of a local network of services to help drug misusers.

HOSTEL PROVISION

14. Some voluntary organisations provide hostels for addicts withdrawn from drugs who need a period of supportive care and help in establishing new patterns of life. A variety of accommodation however is required. A start has been made in London with providing accommodation for married addicts. Voluntary organisations have long played a vital part in our society and I would like to pay tribute to the ways they can respond so readily to changing problems and situations. Unfortunately they need regular financial support to continue, and this particularly is difficult during periods of economic constraints. My Department is however continuing to do what it can to encourage and assist the establishment of suitable accommodation and other facilities.

I conclude that experimental developments in pioneering work of various kinds still dominate the field and this continues for the present largely to be centrally financed. New patterns of provision may well emerge, and I doubt if we should yet seek to draw up a definitive policy for future services. What can be seen already is that there is a clear need for a multi-disciplinary approach and for encouraging the sharing of experience between those working with addicts in different settings. This is a role which this Conference is ideally fitted to perform and I wish you well for the discussion of the very wide-ranging agenda you have before you which I and so many in the field will be watching closely and with expectation.

Section I

General

SEXUAL DISORDERS IN MALE ALCOHOLICS

M.J. AKHTAR

WEST CHESHIRE HOSPITAL

CHESTER

INTRODUCTION

Diminished sexual activity among individuals dependent upon alcohol has been noted by various authors, but no-one has improved on Shakespeare's observation of the sexual effects of alcohol - "It provokes and unprovokes; it provokes the desire, but it takes away the performance". (MACBETH Act II Scene III).

In spite of the often quoted association between impairment of sexual functioning and alcohol, there are few scientific studies to substantiate it and unfortunately the literature on this subject appears to be inconsistent. Levine (1955) assessed the sexual adjustment of 79 alcoholics attending the Veterans Hospital in Washington, and concluded that there was a general diminution in heterosexual relationships among his subjects. Zizic et al. (1967) found the frequency of impotence in alcoholics to be unexpectedly small. They concluded that the action of alcohol as an exogenous toxin on the development of impotence is negligible. Johnson (1968) expressed the view that impotence in alcoholics does not occur consistently except in the beta type of alcoholism. On the other hand, Lemere and Smith (1973) observed that impotence resulting from prolonged alcohol abuse may persist even after years of sobriety. They suggested that alcohol induced impotence is due to the destructive effect of alcohol on the neurogenic reflex arc that serves the process of erection. Scheig (1975) noted impairment of sexual performance among chronic alcoholics due to changes, attributable to liver disease.

Reference should also be made to the widely recognised association between alcoholism and "morbid jealousy" or

"Othello syndrome". There are two excellent reviews of the syndrome in current literature by Enoch et al. (1967) and Shepherd (1961). Most workers relate the symptoms to decrease in potency of the male alcoholics, who by a mechanism of projection, become convinced of their wife's infidelity. Henc (1967) studied the manifestations of jealousy and disorders of potency in chronic alcoholics, by prolonged observation of the patients and their domestic environment. The results confirmed an aetiologic dependence of the mental and functional phenomena upon alcohol consumption.

It has been the author's clinical impression that alcohol induced sexual disorders are common. The present study had three aims:

(1) To assess the extent of sexual dysfunction in a group of male in-patients, admitted to an addiction unit, with the diagnosis of alcoholism.

(2) To see if the data reported by other investigators can be confirmed.

(3) To compare the results of this study with an earlier study of sexual dysfunction by Masters and Johnson (1970).

MATERIAL AND METHOD

The present study took place in the Regional Addiction Unit at Moston Hospital, Chester, between January 1975 and March 1976.

The selection policy of the Unit is to exclude as far as possible cases showing psychosis, or psychopathic personality.

Forty-five male in-patients were studied. Although not consecutive admissions, they were not selected on any basis other than having been admitted with the diagnosis of alcoholism, with absence of obvious signs of brain damage, and their willingness to co-operate in this study. Because of the nature of the inquiry and the type of information to be obtained, patients who had "never married" were not included in the sample. As a measure of ensuring reliability of information, patients were examined within one week to one month of admission when they were all "dry". It was hoped that after achieving sobriety the motivation of falsifying information would be less, and external information greater.

Data on patients was collected during individual interviews with the aid of a suitably designed questionnaire and from case-records and social reports. The interviews were administered by the same psychiatrist in all cases. The interviewer was known to the patients, and he had already established a rapport with them within the therapeutic milieu of the Unit. Before conducting the interview, the purpose of the study was fully explained to the patient. Patients were classified according to their age, marital status, social class (by occupation), number of admissions to hospital with alcoholism drinking history and associated behaviour. Social stability was rated according to the "Straus and Bacon" (1951) scale.

Duration of alcoholism was calculated from the time the patient himself admitted having lost the ability to control alcohol intake, or being unable to abstain from drinking or from the time he first experienced withdrawal "shakes", and memory "blackouts".

Information regarding presence or absence of sexual disorder was obtained by asking questions for various items related to difficulties in a broad area of sexual functioning. The sexual ability of these patients, prior to the development of alcoholism was also rated for the same items and this rating was used as a control to compare any change in sexual function, after the development of alcoholism. These patients thus served as their own control group over time. In the interview an attempt was made to distinguish clearly between transient sedative effects of alcohol and persistent sexual dysfunction.

TABLE NO. I

AGE FREQUENCY OF ADMISSION & MARITAL STATUS OF PATIENTS

Age Distribution of Patients	20-30 years	31-40 years	41-50 years	51-60 years
	4	15	18	8
			Mean age 43-7 years S.D. 9.1 years	

Frequency of Admission for Alcoholism	1st Admission	Re-admission
	18	27

Marital Status	Married	Separated	Divorced	Widowed
	28	11	6	-

TABLE NO. II

SOCIAL CLASS & SOCIAL STABILITY OF PATIENTS

Social Class by Occupation	I & II	III	IV & V
	6	17	22

Percentage of Male Patients answering Yes to Individual Questions :-

Patients Social Stability Straus & Bacon Scale 1951	Straus & Bacon figures in brackets	
	I Ever held a steady Job for 3 years	= (56%)
	II Living in Town of present residence for 2 years	= (86%)
	III Living in own home or with relatives or friends	= (74%)
	IV Married and living with wife	= (53%)

PRESENT STUDY:

I	46%	III	51.7%
II	60%	IV	44.4%

TABLE NO. III

DRINKING HISTORY & ASSOCIATED FACTORS

Duration of Alcoholism	2 - 5 yrs	6 - 10 yrs	11 - 20 yrs
	12	27	6

Drinking Pattern	Continuous Drinking	Bout Drinking
	36	9

Recent or Past H/O Delirium Tremens	Yes	No
	13	32

Police Convictions (Other than Drinking and Driving)	Yes	No
	14	31

TABLE NO. IV

BREAKDOWN OF ITEMS RELATING TO SEX DISORDERS

Items	Method of Assessment
Libido	(i) Patient's own admission of change in sexual desire. (ii) Frequency of Sexual thoughts (Self-Rating Scale Tennents et al. 1974) (iii) Frequency of Sexual intercourse (iv) Frequency of Masturbation
Impotence & Ejaculatory Disorders	Questions Relating to Presence or Absence of Erectile and Ejaculatory dysfunction
Deviant Sex Behaviour	Questions Relating to Presence or absence of Homo-Sexuality and other Sexual deviations

RESULTS

Although one cannot definitely exclude psychological, social and emotional causes of sexual disorders in these patients, the results do indicate that alcoholism is associated with sexual disorders in males. The disturbance of function seems to affect libido, erection and ejaculation. The data analysis also indicated a relationship between length of alcoholism and the severity of sexual disturbance.

TABLE NO. V

TYPES OF MALE SEXUAL DISORDERS DISCOVERED IN PRESENT STUDY

(N = 45
(S.Disorders = 29 = 64.4%)

Nature of Sexual disorder	No. of Patients	Percentage
Diminished Sexual Desire	24 (D.L. only = 3)	53.3%
Erectile Impotence	14	31.1%
Premature Ejaculation	3	6.6%
Ejaculatory Incompetence	8	17.8%
Sexual deviations	1	2.2%

N: Number
D.L: Diminished Libido
S: Sexual

Primary impotence was not found in any member of the group studied. It was noted that the majority of patients with short periods of dependence had either observed no sexual disturbance or reported diminished libido, with or without ejaculatory dysfunction. Patients with long periods of dependence more commonly complained of sexual disturbance and this consisted of diminished libido with erectile impotence or less commonly ejaculatory incompetence. However, the effect of age in determining the severity of disturbance cannot be ruled out with certainty. The extent of deviant sexual behaviour in this study was small (one patient), and further study on a prospective basis, using a larger sample is indicated.

TABLE NO. VI

COMPARISON OF STUDIES (S.D. IN ALCOHOLICS)

FRIDMAN DESPOTOVIC & ZIZIC	STANKVSHEV, PROTIC, SHISHKOV. (N = 373 MEN)	PRESENT STUDY (N = 45 MEN)	VAN THIEL, SHERINS, LESTER (N = 37 MEN WITH LIVER DISEASE)
IMPOTENCE 3.9%	SEXUAL DISORDERS 51% AFFECTING:- (LIBIDO (ERECTION (EJACULATION (ORGASM	SEXUAL DISORDERS 64.4% ERECTILE IMPOTENCE: 31.1%	L.O.L/& IMPOTENCE 78%

Abbreviations: S.D. Sexual Disorders
N. Number
L.O.L. Loss of Libido

DISCUSSION

The findings seem to contradict the conclusion drawn by Zizic et al. (1967) that the frequency of impotence in alcoholics is small (3.9%), but are in broad agreement with the study by Stankvshev et al. (1974) who investigated the disturbance of sexual function in a Bulgarian sample. They carried out clinical studies on 373 male alcoholics aged from 20-50 years and found sexual problems relating to libido and impotence in 51%. Sexual problems were most frequent in spirit drinkers. They also noted changes in the fertility functions of their patients.

Mello and Mendelson (1973) explored the effects of chronic alcoholism on testosterone levels. They found that chronic ethanol ingestion was consistently associated with decreased serum testosterone levels. It is now known that plasma testosterone levels are low in chronic alcoholism, and this may be caused by the change in the NAD:NADH ratio resulting from ethanol oxidation in testicular cells, where it influences the rate-controlling step in the synthesis of androgens and testosterone.

Van Thiel, et al. (1973) investigated the mechanisms of hypogonadism in 37 men with alcoholic liver disease, ranging from hepatitis to end stage cirrhosis. They found loss of libido and impotence in 78% They also found evidence of feminization (62%), testicular atrophy and sterility (57%) and gynaecomastia (16%). They also confirmed low plasma testosterone levels in 68%. They concluded that (i) Feminization and testicular atrophy observed in alcoholic liver disease cannot be explained by excess oestradiol per se, (ii) The failure to see elevated plasma level of luteinising hormone in response to reduced testosterone and the abnormally low response to clomiphene (clomid) stimulation suggest that the hypogonadism seen in alcohol liver disease is in part or whole due to deficient hypothalamic-pituitary function. There is a high incidence of loss of libido and impotence in the group of patients studied by Van Thiel and his associates, as compared to the findings of the present study. This can be attributed to the fact that all patients in their sample had known moderate to severe liver disease.

There is now work showing the characteristics of sex-hormone-binding-globulin in chronic male alcoholics with liver disease. Galvao-Teles et al. (1973) studied 25 such men. Their results clearly demonstrate a fall in plasma unbound (biologically active) androgens (17-OHA) and suggest that this is due to a rise in plasma sex-hormone-binding-globulin (S.H.B.G.) Unbound oestradiol levels were not altered. They concluded that it may be a fall in unbound androgens, rather than any rise in unbound oestradiol, which leads to clinical hypogonadism and gynaecomastia in men with chronic liver disease, especially those with alcoholic cirrhosis.

TABLE NO. VII

COMPARISON OF TYPES OF MALE SEXUAL DYSFUNCTION REPORT IN TWO STUDIES (IN PERCENT)

Nature of Dysfunction	Masters & Johnson (N =448) Study	Present Study (N = 45)
Primary Impotence	7%	Nil
Secondary Impotence	48%	31.1%
Premature Ejaculation	41%	6.6%
Ejaculatory Incompetence	4%	17.8%

Comparison of the forty-five male patients in my sample to 448 male patients (predominantly non-alcoholic) described by Masters and Johnson was done. The two populations are vastly different in many aspects. Masters and Johnson group consisted of highly selected patients, who often had severe problems. Their age ranged from 23 years to 76 years. Because the groups are so different the comparison is all the more interesting. The extent of ejaculatory incompetence in my sample is especially striking.

In conclusion, the results of this study confirm an association between alcoholism and diminished activity in several areas of sexual functioning. Since it is not possible to exclude the effects of psychological and social factors in producing sexual dysfunction in this group of patients, the findings indicate an important, though not necessarily causal relationship between alcoholism and sexual disorders. The data also points to a probable relationship between length of alcohol abuse and severity of sexual disturbance.

An important implication of the findings is that at educational and clinical levels, emphasis should be placed upon the disturbing influence of alcohol on sexual functions. From the diagnostic point of view, complains relating to sexual functioning should be considered in terms of the possibility of co-existing alcohol abuse, and that it may be worthwhile to ascertain the drinking habits of every sexually dysfunctional patient on initial evaluation. Similarly, this study has revealed that in conventional hospital settings alcoholic patients cannot be relied upon to report sexual disorders, unless they are questioned directly. Thus the need for a routine inquiry about sexual functioning, at the time of initial clinical assessment is clearly indicated. In terms of treatment, it would seem clear that to treat sexual disorders separately from treating the alcohol dependence is contra-indicated.

There are limitations of the research method employed in this study, chiefly small sample size, use of retrospective data, and the possibility of uneliminated bias on the part of the interviewer and informant. It is recognised that the findings cannot be readily generalised, and that one can only draw tentative conclusions. At the same time it is felt that the findings have indicated the need for :

(1) Identification, organisation and correlation of existing knowledge about relationship between alcoholism and sexual disorders.

(2) Further systematic clinical study of these relationships with more open minded inquiry and observation.

(3) Well controlled studies with larger sample of patients on a prospective basis.

(4) Comprehensive and carefully conceived physiological research in the field of alcohol-induced sexual disorders.

Finally, it is also important to recognise that in many alcoholics there is a wide spread basic lack of drive and motivation, and that following abstinence from alcohol the sex drive is likely to reassert itself. This presents the challenge of considering how this re-emergence of sex drive might be harnessed in terms of motivation to lead to a better quality of life, than is provided by alcohol dependent behaviour.

REFERENCES

ENOCH, M.D., THRETHOWAN, W.H. and BARKER, J.C. (1967). "Some Uncommon Psychiatric Syndromes." John Wright & Sons Ltd., Bristol.

GALVAD-TELES, A. et al. (1973). "Biologically Active Androgens and Oestradiol in Men with Chronic Liver Disease". Lancet, 1, 172-177.

HENC, I. (1967). "Jealousy of Alcoholics and Disturbances of Potency". (Abstract No. 2460 in Excerpta Medica, Psychiatry, 1968, 21, 8, 403).

JOHNSON, J. (1968). "Disorders of Sexual Potency in the Male". Pergamon Press Ltd., London/Oxford.

LEMERE, F. and SMITH, J.W. (1973). "Alcohol Induced Sexual Impotence". American Journal of Psychiatry, 130. 2,212 - 213.

LEVINE, J. (1955). "Sexual Adjustment of Alcoholics". Quarterly Journal of Studies on Alcohol, 16. 4, 675 - 680.

MASTERS, W.H. and JOHNSON, V. (1970). "Human Sexual Inadequacy". Little Brown and Co., Boston.

MELLO, N. and MENDELSON, J. (1973). "Androgens and Aggression in Alcohol Addicts." 126th Annual Meeting of the American Psychiatric Association, Honolulu, Hawaii, May 7-11.

SCHEIG, R. (1975), "Changes in Sexual Performance due to Liver Disease". (Abstract No. 3365 in Exerpta Medica, Psychiatry 1975, 32, 10, 577).

SHEPHERD, M. (1961). "Morbid Jealousy - Some Clinical and Social Aspects of a Psychiatric Symptom." _The Journal of Mental Science_, 107. 687-704.

STANKVSHEV, T., PROTIC, M. and SHISHKOV, A. (1974). "Disturbances of Sexual Function in Alcoholics (Bulgarian)". (Abstract No. 1883 in Excerpta Medica, Psychiatry, 1976. 33, 5, 333).

STRAUS, R. and BACON, S.D. (1951). "Alcoholism & Social Stability. A Study of Occupation Integration in 2,023 male clinic patients". _Quarterly Journal of Studies on Alcohol_, 12 231 - 60.

VAN THIEL, D.H., SHERINS, R.J. and LESTER, R. (1973). "Mechanism of Hypogonadism in Alcoholic Liver Disease". _Gastroenterology_ 65. 574.

ZIZIC, V., FRIDDMAN, V. and DESPOTOVIC, A. (1967). "Treatment of Impotence in Alcoholics". (Abstract No. 2127 in Excerpta Medica, Psychiatry, 1968, 21, 7, 349).

VITAMIN B_{12} AND FOLIC ACID IN CHRONIC ALCOHOLICS

E. HEILMANN

MEDIZINISCHE POLIKLINIK DER WESTFALISCHEN
WILHELMS-UNIVERSITAT
MUNSTER, GERMANY

The adverse effects of alcohol on haemopoiesis have been well documented. The most important manifestation of alcohol toxicity is anaemia, but abnormalities in platelets and white blood cell production are also observed. One of the earliest indications of the ability of ethanol to interfere in cellular metabolism was the observation by Gram (1884) of macrocytosis in association with cirrhosis. The nutritional deficit associated with alcoholism was suspected by Bianco and Jolliffe (1938); they regarded the macrocytosis of the alcohol addict not as a manifestation of inability on the part of the liver to store a haemopoietic principle, but as an extrinsic deficiency of some necessary haemopoietic substance required to prevent macrocytosis.

The present study reports on vitamin B_{12} in the serum and on folic acid in the serum and red cells of 20 chronic alcoholics given multiple vitamin therapy and 44 untreated chronic alcoholics who consumed more than 100 mg ethanol daily for more than five years. Determinations of vitamin B_{12} were performed with a radio-absorbent assay, (Heilmann & Poblotzki, 1975). Folic acid in the plasma and in red cells were determined with 131_I as marker (Heilmann & Bönninghoff, in press). Vitamin B_{12} and folic acid levels were correlated with other haematological parameters, which were determined in the usual manner.

RESULTS

The levels of vitamin B_{12} in the group of vitamin treated alcoholics ranged between 380 and 1070 pg/ml. These values lie within the normal range. Two patients with elevated vitamin B_{12} levels had been treated with parenteral vitamin B_{12}. The results of vitamin B_{12} in the group of the untreated patients were reduced significantly ($p<0.0001$). The levels ranged between 100 and 370 pg/ml. There was no significant correlation between vitamin B_{12} and Hb_E levels.

In the group of vitamin treated alcoholics serum folic levels ranged from 2 to 10 ng/ml, with a mean of 4.8 ng/ml (normal 7 ng/ml). In the group of untreated alcoholics the mean did not differ significantly from that in the treated group, but more than 50% of the patients showed reduced levels. There was a distinctly significant ($p<0.001$) correlation between folic acid levels in serum and in red cells. No correlation was found between serum folic acid levels and haemoglobin, haematocrit or the Hb_E values.

DISCUSSION

Anaemia is a frequent finding in alcoholic patients for many reasons. Since hard liquor contains no vitamin B_{12}, or folate, it might be expected that chronic alcoholics would develop deficiencies of these vitamins. Because the body stores of vitamin B_{12} ordinarily will last for a number of years, their reduction could be expected to be a very slow process. However, body stores of vitamin B_{12} are much more rapidly exhausted when the normal circulation of vitamin B_{12} is interrupted by the vitamin B_{12} malabsorption which ethanol may produce. The liver disease frequently associated with alcoholism confuses the picture with respect to the vitamin B_{12} nutritional state: with chronic liver disease although serum vitamin B_{12} levels tend to be normal, liver stores of vitamin B_{12} tend to be sharply reduced.

Herbert (1962) performed the first study of the daily requirements for folic acid in normal man and the rate of induction of megaloblastic erytheropoiesis with dietary deprivation. This demonstrated the importance of normal stores as a buffer against inadequate dietary intake. In a series of studies in severe alcoholic patients a high incidence of folic acid deficiency and haematological abnormalities were found. Although the high incidence of folate deficiency is largly due to inadequate dietary intake of folate, other factors also play a role (Halstead et al. 1967, 1971, Herbert 1965, Herbert et al. 1963, Hillman 1975, Sullivan and Herbert 1964).

Alcohol has major adverse effects on gut function, including interference with the absorption of some dietary constituents; but despite many studies of alcohol toxicity on folic acid absorption the results have been variable and inconclusive. There is no specific technique for demonstrating the potential site of alcohol toxicity in the internal metabolism of folic acid. After ingestion, both methylated and non-methylated mono- and diglutamates are absorbed and delivered to the hepatic and renal tissues for storage. When needed the polyglutamate stores in the liver are deconjugated, and the monoglutamate form of N-5-MTHFA is delivered to the serum for participation in DNA and protein synthesis. Each of the steps of the folic acid cycle is a potential site for alcohol toxicity. The most striking finding has been the demonstration of a rapid suppression of circulating N-5-MTHFA levels with alcoholic ingestion. Within six to eight hours of receiving oral or intravenous doses of alcohol, normal individuals show a dramatic fall of the serum levels of folic acid, which then increase eight to nine hours later (Hillman 1975).

Finally, in alcoholic cirrhosis the complication of gastro-intestinal bleeding may induce hyperactivity in the bone marrow with an increased demand for vitamin B_{12} and folate (Herbert 1965). In conclusion chronic alcoholics should be treated with vitamin B_{12} and folic acid.

REFERENCES

BIANCO, A. and JOLLIFFE, N. (1938). "The Anemia of Alcohol Addicts". American Journal of the Medical Sciences, 196. 414.

GRAM, C. (1884). "Untersuchungen über die Grösse der roten Blutkörperchen im Normalzustande und bei Verschiedenen Krankheiten". Fortschritte der Medizin, 2, 33.

HALSTED, C.H., GRIGGS, R.G. and HARRIS, J.W. (1967). "The Effect of Alcoholism on the Absorption of Folic Acid (^{3}H-PGA) Evaluated by Plasma Levels and Urine Excretion". Journal of Laboratory & Clinical Medicine, 69. 116.

HALSTED, C.H., ROBLES, E.A. and MEZEY, E. (1971). "Decreased Jejunal Uptake of Labelled Folic Acid (^{3}H-PGA) in Alcoholic Patients : Roles of alcohol and Nutrition". New England Journal of Medicine, 285. 701.

HEILMANN, E. and POBLOTZKI, F. (1975). "Radioassay zur Vitamin-B_{12}-Bestimmung im Serum". Blut, 31. 199.

HEILMANN, E. and BÖNNINGHOFF, E. (In press). "Radioassay zur Folsäure-Bestimmung im Serum und in den Erythrozyten". Schweizerische Medizinizche Wochenschrift.

HERBERT, V. (1962). Experimental Nutrional Folate Deficiency in Men. Transactions of the Association of American Physicians, 75. 307.

HERBERT, V. (1965). "Hematopoietic Factors in Liver Disease". Progress in Liver Diseases. H. Popper & F. Schaffner, eds. New York.

HERBERT, V., ZALUSKY, R. and DAVIDSON, C.S. (1963). "Correlation of Folate Deficiency with Alcoholism and Associated Macrocytosis, Anemia and Liver Disease". American Journal of Internal Medicine, 58. 977.

HILLMANN, R.S. (1975). "Alcohol and Hematopoesis". Annals of the New York Academy of Sciences, 252. 297.

SULLIVAN, L.W. and HERBERT, V. (1964). "Mechanism of Hematosuppression by Ethanol". American Journal of Clinical Nutrition, 14. 238.

PORPHYRIA CUTANEA TARDA IN CHRONIC ALCOHOLICS

E. HEILMANN

MEDIZINISCHE POLIKLINIK DER WESTFALISCHEN
WILHELMS-UNIVERSITAT
MUNSTER, GERMANY

According to Watson (1960, 1964) and Waldenström (1963) porphyrias are classified by the location of the disturbance of haemesynthesis into erythropoeitic and hepatic forms.

TABLE I: CLASSIFICATION OF PORPHYRIA

I.	Erythropoetic porphyria	
	1.	porphyria congenita erythropoetica
	2.	protoporphyria erythropoetica
II.	Hepatic prophyria	
	1.	porphyria acuta intermittens
	2.	coproporphyria hereditaria
	3.	porphyria cutanea tarda and chronic hepatic porphyrias
	4.	porphyria variegata
	5.	symptomatic porphyria

While erythropoeitic porphyrias are manifest during infancy, hepatic porphyrias typically become clinically evident after puberty. A more detailed differentiation of the various forms of porphyrias is achieved by biochemical analysis of porphyrin precursors in bone marrow, in circulating red blood cells, in serum, liver tissue and in urine.

According to Waldenström (1937) and Watson (1960) hereditary and constitutional factors play a role in porphyria cutanea tarda (PCT), but Doss et al. (1970) and Doss (1974) consider that in most cases PCT is an acquired disease found in the final stages of chronic liver damage. The predominant clinical findings in PCT are vesicles and hyperpigmentation of light-exposed parts of the body. Chronic alcoholism, certain drugs, sunlight and ultra-violet light can determine the clinical expression. The light sensitivity of the dermis is thought to be due to the deposition of uroporphyrin.

Compared to reports on the analysis of porphyrins in PCT there have been only a few studies of disturbances of iron metabolism in the disease. The present study examined changes in iron metabolism of alcoholics with PCT and at the resultant implications for therapy.

PATIENTS AND METHODS

Ten alcoholic patients (age 29-58, two women and eight men) with clinical manifestations of PCT were investigated. The following parameters were measured: haemoglobin, serum iron, latent and total iron binding capacity, transaminases, and cholinesterase. Haemoglobin was estimated photometrically, serum iron colometrically (according to the method of Heilmeyer). Both blind and open liver biopsies were performed. Histology was carried out by Professor Grundmann, Munster. Analysis of porphyrins in urine was performed by Professor Doss, Marburg.

RESULTS

Figure one shows the results of chemical analysis. Haemoglobin was normal in all patients. The average serum iron concentration was double the concentration of normal subjects (maximum value 288g/100 ml). Latent iron binding capacity was lowered on average to half normal values, while total iron binding capacity was normal. Serum glutamic-oxaloacetic transaminase was minimally elevated, Serum glutamic-pyruvic trasaminase was 30% higher than normal. Cholinesterase was slightly lower than the normal range. The diagnosis was ascertained in all cases by the type of porphyrin metalbolites in the urine, that is, by the predominant excretion of uro- and heptacarboxyphorphyrins along with minimal elevation of coproporphyrins.

Figure two shows the relationship between serum iron level, latent iron binding capacity and liver histology. We found three cases of fatty liver, four cases of chronic hepatitis and three of liver cirrhosis. Excluding one patient with chronic hepatitis, in all cases there was increased iron deposition or hemosiderosis of the liver. The elevation of serum iron concentration was not correlated with the severity of liver damage on biopsy. However, increased severity of morphological changes of liver tissue seemed to correlate with decrease in latent iron binding capacity.

DISCUSSION

The pathogenesis of PCT has not been fully elucidated. Many studies indicate that several factors can make clinically manifest a predisposition to disturbed hepatic porphyrin synthesis. Anomalies of iron metabolism are of decisive importance. Utilising radioactive iron, (Schirren et al. 1966) demonstrated an elevation of iron absorption in the disease; ferrokinetically there was an increased turnover of iron with increased haemolysis. This would explain the elevated serum iron concentrations found in all the patients in the present study. There are no reports in the literature with regard to variations of total iron binding capacity. Its levels in the present investigation were normal.

Disturbances of iron metabolism are of therapeutic importance and for this reason Ippen (1961, 1967, 1971) introduced phlebotomy to reduce total body iron stores. By this means reversal of skin symptoms and decrease in porphyrin excretion can be achieved. Whether this is achieved by lowering the porphyrin pool or by iron withdrawal (500 ml blood contains 250 mg iron) has not yet been elucidated.

REFERENCES

DOSS, M., MEINHOF, W., MALCHOW, H., SODOMANN, K.P. and DÖLLE, W. (1970). "Charakteristische Konstellation der Urinporphyrine als Biochmischer Index der Porphyria Cutanea Tarda". Klinische Wochenschrift, 48. 1132.

DOSS, K. (1974). "Klinische Biochemie und Laboratoriumsmedizin der hepatischen Porphyrien". Diagnostik, 7. 489.

IPPEN, H. (1961). "Allgemeinsymptome der spaten Hautporphyrie (Porphyria Cutanea Tarda) als Hinweis fur deren Behandlugn". Deutsche Medizinische Wochenschrift, 86. 127.

IPPEN, H. (1967). "Kausalfaktoren der Porphyria Cutanea Tarda". Dermatologische Wochenschrift, 153. 1351.

IPPEN, H. (1971). "Hepatische Porphyrie". Therapiewoche, 3. 208.

SCHIRREN, C., STROHMEYER, G., WEHRMANN, R. and WISKEMANN, A. (1966). "Ergebnisse der Aderlassbehandlung bei Porphyria Cutanea Tarda". Deutsche Medizinische Wochenschrift, 91. 1344.

WALDENSTRÖM, J. (1937). "Studien über Porphyrie". Acta Medicinae Scandinavica, 82. 1.

WALDENSTRÖM, J. and HAEGER-ARONSON, B. (1963). "Different Patterns of Human Porphyria". British Medical Journal, 5352. 272.

WATSON, C.J. (1960). "The Problem of Porphyria: Some Facts and Questions". New England Journal of Medicine, 263. 1205.

WATSON, C.J. (1964). "Porphyrin Metabolism". G. Duncan: Disease of Metabolism. Saunders, Philadelphia. S.850.

PHYSICAL COMPLICATIONS OF ALCOHOL EXCESS - METABOLISM OF ALCOHOL

D. OWENS

WALTON HOSPITAL

LIVERPOOL

Alcohols are normally produced in small amounts by the bacteria of the small intestine. This presumably explains why the liver is capable of metabolising such substances, (Krebs and Perkins 1970). In individuals consuming large amounts of alcoholic beverages it has been estimated that the liver is capable of oxidising about 100 grams of alcohol, mainly in the form of ethanol, in 24 hours, (Sherlock 1968).

The ethanol which is consumed is rapidly absorbed from the stomach and upper small intestine. The rate of absorption seems to be directly related to the ethanol concentration, but even at higher concentrations absorption is complete, (Cooke and Birchall 1969). The ethanol that is absorbed diffuses rapidly within body water and some may be excreted unchanged in the urine. The majority however (around 80 - 90 per cent) is oxidised in the liver to acetaldehyde, (Thompson 1956). Two different enzyme systems are probably involved in this process. Firstly there is alcohol dehydrogenase (ADH) a zinc containing cytoplasmic enzyme requiring nicotinamide adenine dinucleotide (NAD) as cofactor, which has a pH optimum of 10.4. It is believed that the oxidation catalysed by this enzyme is the rate limiting step in ethanol metabolism by the liver. However, it is not specific for ethanol and also catalyses the oxidation of methanol and glycols,(von Wartburg, Bethune and Vallee 1964). The second enzyme system - the microsomal ethanol metabolising system (MEOS) differs from ADH in that its pH optimum is 7.2 and it requires nicotinamide adenine nucleotide phosphate as cofactor, (Lieber and De Carli 1970). Moreover, this enzyme is inducible by ethanol, (Rubin and Lieber 1968) so that it may play an important part in the metabolism of ethanol in alcoholics. However this point is controversial.

When ethanol is oxidised by ADH and MEOS acetaldehyde is produced which is in turn rapidly oxidised, a reaction catalysed by aldehyde dehydrogenase requiring NAD as cofactor, (Lubin and Westerfield 1945). The acetyl coenzyme A which is produced in this reaction is then mainly oxidised to carbon dioxide and water in the tricarboxylic cycle although some is converted to fatty acids and cholesterol.

An individual's ability to metabolise alcohol is fairly constant, varying between about 50 and 100 mgm per kg b.w. per hour, (Thompson 1956). The only substance which has been shown to increase this in man is fructose, (Tygstrup, Winkler and Lundquist 1965).

There is no doubt that in some individuals excessive ethanol intake alone can lead to altered function and disease in many of the organs of the body. The exact way in which it does this is poorly understood. However, whatever the mechanism it is believed to be a direct result of ethanol toxicity. In other individuals malnutrition is also present and this may give rise to such conditions as Wernicke's encephalopathy, scurvy and beri-beri heart disease. Malnutrition may also potentiate the toxic effects of ethanol of the different organs. The purpose of this article is to describe some of the physical complications of alcohol excess per se.

METABOLIC COMPLICATIONS OF ALCOHOL EXCESS

Profound hypoglycaemia, hyperlipaemia and hyperuricaemia may occur as a direct consequence of the oxidation of excessive quantities of ethanol. Hypoglycaemia is believed to occur because gluconeogenesis is inhibited by the high hepatic levels of reduced nicotinamide adenine dinucleotide (NADH) which are produced in this situation, (Lieber 1966). The mechanism of the hyperlipaemia is not understood but it may be due to a combination of increased lipoprotein production in the liver and to decreased clearance in peripheral sites, (Lieber 1967). Finally hyperuricaemia is believed to be due to decreased renal clearance of uric acid secondary to increased blood lactate levels which are often present in individuals acutely intoxicated by ethanol, (Lieber, Jones, Losowsky, Davidson 1962).

PHYSICAL COMPLICATIONS OF ALCOHOL EXCESS

Effects of gastrointestinal tract

(a) Stomach Probably the most common physical complication of alcohol excess is acute gastritis which explains the high incidence of nausea, vomiting and anorexia found in alcoholics. There is no evidence, however, that this progresses to chronic gastritis or atrophic gastritis. The Mallory-Weiss syndrome of haematemesis from linear mucosal tears at the oesophago gastric junction following severe retching and vomiting and gastric ulceration are commoner in alcoholics than abstainers, (Mallory and Weiss 1929; Engeset, Lygren and Idsoe 1963), but duodenal ulceration appears to occur with the same frequency in both populations.

(b) Oesophagus Carcinoma of the oesophagus appears to be commoner in alcoholics than in the normal population, (Kamionkowski and Fleshler 1965) but the cause of this increased incidence is unknown. It may be related to disordered oesophageal mobility which has been described in alcoholics with peripheral neuropathy, (Winship, Calflish, Zboralske and Hogan 1968).

(c) Intestine Alcoholic patients frequently present with watery diarrhoea which may be due to impaired sodium and water reabsorption in the small intestine and colon. Malabsorption of xylose, (Mezey, Jouve Slavin and Tobon 1970), amino acids, (Israel, Valenzuela, Salazar and Ugarte 1969), folic acid, (Halsted, Griggs and Harris 1967) and vitamin B_{12} have also been described in alcoholics, (Lindenbaum and Lieber 1969a). However, the causes of the malabsorption have not been elucidated and the true incidence of these abnormalities is unknown.

(d) Liver In the United States of America it has been estimated that there are 4.5 million alcoholics, many of whom have liver disease. It is not surprising, therefore, that alcohol liver disease is the fourth most common cause of death in males in that country. Three main types of liver disease occur in these patients, namely fatty liver, alcoholic hepatitis and cirrhosis. At one time it was believed that these conditions were caused by malnutrition, particularly by the lack of lipotrophic factors, methionine and choline, (Himsworth and Glynn 1944), which is so often present in alcoholic patients. However, not all alcoholics are malnourished and it is well known that these patients also develop liver disease. Recently it has been shown that fatty liver, hepatitis and cirrhosis can be induced by a high alcohol intake in the baboon despite high protein and vitamin supplemented diets. Alcohol, therefore, appears to damage the liver by a direct toxic effect which may or may not be aggravated by malnutrition and vitamin deficiency, (Rubin and Lieber 1974). It is now believed that the role of malnutrition has been overemphasised, (Rubin and Lieber 1968). Epidemiological studies have indicated that alcoholic liver injury begins with an intake of more than 80G of ethanol per day, but that cirrhosis is not usually seen unless the intake is above 160G per day, (Lelbach 1974).

Fatty Liver Fatty liver is caused by an accumulation of fat, mainly in the form of triglyceride, within the hepatocytes. This is generally visible on light microscopy as large fatty vacuoles distending the hepatocytes and displacing the cellular organelles to a peripheral position. Even with gross accumulation of fat the degree of inflammation is usually small, (Scheuer 1973). The pathogenesis of this condition has not yet been fully elucidated. However, ethanol probably reduces fatty acid oxidation in the liver which presumably predisposes to accumulation of fat. Decreased lipoprotein release from the liver, which is present in carbon tetrachloride poisoning and choline deficiency, does not appear to occur, (Baraoan and Lieber 1970). Patients with fatty liver are usually asymptomatic and present because hepatomegaly or abnormal serum biochemistry has been discovered during the course of routine medical examination. Biochemical tests of liver function may show mild non specific changes such as slight increases in plasma bilirubin, alkaline phosphatase and aspartate transaminase concentrations. The serum protein and blood coagulation are usually normal. The prognosis is normally very good as the condition is completely reversible on discontinuation of alcohol abuse. Occasionally, however, jaundice and hepatic failure occur but this is uncommon, (Popper and Szanto 1957).

Alcoholic Hepatitis In some alcoholics fatty liver is complicated by alcoholic hepatitis, a condition characterised by fever, leucocytosis, upper abdominal pain and jaundice. The condition may be mistaken for obstruction to the common bile duct by stone, especially if the serum alkaline phosphatase activity is markedly elevated. However, in alcoholic hepatitis the serum aspartate transaminase activity is also often markedly increased and this together with a history of high ethanol intake and the clinical features outlined above enable a clinical diagnosis of ethanolic hepatitis to be made. Occasionally, however, liver biopsy and cholangiography have to be performed before a definitive diagnosis can be made. In most instances alcoholic hepatitis subsides when alcohol is withdrawn, but the mortality ranges from 10 to 30 per cent, the usual cause of death being liver failure. In the patients who survive, alcoholic hepatitis may predispose towards the development of cirrhosis.

Histologically, the liver in alcoholic hepatitis shows widespread polymorphonuclear leucocytic inflammation. Mallory's hyaline may be present and the portal tracts are usually enlarged with proliferated bile ductules and excess mononuclear cells, (Scheuer 1973). Fibrosis may vary from a mild interstitial fibrosis to a fully developed cirrhosis. Occasionally the fibrosis is maximal in the centrizonal areas of the liver lobule and may "compress" the hepatic vein radicles giving rise to portal hypertension in the absence of cirrhosis. This condition has been called sclerosing hyaline necrosis, (Reynolds, Hidemura, Michel and Peters 1967).

Cirrhosis Generally, this is the most serious hepatic complication of alcoholism. Occasionally it follows alcoholic hepatitis, but this is unusual so that most cases develop insidiously. Some patients with cirrhosis may present for the first time in liver failure whilst others present with gastrointestinal bleeding from oesophageal varices, hepatosplenomegaly or abnormal serum biochemistry found during the course of investigation or treatment of another disease.

Physical examination in such patients may reveal the presence of liver palms, spider naevi, bruising, gynaecomastia, parotid enlargement, Dupuytren's contractures and atrophied testes. Hepatosplenomegaly may be found on abdominal examination and ascites, flapping tremor, foetor hepaticus and depressed level of consciousness are usually seen if liver failure is present. The serum biochemistry may show increased levels of bilirubin, alkaline phosphatase and aspartate transaminase whereas the serum albumin concentration is often reduced. Blood coagulation is often defective. However, none of the abnormalities are diagnostic of cirrhosis. Definitive diagnosis is dependent on hepatic histology.

Histologically, in cirrhosis nodules of regenerated liver cells are surrounded by connective tissue septa. In some patients the edges of the nodules are indistinct as a result of piecemeal necrosis and associated chronic inflammatory infiltration whereas in others the nodules are sharply demarcated by connective tissue which contains little inflammatory infiltrate, (Scheuer 1973). Iron may be deposited in the liver parenchyma in such patients but the amount is rarely as great as that seen in primary haemochromatosis, (MacDonald and Mallory 1960). Another complication of alcoholic cirrhosis is malignant hepatoma which develops in about 15 per cent of patients, (Sherlock 1968). Ethanol is not believed to be carcinogenic and the hepatoma is thought to be a complication of abnormal liver cell turnover. The treatment of alcoholic cirrhosis is the treatment of the complications of the disease although all patients should be advised and encouraged to abstain from alcohol so that the rate of progression of the condition may be retarded.

The treatment of gastrointestinal bleeding from oesophageal varices depends on accurate endoscopic diagnosis of the bleeding site followed by blood transfusion, intravenous pitressin and, if necessary, balloon tamponade by means of the Sengstaken tube. Intra-arterial infusion of pitressin into the coeliac or superior mesenteric arteries may also be used. Failure of "conservative" treatment is an indication for emergency surgery which includes injection of varices with sclerosants, ligation of varices, transection procedures and emergency portal-systemic decompression procedures. In most centres the mortality is so high following portal-systemic shunting that this is not now used in the emergency treatment of bleeding varices. In patients in whom bleeding stops with conservative measures elective portal-systemic anastomosis must be considered to prevent further episodes of variceal bleeding. The operation which has been used most often for this purpose is the end to side portal-caval shunt. However, because of the high operative mortality and high incidence of portal-systemic encephalopathy postoperatively even in specialist units,(Hourigan, Sherlock, George and Mendel 1971), newer operations such as the distal splenorenal shunt have been developed and are now being evaluated, (Warren, Fomon and Zeppa 1969). Generally, bleeding from oesophageal varices carries a very high mortality and is one of the commonest causes of death in the alcoholic with cirrhosis. For this reason prophylactic portal-systemic shunting was at one time advocated. However, recent controlled trials have shown that the overall mortality of patients subjected to surgery is greater not less than in unoperated controls,(Conn, Lindenmuth, May and Ramsby 1972).

Ascites in patients with alcoholic cirrhosis indicates a degree of liver cell failure but its cause is not completely understood. Most patients have portal hypertension, hypoalbuminaemia and renal salt and water retention. There is also some evidence that liver lymph production is excessive and this may contribute towards the development of ascites in some patients,(Zimmon, Oratz, Kessler, Schreiber and Rothschild 1969). It is often difficult to evaluate which factors predominate in one individual but the treatment is directed at removing salt and water from the body. This is done by using diuretics combined with a low sodium (22 meq.) diet and fluid restriction to about one litre a day. Many patients respond satisfactorily to this treatment, but in others many different diuretics and combinations thereof have to be tried before a suitable therapeutic regimen can be developed. It is reasonable to give these patients a high protein diet, if they can be tolerated without the development of encephalopathy, for this may increase the serum albumin sufficiently to make ascites easier to control. Moreover, it would appear logical to reduce the degree of portal hypertension as this is believed to be a major cause of ascites. This can be achieved by means of a side to side portal-caval anastomosis. However, as the operation is usually followed by severe portal systemic encephalopathy in this situation it is rarely performed in clinical practice for the control of ascites.

Portal systemic encephalopathy is often present in patients with liver failure and following portal-systemic anastomosis it consists of reduced level of consciousness together with foetor hepaticus and a flapping tremor of the patient's outstretched hands. The cause of the condition is unknown but it may be related to altered carbohydrate metabolism in the brain secondary to accmumulation of ammonia and amino-acids in the central nervous system, (Zieve 1966). These compounds are normally produced in the intestines by the effect of bacterial enzymes on urea and dietary protein, but in the presence of liver disease they are poorly detoxicated so that they accmumulate in the body. The treatment of the condition consists of a high carbohydrate low protein diet combined with purgatives, neomycin and lactulose. The neomycin and lactulose eliminate ammonia producing organisms from the colon and lactulose also reduces ammonia absorption from this organ.

The prognosis in patients with alcoholic liver disease is poor but improves if alcohol is avoided. The five year survivial rate in those who continue to drink is about 34% but 69% in those who abstain, (Brunt, Kew, Scheuer & Sherlock 1974). Unfortunately, hepatoma develops in approximately 15 per cent of patients with alcoholic cirrhosis and these patients are often those who have stopped drinking and who survive for long periods.

Pancreatitis Another common condition seen in alcoholics is chronic pancreatitis and in the United States of America and France alcoholism is the commonest cause of this condition. In Britain however, it seems to be less common, (Dreiling, Janowitz and Perrier 1964).

The pathogenesis of the condition is not clear but it may be due to a direct toxic effect of ethanol on the pancreas. It has been shown that following chronic ethanol ingestion ultrastructural changes develop in acinar mitochondria and fat droplets may be seen Orrego-Matte, Navia, Feres and Costamaillere 1969). Moreover, Sarles, Lebreuil, Tasso, Figarella, Clemente, Devaux, Fagonde and Payan (1971) have produced chronic pancreatitis in rats following 20 to 30 months of ethanol administration and this may be potentiated by a protein deficient diet which is known to cause chronic pancreatitis per se, (Shaper 1965). Another theory is that chronic pancreatitis in the alcoholic is a result of increased pancreatic secretion in association with increased tone of the sphincter of Oddi leading to pancreatic damage. Walton, Shapiro, Yeung and Woodward (1965) have shown that pancreatic duct pressure rises in dogs given ethanol but this, in the acute experiment, does not lead to pancreatitis.

Alcoholic pancreatitis usually occurs after many years of alcoholism. The patients usually present with recurrent attacks of upper abdominal pain of increasing severity which may be associated with weight loss, steatorrhoea and carbohydrate intolerance. Radiological investigation of these patients may show pancreatic calcification and pancreatic secretion tests using secretin and pancreozymin usually reveal that the volume of juice secreted is reduced and that its bicarbonate concentration is low (Wormsley 1969). The treatment of chronic pancreatitis is unsatisfactory. Pancreatic extracts and atropine-like drugs are given in an attempt to control pain but they are rarely satisfactory. Pancreatic insufficiency is usually treated with pancreatic extracts, low fat diets often supplemented with medium chain triglycerides and insulin if diabetes is present.

Haemopoietic System

Haematological abnormalities are frequently encountered in chronic alcoholics and perhaps the commonest abnormality is vacuolation of white and red cell precursors in the bone marrow, (Waters, Morley and Rankin 1966). Iron deficiency, leucopenia, thrombocytopaenia, megaloblastic anaemia and sideroplastic anaemia are also seen. Some of these abnormalities may be due to deficient dietary intake of haematinics or they may be a consequence of alcoholic liver disease. However, there is also good evidence that ethanol is directly toxic to the marrow (Lindenbaum and Lieber 1969b) and that it depresses both the absorption and utilisation of haematinics, (Halsted, Griggs and Harris 1967; Lindenbaum and Lieber 1969b; Sullivan and Herbert 1964).

Nervous System

The neurological syndromes associated with alcoholism may be of four types. Firstly, is the syndrome of alcohol excess leading to acute intoxication and associated disturbance of balance, gait, speech and level of consciousness. Secondly, is the syndrome caused by withdrawal of alcohol in an alcohol dependent subject. The commonest manifestation of this condition is the "morning shakes" whereas the full-blown syndrome consists of an acute confusional state - delirium tremens. The third group of conditions seen in alcoholic patients are those which result from an associated nutritional deficiency. Wernicke's encephalopathy and the Korsakov psychosis are examples of this type of condition. Finally, the fourth group of abnormalities affecting the central nervous system in the alcoholic are of unknown aetiology but possibly caused by a direct toxic effect of ethanol on the central nervous system. The commonest abnormality of this type is peripheral neuropathy usually the result of primary axonal degeneration. Primary cerebellar degeneration, degerneration of the corpus callosum, central pontine myelinolysis, epilepsy and myopathy are also well recognised complications of chronic alcoholism, (Victor 1962).

Cardiovascular System

Alcoholics may present with four types of heart disease. Firstly, they may suffer from beri-beri heart disease due to thiamine deficiency which improves rapidly when the vitamin deficiency is corrected. The second type of problem is one of recurrent arrhythmias (Brigden and Robinson 1964). The third condition is alcoholic cardiomyopathy giving rise to heart failure, cardiomegaly and electrocardiographic evidence of severe myocardial damage. In this situation patients respond poorly to digitalis and diuretics and gradually deteriorate, although the rate of progression of the disease may be retarded by prolonged bed rest, (Burch and DePasquale 1969). Finally, the fourth type of alcoholic heart disease which has been described in alcoholics is that caused by cobalt toxicity. Most reports of this condition have come from Canada where cobalt was added to beer to improve the quality of its foam,(Sullivan, Egan and George 1969). This practice has since been discontinued.

From the preceding account it can be seen that chronic alcoholism may be associated with disordered function and disease in most systems of the body. Some of the abnormalities such as acute gastritis, peripheral neuropathy and cirrhosis are very common, whereas conditions such as central pontine myelinolysis and degeneration of the corpus callosum are rare. For this reason the psychiatrists and social workers who often treat alcoholics must be aware of these physical complications of alcoholism. Moreover, the general physician confronted by a patient with neuropathy, diarrhoea or anaemia of unknown aetiology must say to himself - could it be due to excessive alcohol consumption ?

REFERENCES

BARAONA, E. and LIEBER, C.S. (1970). Journal of Clinical Investigation, 49, 769.

BRIGDEN, W. and ROBINSON, J. (1964). British Medical Journal, 2, 1283.

BRUNT, P.W., KEW, M.C., SCHEUER, P.J. and SHERLOCK, S. (1974). Gut, 15,52.

BURCH, G.E. and DePASQUALE, N.P. (1969). American Journal of Cardiology, 23, 723.

CONN, H.O., LINDENMUTH, W.W., MAY, C.J. and RAMSBY, G.R. (1972). Medicine (Baltimore), 51, 27.

COOKE, A.R. and BIRCHALL, A. (1969). Gastroenterology, 57, 269.

DREILING, D.A., JANOWITZ, H.D. and PERRIER, C.V. (1964). "Pancreatic Inflammatory Disease", Hoeber, New York.

ENGLESET, A., LYGREN, T. and IDSOE, R. (1963). Quarterly Journal Studies on Alcohol, 24, 622.

HALSTED, C.H., GRIGGS, R.C. and HARRIS, J.W. (1967). Journal of Laboratory and Clinical Medicine, 69, 116.

HARTCROFT, W.S. (1967). Federation Proceedings, 26, 1432.

HIMSWORTH, H.P. and GLYNN, L.E. (1944). Clinical Science, 5, 93.

HINES, J.D. (1969). British Journal of Haematology, 16, 87.

HOURIGAN, K., SHERLOCK, S., GEORGE, P. and MENDEL, S. (1971). British Medical Journal, 4, 473.

ISRAEL, U., VALENZUELA, J.E., SALAZAR, I. and UGARTE, G. (1969). Journal of Nutrition, 98, 222.

KAMIONKOWSKI, M.D. and FLESHLER, B. (1965). American Journal of Medical Science, 249, 696.

KREBS, H.A. and PERKINS, J.R. (1970). Biochemical Journal, 118, 635.

LELBACH, W.K. (1974). "Biochemical, Epidemiological and Clinical Aspects of the Etiology of Alcoholic Liver Pathology" Israel, U. (ed.).

LIEBER, C.S., JONES D.P., LOSOWSKY, M.S. and DAVIDSON, C.S. (1962). Journal of Clinical Investigation 41, 1863.

LIEBER, C.S. (1966). Gastroenterology, 50, 119.

LIEBER, C.S. (1967). Annual Review of Medicine, 8, 35.

LIEBER, C.S. and De CARLI, L.M. (1970). Journal of Biological Chemistry, 245, 2505.

LINDENBAUM, J. and LIEBER, C.S. (1969a). New England Journal of Medicine, 281, 333.

LINDENBAUM, J. and LIEBER, C.S. (1969b). Nature, 224, 806.

LUBIN, M. and WESTERFIELD, W.W. (1945). Journal of Biological Chemistry, 161, 503.

MACDONALD, R.A. and MALLORY, G.K. (1960). Archives of Internal Medicine, 105, 686.

MALLORY, G.K. and WEISS, S. (1929). American Journal of Medical Science, 178, 506.

MEZEY, E., JOWE, E., SLAVIN R.E. and TOBON, F. (1970). Gastroenterology, 59, 657.

ORREGO-MATTE, H., NAVIA, E., FERES, A. and COSTAMAILERE, L. (1969). Gastroenterology, 56, 280.

POPPER, H. and SZANTO, T.B. (1957). Journal of the Mount Sinai Hospital, 24, 1121.

REYNOLDS, T.B., HIDEMURA, R., MICHEL, H. and PETERS, R. (1967). Annals of Internal Medicine, 70, 497.

RUBIN, E. and LIEBER, C.S. (1968). American Journal of Medicine 45, 1.

RUBIN, E. and LIEBER, C.S. (1974a). New England Journal of Medicine, 290, 128.

RUBIN, E. AND LIEBER, C.S. (1974b). Clinics in Gastroenterology, 4, 247.

SARLES, H., LEBREUIL, G., TASSO, F., FIGARELLA, C., CLEMENTE, F., DEVAUX, M.A., FAGONDE, B. and PAYAN, H. (1971). Gut, 12, 377.

SCHEUER, P.J. (1973). Liver Biopsy Interpretation, (2nd edition). Williams & Wilkins, Baltimore.

SHAPER, A.G. (1965). British Medical Journal, 1, 1607.

SHERLOCK, S. (1975). Diseases of the Liver and Biliary System, (5th edition). Blackwell, London.

SULLIVAN, L.W. and HERBERT, V. (1964). Journal of Clinical Investigation, 43, 2048.

SULLIVAN, J.F., EGAN, J.D. and GEORGE, R.P. (1969). Annals of the New York Academy of Science, 156, 526.

THOMPSON, G.N. (1956). Alcoholism, Thomas, Springfield, Illinois.

TYGSTRUP, N., WINKLER, K. and LUNDQUIST, F. (1965). Journal of Clinical Investigation, 44, 817.

VICTOR, M. (1962). Clinical Neurology, A.B. Baker (ed.) 2, Hoeber and Harper.

WALTON, B.E., SHAPIRO, H., YEUNG, T. and WOODWARD, E.R. (1965). Americal Journal of Surgery, 31, 142.

WARREN, W.D., FOMON, J.J. and ZEPPA, R. (1969). Annals of Surgery 169, 652.

WARTBURG, J.P. von, BETHUNE, J.L. and VALLEE, B.L. (1964). Biochemistry, 3, 1775.

WATERS, A.H., MORLEY, A.A. and RANKIN, J.G. (1966). British Medical Journal, 2, 1565.

WINSHIP, D., CALFLISH, C.R. ZBORALSKE, F.F. and HOGAN, W.J. (1968). Gastroenterology, 55, 173.

WORMSLEY, K.G. (1969). Scandinavian Journal of Gastroenterology, 4, 623.

ZIEVE, L. (1966). Archives of Internal Medicine, 118, 211.

ZIMMON, D.S., ORATZ, M., KESSLER, R., SCHREIBER, S.S. and ROTHSCHILD, M.A. (1969). Journal of Clinical Investigation, 48, 2074.

IS ILLEGAL DRUGTAKING A PROBLEM ?

MARTIN A. PLANT

MEDICAL RESEARCH COUNCIL UNIT FOR EPIDEMIOLOGICAL
STUDIES IN PSYCHIATRY
UNIVERSITY DEPARTMENT OF PSYCHIATRY, EDINBURGH

DEFINING "ABUSE"

At the Second International Conference on Alcoholism & Drug Dependence, Liverpool, somebody said that an alcoholic is a person who drinks more than his doctor.

In practical terms that is probably as good a definition as any. Definitions of drug use or abuse are just as relative as that. A drug <u>user</u> gets high with whatever drug you use. An <u>abuser</u> gets high with something different. Most societies favour some form of psychotropic substance which is regarded as acceptable and as useful. Usually the drug they choose is simply whatever is available. In South America coca is abundant, in Mexico peyote, in Britain alcohol. Definitions of use and abuse are partly traditional, partly, and only partly, scientific. Some drugs are viewed as good in some societies yet are regarded as dangerous elsewhere. Often both views are unrelated to the chemical properties or effects of the drug in question.

In Britain it is estimated that there are 350,000 alcohol addicts, yet alcohol is our society's "chosen, acceptable drug", which is woven inseparably into the fabric of our daily lives. In comparison with the enormity of the damage caused by alcohol abuse, the "drug problem" pales into insignificance, and yet the use and abuse of illegal drugs is still sometimes coupled with alcoholism as a major social problem. In some countries it clearly is. Here it is intended to be parochial and to consider the situation in Britain.

Illegal drugtaking by young people is no longer a novelty. Drugs like cannabis (marihuana), L.S.D. and heroin have been around now for quite a while. Now that much of the fuss has died down, in this paper the evidence will be reviewed and some of the issues that still need to be faced, discussed.

When, during the 1960's and 70's non-medical drugtaking gained popularity amongst young people there was a very real concern that an unbridled epidemic of addiction was beginning. Our fears in Britain were partly founded upon the emergence of the huge narcotics problem in the U.S.A. To some extent the wave of public horror about drugtaking was due to a widespread equation of drug use with drug addiction. Even today there remains widespread ignorance about drugs, and there exists a general failure to distinguish either between drugs or between those who use them.

Public opinion about illegal drugtaking is both reflected and influenced by the mass media which frequently over-simplify issues and generalize from the least favourable examples of illegal drug use. If one reads back through newspaper and other popular coverage of drug use one finds a general assumption that individuals who use illegal drugs do so because they are psychologically disturbed, or that such drug use is intrinsically harmful or likely to develop into addiction.

The impression one forms of illegal drug use largely depends upon which group of drugtakers one considers. The exact prevalence of drug use in the community is unknown, and suggested rates of the use of specific drugs are at best inspired guesses. Epidemiologists therefore, face considerable problems in attempting to assess the extent of drug use. Any kind of illegal behaviour is difficult to investigate thoroughly, particularly if those who infringe the law risk severe penalties if detected, or if their behaviour incurs the great social stigma attached to both illegal drug use and to alcoholism.

The popular impression of illegal drugtaking as symptomatic of disturbance and necessarily harmful is probably due to the fact that much of the available information relates to drugtakers known to "official agencies". These include people convicted of drug offences and those who have received medical treatment in connection with the abuse of drugs. These are the casualties of the drug scene, and are by no stretch of the imagination a representative group of those in the community who use drugs, but who generally pass unnoticed.

It would be as misleading to assume that all drugtakers resemble either the convicted offenders or those needing medical aid as it would be to assume that hospitalized alcoholics typify all those who sometimes take a drink. Just as one may trace a broad continuum between moderate drinking and chronic alcoholism, so there is a great gulf between the extremes of the experimental or controlled use of illegal drugs and drug addiction involving physical drug dependence.

One very real problem is that public attitudes to both alcoholism and drug dependence are dominated by over-simplified stereotypes. Many people sincerely believe that to be an alcoholic, a person must drink at least spirits if not meths or Brasso, and must resemble the social derelict on Skid Row. In fact, few alcoholics are on Skid Row. Many are young, highly successful and attractive. In fact many are the exact opposite of the popularly accepted stereotype.

Both the law and public opinion relating to illegal drug use are similarly bedevilled by an equally potent stereotype, that of the junkie. The junkie is portrayed as a youthful counterpart of the Skid Row wino, as physically and emotionally dependent upon regular supplies of a drug such as heroin sold him by the real villain of the piece, the pusher. This is the commercially motivated peddler who is widely blamed as the source of supply and likely reason for any young person starting to use drugs in the first place.

To get a broader view of drug use as it really is in the community one needs to piece together a jigsaw. There have been hundreds of studies on the social aspects of drugtaking which now enable an impression to be formed of the nature and approximate extent of different types of drug use. There is no lack of information.

Apart from numerous clinical studies, (Ritson et al 1973, Stimson 1973, Woodside 1973, Plant 1976, Plant and Ritson 1976), and studies of convicted offenders, (Noble 1970 and 1971, Cockett 1971, Mott and Taylor 1974, Bean 1974), there have been many sample interview surveys of drugtaking amongst young adults in various social settings, (Binnie and Murdock 1969, Hindmarch 1970, Kosviner et al 1973, Mackay et al 1973, Hindmarch et al. 1975). Most drug surveys have been confined to schools and universities. Even so, they provide a very good idea of drugtaking amongst young people most of whom have neither been convicted of drug offences nor required medical help for drug abuse. There have been participant observation studies of small numbers of drug users in their natural settings, (Reeves 1973, Plant 1975), and useful autobiographical accounts by drug users, (Trocchi 1966, Crowley 1972, Huxley 1972). In addition there are several highly perceptive general books such as Andrew Weil's (1975) classic work The Natural Mind, see also, (Goode 1970, Johnston 1973, Wells 1973, Judson 1974, Einstein 1975, Inglis 1975).

From this evidence, it is apparent that far more people indulge in illegal drug use than ever attract the attentions of official agencies. Most surveys have shown that at least a minority of young adults in secondary schools, colleges and universities have experimented with cannabis, L.S.D. and other proscribed drugs. It is evident that most of those who use such substances do not conform to the unattractive stereotype of the junkie any more than most alcohol users resemble Skid Row alcoholics.

PREVALENCE

Home Office figures show that during the past two decades there has been a meteoric increase in the numbers of people convicted annually of drug offences, mainly related to cannabis. Between 1958 and 1972 there were 57,963 cannabis convictions. These figures do not necessarily represent the growth of illegal drug-taking in the community. They are influenced by the level of police activity in enforcing the drug control laws, and by the concomitant formation during this period of specialist police drug squads in many towns and cities. During the same period Home Office figures show that there was a steep rise in the number of registered addicts until 1969. Since then numbers of known addicts have remained relatively stable, at around 3,000.

TABLE I: TRENDS IN DRUG ABUSE - GREAT BRITAIN *

		Registered Addicts	
Year	Cannabis Convictions	Heroin	Methadone
1958	99	62	12
1959	185	68	60
1960	235	94	68
1961	288	132	59
1962	588	175	54
1963	663	237	55
1964	544	342	61
1965	626	521	72
1966	1119	899	156
1967	2393	1299	243
1968	3071	2240	486
1969	4683	1417	1687
1970	7520	914	1820
1971	9219	959	1927
1972	12611	868	2171
1973	14119	866	2247

* Derived from Home Office Press notes relating to Drug Addiction and Drug Offences.

Most surveys of student populations at British colleges and universities have indicated that 5-30% of those interviewed have used cannabis and other drugs at least experimentally. Prevalence rates vary and none can be relied upon. A large scale survey, (Treisman, unpublished, 1973) suggested that approximately 3.8 million people in Britain have used cannabis. Whatever the real number of drug users in the population is, there can be little doubt that it greatly exceeds the "tip of the iceberg" represented by Home Office figures. A number of studies, (Plant & Reeves 1973), have shown that even amongst those regularly injecting heroin and other drugs, there are many who pass unrecorded by the Home Office. So the numbers of those using drugs is open to conjecture. In spite of this, many studies have produced consistent information about the types of people who use illegal drugs, and about their patterns of use.

WHO USES DRUGS ?

There is broad agreement from current surveys that people who report having used illegal drugs differ in the following respects from people who report not having used them;

Age: Many studies indicate that most illegal drugtakers are young, either teenagers or in their twenties or early thirties.

Sex: Males are more likely than females to experiment with drugs. This is consistent with male over-representation in relation to many other forms of illegal behaviour.

Alcohol and Tobacco Use: Those who report using illegal drugs are more likely than others to report that they also use alcohol and tobacco. In general the young people who experiment with proscribed drugs are recruited from amongst those who also use legal drugs.

Separation from Parents: Drugtakers are more likely than others to be living apart from their parents, in flats, bedsitters or other accommodation of their own. In such circumstances, young people have more opportunity to lead their own lives and to indulge in behaviour of which their parents would disapprove, including illegal drug use.

Sexual Experience: Consistent with the greater freedom of young people living apart from their parents, drugtakers are more likely than others to report that they are sexually experienced.

Education: Drugtakers are more likely than non-drugtakers to report that they have stayed on at school after the minimum leaving age, or that they have had some form of higher or further education. This is consistent with the popular view that drugtaking is especially commonplace amongst college and university students.

Social Class: Drugtakers are more likely than non-drugtakers to come from middle-class homes; that is, to have fathers who are non-manual workers.

Ideology: As discussed by Dr. Jock Young (1971), drugtakers are especially likely to state that they regard themselves as politically radical rather than conservative, and that they have religious views which differ from their parents' traditional beliefs. Many drugtakers identify themselves as agnostic, atheist or as having adopted Oriental forms of belief such as Bhuddism, or practicing some form of meditation. Many belong to recently formed sects such as the Divine Light Mission or to some of the so-called "Jesus Freak" groups originating in the U.S.A. In summary, drugtakers are more likely than others to report that their ideology, whether political or religious, is opposed to the established beliefs and organisations of their society.

WHY DO PEOPLE USE DRUGS ?

As already observed, it is a fairly popular belief that many young people are "seduced" into drugtaking by so-called pushers. In fact the majority of those who do experiment with illegal drugs begin in much the same way as most young people begin to drink alcohol or to smoke cigarettes.

Davies and Stacey (1972), investigated the use of alcohol by Glasgow adolescents aged 14 - 17. They found that most of the young people in their study began to drink because they were encouraged to do so by friends of the same age. The described their findings in these words;

" ..... young people drink, and possibly smoke too, because they believe that alcohol and tobacco confer upon the user certain qualities, which include 'toughness', 'maturity', 'sociability' and 'attractiveness' to the opposite sex, and may for some individuals be symbolic of their rejection of adult authority and/or control ..... A decision about drinking or smoking is therefore likely to be mediated by considerations of how favourably one will be perceived by one's peers."

In just the same way, the great majority of young people who try illegal drugs do so largely because of strong peer group pressure and because they are curious about the effects of drugs. Most people begin to use drugs only after a period of encouragement to do so by friends or others whom they like or regard as exemplars. The first step is contact with people who have already used drugs who assure them that cannabis, or whatever, is pleasant and harmless. Initial doubts are allayed, and the individual is reassured and convinced that any initial fears or reservations are unfounded. Most people do not begin to use drugs until they accept that it is safe to do so. In addition, drug use invariably occurs when it is clearly defined as prestigious; when a person's friends make it clear that this behaviour is approved of and desired.

Reports from drugtakers invariably emphasise the strong social pressure to use drugs. Usually drug use is prompted not by strangers, but by friends, and is seen by those who indulge in it as a token of intimacy and a bond.

Most of those who use illegal drugs do so in concert with others, and regard their drugtaking as a facet of their leisure, their friendship and beliefs; "I wanted to be part of the scene. Smoking shows where you're at". Whether or not young people do use drugs largely depends upon those with whom they associate. Like under-age smoking and drinking, illicit drug use carries great prestige in some circles, and is sometimes seen as a potent way of rebelling. Most young people drift into drugtaking because of their friendships at a given time. Some get more encouragement to try drugs than others. Just as people drift into drug use, so many drift away again, either because they lose interest, or because they find new friends or fall in love.

This is all very different from being lured into drug use and then into addiction by the pusher. Clearly some people do import and traffic in large amounts of illegal drugs. The chains of supply between the big dealer and the majority of users are very long. Most users buy and sell small amounts to friends, and very few drugtakers would risk buying drugs from people they do not know well. In addition, most sellers also use drugs themselves and handle only small amounts; so the pusher is a very remote figure to most young people who take drugs. Certainly availability is an important consideration, but people who still see the drug scene as primarily a dialogue between user and pusher miss the whole point. It is often hard to distinguish between the two, and the drug scene is essentially a network of friends, not business associates.

PATTERNS OF USE

Most British studies indicate that cannabis is the most commonly used of the illegal drugs. In addition, a perplexing array of other substances have been tried by young people in search of some form of "buzz" or pleasant experience. These include amphetamines, L.S.D. and other hallucinogens, barbiturates, methaqualone, opiates, cocaine, cough linctuses, tranquillizers, sedatives, herbs, inhalers, even glues and solvents, (Watson 1975). There is almost no limit to the things people will use in order to attain a changed form of consciousness. Many young people experiment with a range of drugs, depending upon which are available and recommended by their friends. Most studies show that poly drug use - the use of several different types of drug - is commonplace.

In most cases a user will experiment for a while, normally with cannabis, then with a few other drugs. Thereafter interest generally wanes and most people either cease their drug use or continue to smoke cannabis on occasion, invariably with friends. There is a great deal of evidence that only a small number of those who use illegal drugs ever inject drugs or become dependent on them. Most surveys have shown that the great majority of those who use drugs such as cannabis are hostile to the use of opiates and other injected drugs, and it is clear that amongst most groups of drug-takers, certainly amongst most students and non-manual workers there is little support for such practices. There is no inevitable escalation from cannabis to heroin addiction any more than there is from beverage alcohol to meths drinking.

WHO HAS PROBLEMS ?

So far, a rosy picture has been painted. Probably some of you will feel the problems have been ignored. Clearly, there are problems connected with illegal drug use. Any psychotropic drug, legal or illegal, can be abused and most of them are.

For the majority of users cannabis and other drugs provide an enjoyable facet of their leisure. Most young people are not unduly preoccupied with drugs, since they have other more pressing concerns, and tend to be at most "weekend hippies".

Those who do become very involved with the lifestyle of the drug scene are usually those who lack many alternative interests; those who have least stake in the workaday world of jobs, education and families; those who have the greatest need to create for themselves some form of social status and prestige. Most of those who become drug casualties, either as offenders or patients, are very different from the majority of those who use drugs. Most of these casualties are multi-problem individuals, commonly from social class V backgrounds and from disturbed families. Many are unusual in other respects. Often they abuse alcohol as well as drugs, are unemployed through choice, homeless and had disturbed biographies prior to their drug abuse. In general a person's drugtaking reflects his or her overall wellbeing. Those whose drug-taking is extreme and causes problems are invariably disturbed in many other ways. The drug scene provides a lifestyle which is very free and easy, which is a haven for the eccentric, unhappy or disturbed young person who cannot or will not conform to the demands of the mainstream of society. Within the hard core of the drug scene disaffected and deprived young people create their own status system partly based upon drug use, so that the person who uses the widest range of drugs has the greatest prestige. Even so, physical dependence upon a single drug is rare. Most of those needing medical aid do so because of an accident; an overdose, bad trip or a social crisis (such as being "busted"), not because they are addicted to any one drug. Dependence, when it occurs, is more often upon the lifestyle of being a drugtaker than upon a particular drug.

So, while most who use illegal drugs do not approach the medical services or other agencies for help, a minority of individuals do. A few are genuinely drug dependent, and some cause themselves physical damage through drug abuse. Such individuals do present real medical and psychiatric difficulties which will be well known to anyone who has worked in a treatment centre or similar agency. Young drug abusers are not easy to help. Many are hostile to medicine and social work, and most have few ties outside the drug scene. Such individuals are unlikely to relinquish their allegiance to drugs until an equally attractive alternative becomes available. Few of those working in treatment agencies can conjure up new lifestyles for their patients, and so drug abusers are often found to be unrewarding patients, many of whom regard the clinic not so much as a way out of drugtaking, but as a source of supply.

IS THE BRITISH SYSTEM DEAD ?

In 1933 the Rolleston Committee established the so-called "British System" of managing drug addiction as a medical issue. At that time there were relatively few addicts, and drug abuse was not viewed as a social threat. Since the Second World War things have changed. We have built up an elaborate system of drug controls covering a wide range of behaviour. Today far more people are convicted of drug offences than are recorded as drug dependent. The "British System" as originally conceived, is dead.

Most of those convicted of drug offences are not drug addicts, neither are most of them large scale drug traffickers. On the contrary, the great majority are convicted of possessing small amounts of cannabis. The author believes that present drug laws were introduced in order to ward off an epidemic which never really arrived. A large number of people are penalised who are not really a problem. The penalties attached to many trivial drug offences are out of all proportion with the offences concerned. A sledge hammer is being used to crack a nut.

A society may realistically adopt either of two approaches to drug use and drug abuse. Firstly, it may decide to enforce prohibition. This has been applied in the past, and has generally caused as many problems as it has solved. Secondly, it may accept that some form of drug use and abuse will occur and make provisions to help the minority of people who abuse drugs and who need help. This second alternative is in keeping with the spirit of the original Rolleston recommendations. It is also roughly what is done in relation to alcohol. Some drugtakers need help. Most do not. It is time that Britain reviewed the situation and altered course.

Presumably the drug control laws were introduced to minimise human misery. They do not achieve this objective. On the contrary, they add to human misery. Would it not be more constructive to help the minority of drugtakers who need it than to waste police time and public money, and to alienate thousands of young citizens by continuing with a witch hunt ?

REFERENCES

BEAN, P. (1974). The Social Control of Drugs, Martin Robertson, London.

BINNIE, H.L. and MURDOCK, G. (1969). "Attitudes to Drugs and Drugtakers of Students at the University and Colleges of Higher Education in an English Midland City". Vaughan Papers, 14, University of Leicester.

COCKETT, B. (1971). Drug Abuse and Personality in Young Offenders, Butterworths, London.

CROWLEY, A. (1972). Diary of a Drug Fiend, Sphere, London.

DAVIES, J. and STACEY, B. (1972). Teenagers and Alcohol : A Development Study in Glasgow, Vol. II, H.M.S.O.

EINSTEIN, S. (1975). Beyond Drugs, Pergamon, London.

GOODE, E. (1970). The Marijuana Smokers, Basic Books, New York.

HINDMARCH, I. (1970). "Patterns of Drug Use in a Provincial University". British Journal of Addiction, 64, 395-403.

HINDMARCH, I., HUGHES, I. and EINSTEIN, R. (1975) "Attitudes to Drug Users and to the Use of Alcohol, Tobacco and Cannabis on the Campus of a Provincial University". Bulletin of Narcotics, XXVII, 1, 27-36.

HUXLEY, A. (1972). The Doors of Perception and Heaven and Hell, Harmondsworth, Penguin.

INGLIS, B. (1975) The Forbidden Game: A Social History of Drugs, Hodder and Stoughton, London.

JOHNSTON, B.D. (1973). Marihuana Users and Drug Subcultures, John Wiley, London.

JUDSON, H.F. (1974). Heroin Addiction in Britain, Harcout Brace Jovanovich, London.

KOSVINER, A., HAWKS, D., and WEBB, M.G.T. (1973). "Cannabis Use Amongst British University Students". British Journal of Addiction, 69, 35-60.

MACKAY, A.J., HAWTHORN, V.M., and McCARTNEY, H.N. (1973). "Drug Taking Among Medical Students in Glasgow". British Medical Journal, 1, 540-543.

MOTT, J. and TAYLOR, M. (1974) "Delinquency Amongst Opiate Users". Home Office Research Unit Report, H.M.S.O., London.

NOBLE, P.J. (1970). "Drugtaking in Delinquent Boys". British Medical Journal, 1, 102-105.

NOBLE, P.J. (1971). "Drugtaking in Adolescent Girls: Factors Associated with Progression to Narcotics Use". British Medical Journal, 1, 620-623.

PLANT, M.A. and REEVES, C.E. (1973). "Social Characteristics of Drugtakers in Two English Urban Areas". Drugs and Society, 2, 11, 14-18.

PLANT, M.A. (1975) Drugtakers in an English Town, Tavistock, London.

PLANT, M.A. and RITSON, E.B. (1976). "The Scottish Drug Scene". Health Bulletin, (Scotland), 18, 1, 12-15.

PLANT, M.A. (1976) "Young Drug and Alcohol Casualties Compared: Review of 100 Patients at a Scottish Psychiatric Hospital". British Journal of Addiction, 71, 13-20.

REEVES, C.E. (1973). "Sociological Aspects of Drug Taking in Southern Hampshire". Paper presented at Second International Conference on Alcoholism and Drug Dependence, Liverpool

RITSON, E.B., TOLLER, P. and HARDING, F. (1973). "Drug Abuse in the East Midlands: A Study of 139 Patients Referred to an Addiction Unit". British Journal of Addiction, 68, 209-214.

STIMSON, G.V. (1973). Heroin and Behaviour, Irish University Press, London.

TROCCHI, A. (1966). Cain's Book, Calder, London.

WATSON, J.M. (1975) "A Study of Solvent Sniffing in Lanarkshire 1973/74". Health Bulletin, XXXIII, 1-3.

WELLS, B. (1973). Psychedelic Drugs, Harmondsworth, Penguin.

WEIL, A. (1975). The Natural Mind, Harmondsworth, Penguin.

WOODSIDE, M. (1973). "The First 100 Referrals to a Scottish Drug Treatment Centre". British Journal of Addiction, 68, 231-241.

YOUNG, J. (1971). The Drugtakers, Paladin, London.

ALCOHOLISM AND DRUG DEPENDENCE - A MULTIDISCIPLINARY PROBLEM: THE SOCIOLOGIST'S VIEWPOINT

DAVID ROBINSON

ADDICTION RESEARCH UNIT

INSTITUTE OF PSYCHIATRY, UNIVERSITY OF LONDON

Whenever alcoholism or drug-dependence are being discussed it can confidently be expected that someone will call them "multidisciplinary" problems. David Pittman (1967), for example, speaks of : ".....a multidisciplinary problem which involves researchers and practitioners from the biological and natural sciences, the psychological sciences and the social sciences", while Catanzaro (1968) casts his net even wider : "Professionals from a wide variety of disciplines must be involved" he says, "including physicians, social workers, psychiatrists, psychologists, nurses, clergymen, vocational rehabilitation counsellors, lay counsellors, judges, educators, sociologists and many others". Significantly, perhaps, sociologists come immediately before the unspecified "many others" at the end of Catanzaro's list. Significantly, because people who write about alcoholism and drug dependence often include sociologists in their list of people who might be "involved" and yet rarely give any indication of what sociologists actually do, or why what they do might be of interest, much less of use.

Outlined are some of the questions and areas of enquiry with which the sociologist might be particularly interested in relation to alcoholism and drug dependence. Most of the issues will be discussed in much greater detail, of course, during other phases of the Conference. This is merely by way of setting the scene.

At heart, the sociologist is concerned with "context". He will always be trying to drag attention away from individuals: in this setting, from individuals who are taken to have some drug-related problem; alcoholics, drug addicts, people who are dependent, call them what you will. So, if the sociologist has anything to contribute to this particular problem area it must be through his attempts to place individuals in their wider social context; in the context of everyday ideas, activities and relationships, which themselves change over time and from place to place.

The sociologist obviously has nothing to contribute to the chemical analysis of particular drugs, or to the intricate arguments about the bio-chemical effects on those who consume them. On the other hand, he may be very interested in several other questions concerning those substances.

At a basic level there is the matter of who uses, how much, of what substances, in what way, when, where and with whom. This may seem pretty simple-minded. But these are not merely extremely difficult questions to answer - as anyone who has, for example, attempted to conduct a general population drinking survey will readily acknowledge - but also extremely important questions which are too often neglected in the mad rush toward "problems" and "patients". In his introduction to Cahalan and associates' book American Drinking Practices, Selden Bacon (1969) puts it this way: "..... there are studies and books and pamphlets, there are conferences and national organisations and films, there are newspaper reports and laws and hand-books on procedures... One is concerned with disease, another with accidents, a third with sales control, a fourth with criminal justice. One is based on biochemistry, another on personality, a third on political art or science and a fourth on community organisation and public health. Cahalan, Cisin and Crossley have centred attention on the phenomenon central to all these approaches and central to all the problems no matter what their form, no matter what discipline or language is employed: namely man using alcoholic beverages ... All too frequently" Bacon says, "this core, this essential and crucial precondition to all questions and answers, has been forgotten".

Closely related to this essentially descriptive material are questions about the place of particular drugs in everyday life: "what is the relation of drinking, smoking, opium eating, or whatever, to other activities?", "what social, cultural, religious or other meanings does it have for those involved?". Sociological studies of drinking, for example, range from the ambitious, but no longer fashionable, "supra-cultural orientation" of people like Horton (1943), Roebuck & Kessler (1972), Field (1962) and Bales (1959), who tried to identify the function of alcohol in all cultures, via the many studies of the place of alcohol in the lives of particular groups or sections of one society, (Snyder 1958, Maddox & McCall 1964) to the small but growing body of micro-sociological work like that of Sherri Cavan (1966) who was content to analyse the complex social rules underpinning cocktail-bar behaviour. But, at whatever level, the core aim of all this sociological work is the same: to try to understand the social context in which a drug is produced, distributed and used.

The author says produced and distributed as well as used because, of course, drugs do not just appear as if by magic, in a glass, capsule or syringe. Drug production and distribution are parts of everyday agricultural-industrial-commercial life. Listen to Griffith Edwards (1971) as he elegantly makes that point at the start of his 1971 Edwin Stevens Lectures: "If you were to fly low over any part of the earth's land surface, you would have a fair chance of before long seeing below you some process of drug cultivation. You would see vineyards, coffee plantations, wide areas where the tobacco crop shaped the economy, fields of opium poppy, patches of lank Indian hemp flowering even on vacant city lots ... As you look down on the great industrial countries it would be the factories, however, rather than the fields which would probably catch your attention: modern technology spills out its tranquillisers, its stimulants analgesics, and anti-depressants by the billion, and the chemist much improves on cactus and mushroom ... Sometimes the state would sell a drug and take the profit, sometimes the profit would go to the legally operating entrepreneur, sometimes that role was played by the man with the mule train who makes his way over mountain paths ... You could conclude" he says, "that one of the main businesses of the world was to cultivate, manufacture, advertise, legislate on, tax, consume, adulate and decry mind-acting substances".

No surprisingly then, sociologists are interested in 'drugs' since, clearly, they are so tightly woven into the fabric of everyday social life.

But what is a 'drug'? It might seem rather fatuous to raise the question at this late stage, but do we perhaps lose something with all the talk of substances ? A drug, says the Concise Oxford Dictionary is : "an organic simple medicinal substance organic or inorganic used alone or as an ingredient". The Penguin Dictionary, (Garmondsway 1965) defines it as : "a chemical substance used in medicine" while Young (1971), a sociologist who should know better, begins his book, The Drug takers, by saying that : "a drug is any substance which has an effect on the metabolism of the body".

No one could violently disagree with those three definitions. On the other hand, they hardly convey the flavour of what many people take "drugs" to be. Perhaps "portable ecstacies", De Quincey's (1932) description of laudanum, or " a stimulator of the mystical consciousness" William James (1969) description of ether, convey the notion better. Or maybe the word drug has quite different associations. William Burroughs (1961) suggested, in his address to the American Psychological Symposium in 1961, that : "... the word 'drug' activates a relfex of fear, disapproval and prurience in Western nervous systems". He, thus, takes us beyond dictionary definitions to a consideration of those real, everyday definitions in terms of which people structure the social world about them. In short, he reminds us, if such was needed, that 'drugs' are no more and no less than what people categorise as such in the normal business of their lives.

Not long ago for example, the popular press was full of the reports of people becoming "addicted" to peanut butter, and glue-sniffing was a common-place. While a paper in the British Journal of Addiction, (Beattie 1968) began as follows : "There is an increasing volume of casual evidence that nutmeg is being consumed in some circles for its supposed psychological effects .. Users claim that nutmeg causes arousal or elation ... Students have suggested that nutmeg makes exam questions look easy ..." Mr Beattie then went on with admirable scientific zeal to say that: "... a controlled trial using a group of young people might be timely, since nutmeg-taking is as yet unpractised here to the same extent as in America".

But after investigating whether nutmeg ingestion affected ease of problem solving, the speed and accuracy of problem solving, or was associated with "unusual changes in experience" our intrepid researcher was able to report, to a relieved world, that: "Nutmeg has no psychological effect in the areas of behaviour sampled". It is clear that it would be difficult to say with certainty of any substance that it never has been, or never would be, classed by some people as "a drug" - with all that that to them implies - and likely to be "a problem".

Once we move from "substances" to "problems" - with "drugs" as the link word - a whole new set of questions are raised: "why, and by whom, is some drug-related response or activity taken to be "a problem'?", "whose problem is it anyway?", and "who says so?". There is no need to tell this audience that new and sophisticated definitions of alcoholism and drug-dependence drop "like the gentle rains from heaven". Embedded in most of them however, are two core themes: the notion of dependence and the notion of cultural appropriateness. The W.H.O. definition of alcoholism, for example, contains both, as does, particularly clearly, Cohen's (1969) definition of drug abuse as: "The ... self-administration of any drug which has resulted in psychological or physical dependence or which deviates from approved social patterns of the culture". Here, bound definitionally together, are: the drug, its effect on the individual and the cultural context of its use. The interconnection between them is not merely in the eye of the experts, but reflects the way in which the situation is conceptualised by the drug user, on the one hand, and by those who react to him, on the other. Listen to a young R.A.F. man on the subject of cannabis, (Glatt 1968) "What is wrong with pot is not me, is not the drug, but society's attitude to it ... I could give up pot ...", he continued, "... any time I want to, but rather than accept this country's bad law on the drug I prefer going abroad".

Others have discussed the relationship between the drug, the user and society in rather more flamboyant terms. Robert Maxwell (1969), speaking at a press conference to raise money for a film about drug addiction asserted that : "Individuals indulging in drug taking, including cannabis, so weaken themselves that they cease to be useful members of society. If sufficient numbers of individuals were to follow suit then the whole of our society would be weakened and endangered". He also felt that: "Information about ... drugs prompts the desire in many adolescents to experiment with them". It's interesting then, that Maxwell was so keen to raise £100,000 to finance a film about drug addiction as part of "a competent educational programme for our children".

But just as "drugs" are no more and no less than what people categorise as such, so the rhetoric of the debate about drug use serves to remind us, again if such was needed, that "alcohol and drug abuse" are no more and no less than a sub-class of drug use which people categorise as such in the normal business of their lives. We all know, in our bones, which instances of "drug use" deserve the accolade "drug abuse". The difficulty is of course, that just as we all have our own bones so we tend to "know" different things. Not only that but our ideas change.

One area in which ideas have fluctuated over the past hundred years is on the question of the relationship between alcoholism and disease. Jellinek's (1960) classic discussion, placed the debate in its historical and cultural context while, at a more day-to-day level, Griffith Edwards (1970) reminds us that : "... when all the relevant information on the causes of abnormal drinking has been gathered in, the decision as to alcoholism being a disease will still rest .. on the definition of "alcoholism" on the one hand and of "disease" on the other". In some quarters, however, the question of the relationship between alcoholism and disease is not taken to be one of debate or definition but one of decree or declaration. Moser (1974) for example, in a recent W.H.O. survey of Problems and Programmes Related to Alcohol and Drug Dependence in 33 Countries, reports that : "... almost all the respondents affirmed that dependence on alcohol and other drugs is recognized as illness in their country ..." and, as an example, cites: "Costa Rica, where alcoholism was officially declared as a disease according to presidential decree issued in 1954". Gitlow (1973) also takes the matter to be settled and beyond debate on the grounds that: "the American Medical Association, American Psychiatric Association, American Public Health Association, American Hospital Association, American Psychological Association, National Association of Social Workers, World Health Organisation and the National College of Physicians have now each and all officially pronounced alcoholism a disease". In the light of this formidable array of official pronouncements that alcoholism is a disease he feels that "the rest of us can do no less". In fact, he goes further and suggests that "we must now question the motives of those who would strip the alcoholic of his 'disease' label".

The sociologist's job however is neither to settle the debate, nor to support or oppose the decrees and declarations. He is much more concerned with the use of the idea that alcoholism is a disease, (Robinson 1976). For as Robin Room (1972) puts it : "The necessities that nurtured the disease conception ... were practical rather than theoretical ... brought about essentially as a means of getting a better deal for the 'alcoholic' rather than as a logical consequence of scholarly work and scientific discoveries". The logical conclusion to this using of the disease concept as an administrative or therapeutic strategy is the position of "humane cynicism" as Room calls it, Room, Ibid. exemplified by Pattison (1969) who felt that: "The major advantage of defining alcoholism as a disease is social. It legitimates social support for rehabilitation of the alcoholic rather than punishment. To say that alcoholism is an illness should not be construed as a statement about the aetiology or treatment of alcoholism". D.L.Davies (1974) has gone further and suggested that "the concept of alcoholism has outlived its usefulness". It is understood why he thinks that, and he's right, but unfortunately it is not so easy to legislate for what people will take words and phrases to really mean. For while Pattison, for example, may be quite clear in his own mind, that the phrase "alcoholism is a disease" should not be "construed as a statement about etiology and treatment" it certainly may not be so clear either to his colleagues or to members of the general population; especially as they are constantly being told through health education campaigns and the popular press merely that alcoholism is a disease, with all that that to them implies, such as, for example, that the helping professions have something to offer - perhaps even "a cure" - with all that that to them implies.

It is not the intention to end up however, by giving the impression that all is unhelpful confusion. Far from it. Over the past few years there have been a number of significant developments; among the most promising of which has been the much greater willingness - encouraged by government interest, policy and finance - to divert attention from "alcoholics" and "drug addicts" to a consideration of the whole range of relationships between drinking, or the use of any other drug, and the development, recognition, course and handling of "problems" which one may or may not want to argue, constitute "alcoholism" or "drug dependence".

Symptomatic of the greater readiness to place particular issues in their broader social context has been the recent discussion about the Erroll committee report on liquor licensing. Mr Maudling and the Home Office clearly imagined that the question of liquor licensing would merely set off the same old tired debate between the trade, on the one hand, and the temperance organisations, on the other, with the general public - tongues hanging out - somewhere in the background. In the event of course, the medical profession waded in with the strongest evidence, most of the trade organisations wanted to maintain the status quo while the "heavy" press, the "popular" press and the general public were all against the committee's main proposals to relax controls. In fact the members of the Erroll committee appeared to be almost the only people in England and Wales who didn't understand the implications of the relationships between alcohol consumption, alcohol problems and everyday social life.

As suggested earlier, in the paper, if the sociologist has anything to contribute to discussions about alcoholism and drug dependence it is through his constant attempts to place particular issues in their everyday social context: in the ever-changing context of ideas, activities and relationships.

REFERENCES

BACON, S. (1969). introduction to Cahalen, D., Cisin, I. and Crossley, M. American Drinking Practices: a National Study of Drinking Behaviour and Attitudes, Rutgers Centre of Alcohol Studies, New Brunswick.

BALES, R.F. (1959). "Cultural Differences in Ideas of Alcoholism", Drinking and Intoxication, McCarthy, R.G. (ed) Free Press, New York.

BURROUGHS, W.S. (1961). Points of Distinction between Sedative and Mind-expanding Drugs". Paper to American Psychological Symposium.

BEATTIE, R.T. (1968). "Nutmeg as a Psycho-active Agent". British Journal of Addiction, 63. 105-109.

CATANZARO, R.J. (1968). Alcoholism: the Total Treatment Approach, Charles C. Thomas, Springfield, Illinois.

CAVAN, S. (1966). Liquor License: an Ethnography of Bar Behaviour, Aldine, Chicago.

COHEN, D. (1969). The Drug Dilemma, McGraw-Hill Inc. New York.

CONCISE OXFORD DICTIONARY

DAVIES, D.L. (1974). "Is Alcoholism Really a Disease?" Contemporary Drug Problems.

DeQUINCEY, T. (1932). Confessions of an English Opium Eater, Three Sirens Press, New York.

EDWARDS, G. (1971). Unreason is an Age of Reason, The Edwin Stevens Lectures.

EDWARDS, G. (1970). "The Status of Alcoholism as a Disease". Modern Trends in Drug Dependence and Alcoholism, Phillipson, R.V. (ed) Appleton-Century-Crofts, New York.

FIELD, P.B. (1962). "A New Cross-cultural Study of Drunkenness". Society, Culture and Drinking Patterns, Pittman, D.J. and Snyder, C.R. (eds), Wiley, New York.

GARMONDSWAY, G.N. (1965). The Penguin English Dictionary, Penguin, Hammondsworth, Middlesex.

GITLOW, S.E. (1973). "Alcoholism a Disease". Alcoholism: Progress in Research and Treatment, Bourne, P.G. and Fox, R. (eds), Academic Press Inc. New York.

GLATT, M.M. (1968). "Recent Patterns of Abuse of and Dependence on Drugs". British Journal of Addiction, 63, 111-128.

HORTON, D. (1943). "The Functions of Alcohol in Primitive Societies". Quarterly Journal Studies on Alcohol, 4, 199.

JAMES, W. (1969). "Varieties of Religious Experience". The Marijuana Papers, Soloman, D. (ed) Panther Books, London.

JELLINEK, E.M. (1960). The Disease Concept of Alcoholism, Hillhouse Press, New Haven.

MADDOX, G.L. and McCALL, B.C. (1964). Drinking Among Teenagers, Rutgers Centre of Alcohol Studies, New Brunswick, New Jersey.

MAXWELL, R. (1969). Cited in notice about a fund for a film on drug addiction. British Journal of Addiction, 64, 149-150.

MOSER, J. (1974) Problems and Programmes Related to Alcohol and Drug Dependence in 33 Countries, World Health Organisation, Geneva, Offset Publication No. 6.

PATTISON, E.M. (1969). "Comment on the Alcoholic Game". Quarterly Journal Studies on Alcohol, 30, 953-956.

PITTMAN, D.J. (1967). Alcoholism, Harper and Row, New York.

ROBINSON, D. (1976), From Drinking to Alcoholism: A Sociological Commentary, Chapter 4, Alcoholism and Disease, Wiley and Sons, London.

ROEBUCK, J.B. and KESSLER, R.L. (1972). The Etiology of Alcoholism: Constitutional, Psychological and Sociological Approaches, Charles C. Thomas, Springfield, Illinois.

ROOM, R. (1972). "The Alcohologist's Addiction: Some Implications of Losing Control over the Disease Concept of Alcoholism". Quarterly Journal Studies on Alcohol, 33.4.72.

SNYDER, C.R. (1958). Alcohol and the Jews, Free Press, Glencoe, Illinois.

YOUNG, J. (1971). The Drug takers: the Social Meaning of Drug Use, MacGibbon & Kee, London.

ALCOHOL AND THE GENERAL PHYSICIAN

ROBIN WALKER

WALTON HOSPITAL

LIVERPOOL

It is unfortunate that physicians have not shown an interest in the problem of alcohol abuse. Most doctors do not view alcohol and its related problems as a therapeutically rewarding field. This is mainly engendered by the high readmission rates which occur with this problem and perhaps also because of the inordinate amount of time which such patients tend to consume. Whilst there may be advantages to treating certain patients in a general medical ward environment, it only takes one recalcitrant and recidivist alcoholic to dampen the interest of physicians, nursing staff and others in the medical field. This is perhaps easy to understand, but the vast majority of patients with alcohol-related problems do not fit into this category. Those physicians who have for one reason or another developed an interest in the problem may derive some satisfaction from a recent article from Glasgow surveying the attitudes towards alcoholism amongst psychiatrists, (MacDonald & Patel 1975). It might be supposed that the role of the psychiatrist in the treatment of alcoholism was well established and in the particular hospital from which this account came, up to 40% of their acute male psychiatric admissions were alcoholics. The paper established on the basis of a questionnaire that at least in Scotland this problem is regarded less favourably than any other psychiatric or organic illness with the exception of self-poisoning and drug-dependence. If this attitude is prevalent amongst psychiatrists despite their presumed orientation towards the social and emotional aspects of illness, then what hope is there for physicians, general practitioners and other medical specialists ?

It is surely relevant that in this same study the organic group least favoured amongst psychiatrists was neoplasia. As with alcohol-related problems there has been an attitude that neoplasia is a largely untreatable condition, so that it has tended to become the pariah amongst organic diseases. There is some recent evidence that an interest in the treatment of neoplasia is being kindled by the successes scored in the field of chemotherapy. There seems little doubt that if we can show that the treatment of alcohol-related problems is therapeutically and perhaps emotionally rewarding, then we will gain the support and enthusiasm of all those who have to treat such patients. Much of this must come through education, but perhaps of equal importance is that as a profession we must examine our attitudes to the problem. A cynic has remarked that "an alcoholic is a person who drinks more than his doctor", (Shropshire 1975). This half-truth highlights the puritanical streak which is prevalent amongst the medical profession. That it is an attitude which we can no longer afford is supported by the suggestion that in as many as 10-20% of acute medical admissions to hospital alcohol may be playing a significant part. Wilkins (1974) has suggested that as many as 100 patients may be affected by the problem in each general practice, either directly or indirectly through a relative.

In the light of such figures we must as physicians recognise that we are likely to meet with this problem nearly every day of our working lives. It would be no exaggeration to state that alcohol has become the great mimic of modern medicine. Virtually every organ in the body may be affected directly or indirectly. The long list of possible physical consequences of alcohol abuse can be obtained elsewhere (Seixas et al 1975). These will not be reiterated as whatever the specialty we should ask ourselves with every patient seen whether alcohol may be playing a part in the illness.

It is pertinent to consider the damage caused to the liver by alcohol in a little more detail. Damage caused to this organ has been used as an index of the extent of the alcohol problem in a community and it is perhaps the organ damaged by alcohol about which our knowledge is most complete. In this country it is likely that about 70% of cases of cirrhosis are now due to alcohol, (Blendis et al 1975), and this is certainly the case for Liverpool, (Forshaw 1972). Only a few years ago we were claiming that we did not have alcohol-induced liver disease as a significant problem in this country. The lesions which may occur in the liver are three in number: (a) fatty liver, (b) hepatitis, (c) cirrhosis. Fatty liver probably occurs in all people taking an excess of alcohol. The triglyceride content in the liver of non-alcoholic volunteers may double during the course of a weekend of heavy drinking and perhaps quadruple over a period of ten days despite taking a high protein/low fat diet, (Rubin & Lieber 1968). This lesion may be associated with hepatomegaly and minor abnormalities in liver function tests, but in the absence of necrosis or inflammation would appear to be entirely reversible and there is no good evidence that in itself it is a pre-cirrhotic lesion.

Increasing use of the percutaneous liver biopsy has demonstrated that a significant proportion of a population of heavy drinkers will show the histological picture of alcohol-induced hepatitis. In this lesion ballooning and necrosis of hepatocytes, infiltration of the lobule with segmented leucocytes, hyaline of Mallory, variable steatosis and centrilobular perivenous fibrosis are seen. In one series, (Ugarte et al 1970), this lesion was observed in 11% of cases. This lesion was formerly thought to be associated with an extremely high mortality. In the patients with a normal prothrombin time the mortality rate is probably in the region of 7% but is very much higher where the prothrombin time is prolonged and where liver biopsy cannot be performed, (Galambos 1972).

There is a growing body of evidence that the lesion of alcohol hepatitis is pre-cirrhotic and that continued drinking in those showing this lesion will inevitably lead to the development of cirrhosis. Unfortunately it may be that many of those who abstain will also develop cirrhosis. The latter is a highly malignant condition and clearly our efforts should be directed towards preventing its development. Even so the importance of maintaining abstinence is emphasised by the survival figures for patients with established cirrhosis. Overall about 50% will be alive at five years,(Brunt et al 1974). In those that abstained in this series the survival rate was 69% whilst those who continued to drink had a survival of only 34% at five years.

Many physicians have still continued to deny the importance of alcohol as a direct toxin in the aetiology of cirrhosis. We do not yet have sufficient evidence to deny the possible importance of nutritional deficiency and the recent epidemiological survey of 304 alcoholic patients suggested that non-cirrhotics have a higher food caloric intake and higher protein intake than patients who develop cirrhosis,(Parek et al 1975). It is equally possible that nutritional deficiencies may be more important in determining the prognosis in individual patients than in being an aetiological factor in the development of liver disease. Patients should certainly not be given the impression that if they maintain an adequate nutrition they will not develop liver damage if they continue to take excessive amounts of alcohol, (Lieber 1975 a).

There is an impressive body of evidence to suggest that alcohol itself is a direct toxin to the liver. Much of the evidence is epidemiological, the presence of cirrhosis being much more frequent in alcoholics with more than 15 years of excessive drinking (4.8). The development of cirrhosis is related to the amount of alcohol taken and the duration of time over which it is taken. Until recently a similar lesion could not be produced in animals without manipulation of the diet. Fortunately an experimental model has now been found in the baboon which has a liver morphologically similar to that of man, (Lieber 1975 a). Baboons fed alcohol over a number of years developed similar lesions to those seen in man, namely fatty liver, hepatitis and cirrhosis. This occurred despite an adequate diet and a control group of animals developed no such lesions.)

Finally it would be appropriate to comment on the individual variation which exists in the development of organic disease. Some patients remain immune to such damage after many years of heavy drinking. Some develop damage in one organ and not in others. Even in heavy long term drinkers only about 10% develop cirrhosis. Chronic calcific pancreatitis is exceptionally rare in this country, but common in France and other countries. Whilst environmental factors such as diet have not been entirely ruled out as important in explaining some of these anomalies, there is suggestive evidence that constitutional, possibly even genetic, factors may play a large part. Although prevention should be the main aim of people working in this field, there can be no doubt that the physical consequences of excessive alcohol intake will remain for the foreseeable future. This being the case, if we could develop the ability to weed out the sensitive individuals in our midst, we might be able to halt the inexorable increase in diseases caused by alcohol. This of course assumes that our alcohol treatment services are such as to allow us to persuade a sizeable proportion of our patients with physical disease to abstain completely from alcohol.

REFERENCES

BLENDIS, L.M., LEEDS, A.R., and JENKINS, D.A.J. (1975). "Incidence of Alcoholic Liver Disease". Lancet, i, 499.

BRUNT, P.W., KEW, M.C., SCHEUER, P.J., and SHERLOCK, S. (1974). "Studies in Alcohol Liver Diseases in Britain". 1. Clinical and Pathological Pattern Related to Natural History. Gut, 15, 52.

FORSHAW, J. (1972). "Alcoholic Cirrhosis of the Liver". British Medical Journal, 4, 608-9.

GALAMBOS, J.T. (1972). "Alcoholic Hepatitis: its Therapy and Prognosis". Progress in Liver Diseases, Popper H. & Schaffner, F.(eds) IV, 567-588.

LIEBER, C.S. (1975a). "Alcohol and Malnutrition in the Pathogenesis of Liver Disease". Journal of the American Medical Association, 233, 1077.

LIEBER, C.S., DeCARLI, L.M., and RUBIN, E. (1975b). "Sequential Production of Fatty Liver, Hepatitis and Cirrhosis in Sub-human Primates fed Ethanol with Adequate Diet". Proceedings of the National Academy of Sciences of the United States, 72, 437-441.

MacDONALD, E.B. and PATEL, E.R. (1975). "Attitudes Towards Alcoholism". British Medical Journal, 2, 430-31.

PAREK, A.J., IMRE, G.T., SAUNDERS, M.G., CASTRO, G.A., and ENGEL, J.J. (1975). "Alcohol and Dietary Factors in Cirrhosis. An Epidemiological Study of 304 Alcoholic Patients". Archives of International Medicine, 135, 1053-1057.

SEIXAS, F.A., WILLIAMS, K, and EGGLESTON, S. (eds). (1975). "Medical Consequences of Alcoholism". Annals of the New York Academy of Sciences, 252, 1-399.

SHROPSHIRE, R.W. (1975). "The Hidden Faces of Alcoholism". Geriatrics, 30, 99-102.

RUBIN, E., and LIEBER, C.S. (1968). "Alcohol Induced Hepatic Injury in Non-alcoholic Volunteers". New England Journal of Medicine, 278, 869-876.

UGARTE, G., ITURRIAGA, H., and INSINZA, I. (1970). "Some effects of Ethanol on Normal and Pathological Livers". Progress in Liver Diseases, Popper, H. & Schaffner, F. (eds) III, 355-370.

WILKINS, R.H. (1974). "The Hidden Alcoholic in General Practice". Elek Science, London.

DENIAL OF THE HIDDEN ALCOHOLIC IN GENERAL PRACTICE

RODNEY H. WILKINS

DEPARTMENT OF GENERAL PRACTICE

MANCHESTER UNIVERSITY

Denial is a glue choking the detection mechanism leading to successful identification and management of the alcoholic. It is a feature not only of the alcoholic, but often of the family, workmates, society, and sometimes the medical profession (Wilkins 1973). Numerous questionnaires have been devised to overcome this fact that "the very nature of the condition has written into it the probability of the patient's often attempting to deny his problem (Hensman et al, 1968). This aspect of denial in relation to a study carried out by a group of general practitioners to detect the hidden alcoholic in general practice will be discussed.

For a period of one year, the general practitioners of the Manchester University Department of General Practice administered questionnaires to patients, aged 15-65 years, attending a Health Centre who were considered "at risk" to be alcoholics. An "at risk" patient was one identified as possessing one or more of the 63 "at risk" factors listed on an Alcoholic At Risk Register (Wilkins 1974). This list included certain physical and mental diseases, occupations, types of marital status, work problems, family disharmony, families with children suffering from psychological or psychosomatic disturbance, accidents, criminal offences, living in a hostel for the homeless, and smelling of drink.

A short disguised questionnaire, the Spare Time Activity Questionnaire (STAQ) was devised for use in general practice. Apart from the usual problems of questionnaire design, we had the additional problem of the limited amount of time available. On the basis of pilot studies, a questionnaire was developed which had two cut-off points; if the patient answered negatively to the early questions, the interview was concluded. In such cases, completion took only about three minutes compared with seven to eight minutes for the unabridged version. Because of this time factor the questions used had to be very selective. Four types of question relating to alcohol abuse were asked: quantity-frequency of drinking alcohol, the attitude of the patient and his family to drinking, the acknowledgement of problems from heavy drinking in the areas of health, family arguments, work, finance and criminal proceedings, and, finally, questions relating to symptoms of alcohol addiction. These questions were embedded amongst others about cigarette smoking, television viewing, etc.

A total of 546 questionnaires were analysed (practice population, 12,000) and 28.4% of the "at risk" patients were classified as present or past alcoholics. The prevalence rate for all alcoholics was estimated to be 18.2 per 1,000 of the practice population aged 15-65 years. Discussion of the prevalence rates (Wilkins 1972) and the discriminating value of the different "at risk" factors (Wilkins 1976) have been described elsewhere. This paper highlights the problem of denial as illustrated by a reliability study of the questionnaire.

The author's remarks must be prefaced by a comment on definitions. Keller (1962) has pointed out that "the definitions of alcoholism (and alcoholics) have long been marked by uncertainty, conflict and ambiguity". When one remembers that Jellinek (1960) has written 246 pages of detailed analysis of over 100 definitions, clearly every research worker must explain the terms he uses. In this study there are classified three types of abnormal drinker based on a scoring system in which one or more points were allocated to certain answers depending on the significance attached to them. In essence, the "alcohol addict" admitted at least two symptoms of alcohol addiction, and the "problem drinker" acknowledged at least one problem from alcohol abuse, but no symptoms of physical dependence. These two categories comprised the "alcoholic". The third variety of abnormal drinker, the "heavy drinker" was the patient who scored at least two points but did not admit to any problems, or symptoms of addiction. Patients were also classified into "present" or "past" abnormal drinkers.

Does alcoholic denial affect the reliability of the questionnaire ? In the opinion of Bailey (1967) "the reliability of alcoholism survey data leaves a great deal to be desired" and indeed only five such studies have been reported. It is not difficult to understand why when one examines some of the results. For example, Guze and Goodwin (1971) interviewed 176 male criminals and asked the same 17 questions as seven to eight years previously with the aim of eliciting a "lifetime prevalence of alcoholism. Of those classified as "definite" or "questionable" alcoholics at the first interview, 27% were classified as non-alcoholics at the second interview. Of those originally identified as non-alcoholics, 32% were considered to be "definite" or "questionable" alcoholics at follow-up. Summers (1970) interviewed 15 males on their entry into an in-patient alcoholism programme. Two weeks later on re-interview, 14 of them had changed their responses to 50% of the questions. About equal numbers admitted to more or less serious problems. Similar trends were shown in studies by Bailey et al (1966), Edwards et al (1973) and Bowden (1975). All these investigations adopted the test-retest technique, although it is usually considered that there are better methods (Moser 1958).

At the conclusion of the survey year, 41 patients who had previously completed a questionnaire were invited to answer another one (Wilkins 1974). The varying timespan (one to 12 months) between administering the original and reliability Spare Time Activity Questionnaire, and the small numbers, gave rise to difficulty in assessing what meaning to assign to the "present" and "past" categories of abnormal drinker. The results for both categories were therefore combined. Comparisons between the two Spare Time Activity Questionnaire classifications were made by calculating the Serious Disagreement Rates, adapted from Kendell et al (1968). This was defined as the proportion of alcoholics classified in the original questionnaire who were re-classified as nonalcoholics in the reliability questionnaire, and vice-versa. An "alcoholic" was defined as a problem drinker or alcohol addict, and a "nonalcoholic" as a social drinker or heavy drinker.

The results showed that 67% (six in nine) of the alcoholic patients were re-classified as non-alcoholics in the reliability study compared with only 6% (two in 32) of the non-alcoholics who, on re-interview, were classified as alcoholics. This difference was statistically significant (exact test: $p = 0.0009$; significant at 0.1% level).

The questionnaire comprised two sections, one relating to the informant and the other to the immediate family. We found that 36% (four in 11) of the alcoholic family members were re-classified as non-alcoholics in the reliability study. Only 3% (one in 30) of the originally classified non-alcoholics were redefined as alcoholics on re-interview. This difference was statistically significant (exact test: $p = 0.028$; significant at 1% level).

There was no significant difference in the proportions of alcoholics to non-alcoholics (original Spare Time Activity Questionnaire) between the informants and family members ($X^2 = 0.066$; d.f. = 1; $0.80 > p > 0.70$; not significant at 5% level). It is therefore permissible to pool the non-alcoholics and alcoholics in the two groups, and compare the total serious disagreement rates. The difference was not statistically significant. (Yates correction: $X^2 = 0.365$, d.f. = 1, $0.60 > p > 0.50$; not significant at 5% level).

What meaning can be attached to these findings ? Firstly, there is the possibility of observer bias. It is likely that in a number of cases the original classification was remembered. However, this might have tended to increase the number of agreements between the original and reliable Spare Time Activity Questionnaire classifications, and does not explain the relatively high proportion of disagreement found.

Secondly, as in any test-retest technique, the possibility of respondent bias has to be considered. It has been suggested that if respondents are re-interviewed after a comparatively short time lapse they may remember their original answers, whereas with a longer time interval, there may have been a real change in drinking behaviour (Edwards et al 1973). However, Bowden refutes this opinion by showing that the same degree of reliability of 70-75% was found in four studies with a time span of one month (Bowden 1975), two to three months (Edwards et al 1973), two to 3 years (Bailey et al 1966) and eight years (Guze and Goodwin 1971). Chalke and Prys Williams (1971) believe that "re-approach at a later date is scientifically useless, because there is no means of knowing if, and how far, the respondent has been conditioned - in memory and even as to performance - by the process of completing the first interview". It should be noted that the survey was concerned with a lifetime experience and not just the current drinking status. Even if a drinking alcoholic had become sober in the intervening few months, the reliability Spare Time Activity Questionnaire should have indicated that the patient was a "past" alcoholic, and so shown no disagreement. It is suggested that some alcoholics, realising the purpose of the questionnaire, deliberately falsified their responses on re-interview. It is possible that the two patients originally classified as non-alcoholics had become alcoholics later. However, it is more probable that they had decided to reveal their true drinking behaviour at the second interview.

In the study, the more serious disagreement was found amongst the alcoholics, with a far higher degree of reliability amongst the non-alcoholics. A similar finding was reported by Bailey et al (1966). However Guze and Goodwin (1971) observed the opposite; reliability was more likely in patients with the more serious drinking problem. Bowden (1975) has compounded this confusing picture in his study of the reliability of prisoners' responses to drinking questions. Over 600 prisoners in the South East of England completed a self-administered questionnaire, and one month later participated in a structured interview. The overall reliability was 69%, but he found that the two extremes of non-heavy drinkers and the heaviest drinkers were the most reliable, and that an intermediate group were the least reliable. One could postulate that social drinkers and established, confirmed admitting alcoholics have nothing to hide, and it is the in-between group of problem drinkers who are still denying, either consciously or subconsciously that they are abusing alcohol. The conflicting findings of these different studies may be more apparent than real, and in fact reflect a difference of definition of the terms used to describe the severity of the drinking problem.

What conclusions can be drawn ? The results illustrate one of the difficulties in calculating accurate prevalence rates. A total of 41 patients were interviewed on two occasions. Only three alcoholics were identified from both questionnaires, and eight from one of them. The original Spare Time Activity Questionnaire suggested a total of nine alcoholics, and the reliability Spare Time Activity Questionnaire indicated only five. It is considered unlikely that individuals would admit to being alcoholics if they were not. A validation study of the questionnaire by two psychiatrists confirmed that no false positives were reported (Hore et al 1976). Thus any prevalence rate obtained is almost certainly an underestimate.

Despite the difficulties, general practitioners can detect a considerable number of alcoholics by asking the right questions to the right patients. Of the 155 alcoholics identified in the survey, only eight consulted specifically for advice about their drinking problem, and eleven had previously received psychiatric treatment for alcoholism. Because of denial, some will be missed at the first interview. But, once suspicions are aroused the moment of recognition is likely to be brought forward. It is not believed that the problem of denial can be defeated, but a hole can be knocked in this wall which the alcoholic often builds around himself.

REFERENCES

BAILEY, M.B. (1967). American Journal of Public Health, 57, 987.

BAILEY, M.B., HABERMAN, P.W. and SHEINBERG, J. (1966). Quarterly Journal Studies on Alcohol, 27, 300.

BOWDEN, P. (1975). Psychological Medicine, 5, 307.

CHALKE, H.D. and PRYS WILLIAMS, G. (1971). "Alcohol and the Family". Christian Economic and Social Research Foundation, Priory Press, Royston, Herts.

EDWARDS, G., HENSMAN, C. and PETO, J. (1973). Quarterly Journal Studies on Alcohol, 32, 1244.

GUZE, S.B. and GOODWIN, D.W. (1971). Quarterly Journal Studies on Alcohol, 32, 808.

HENSMAN, C., CHANDLER, J., EDWARDS, G., HAWKER, A. and WILLIAMSON, V. (1968). Medical Officer, 120, 215.

HORE, B., ALSAFAR, J. and WILKINS, R.H. (1976). British Journal of Addiction. (In Press).

JELLINEK, E.M. (1960). The Disease Concept of Alcoholism, Hillhouse Press, New Haven, Connecticut.

KELLER, M. (1962). In Society, Culture and Drinking Patterns D.J. Pitman and C.R. Snyder (eds.) Wiley, New York.

KENDELL, R.E., EVERETT, B., COOPER, J.E., SARTORIUS, N. and DAVID, M.E. (1968). Social Psychiatry, 3, 123.

MOSER, C.A. (1958). Survey Methods in Social Investigations, Heinemann, London.

SUMMERS, T. (1970). Quarterly Journal Studies on Alcohol, 31, 972.

WILKINS, R.H. (1976). "An Alcoholic at Risk Register in General Practice - Why Bother?" 20th International Congress of General Practice, Stratford-upon-Avon.

WILKINS, R.H. (1972). "A Survey of Abnormal Drinkers in General Practice". 30th International Congress on Alcoholism and Drug Dependence, Amsterdam.

WILKINS, R.H. (1973). Proceedings of the First International Medical Conference on Alcoholism, London, Edsall and Co. Ltd., London.

WILKINS, R.H. (1974). "The Hidden Alcoholic in General Practice". Elek Science, London.

Section II

Aetiology and Epidemiology

THE VALIDITY OF PER CAPITA ALCOHOL CONSUMPTION AS AN INDICATOR OF THE PREVALENCE OF ALCOHOL RELATED PROBLEMS: AN EVALUATION BASED ON NATIONAL STATISTICS AND SURVEY DATA

A.K.J.CARTWRIGHT, S.J.SHAW and T.A.SPRATLEY

MAUDSLEY ALCOHOL PILOT PROJECT

THE BETHLEM ROYAL & MAUDSLEY HOSPITAL, LONDON

The French alcohologist Ledermann (1964) was the first to use per capita consumption of alcohol as an indicator of the prevalence of alcohol related problems. Following his work, rises in per capita consumption of alcohol has frequently been used as evidence that the prevalence of alcohol related problems has been increasing.

This paper considers the validity of predictions based upon Ledermann's work, in the light of both national statistics and survey data.

Ledermann postulated that in any homogeneous population the distribution of alcohol consumption followed a special case of the log normal curve. Such a curve is depicted in Figure 1.

This distribution has three notable characteristics. Firstly, the curve is smooth, describing a mathematically definable path. Secondly, in the higher consumption levels, each consecutively higher consumption category contains a smaller proportion of the population. Thirdly, then, the mean of the distribution lies to the right of the mode and median.

This consumption curve differs from other log normal curves in that its end points are considered fixed. Ledermann believed that the upper limit of alcohol consumption could be defined by fatal dose levels of alcohol, whilst the lower limit must of course be zero. Therefore he could argue that an increase in per capita consumption could only be explained by a general move within the population towards heavier drinking. Thus, the implication of a rise in per capita consumption within a population is not simply that there are more excessive drinkers, but that there is a general pattern of heavier drinking throughout the population as a whole.

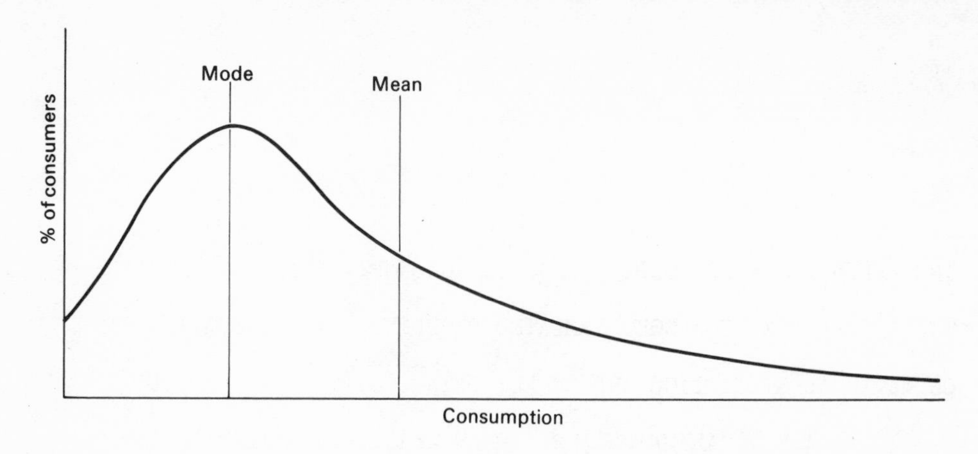

FIGURE 1. A Hypothetical Log Normal Curve

Aetiologically, Ledermann believed that the most significant determinant of problems was the quantity of alcohol consumed. The greater the quantity consumed the greater the risk of experiencing a variety of problems. Thus, as an individual's consumption rose, so proportionately did his vulnerability to experiencing alcohol related problems.

When this aetiological perspective is combined with the theory of the distribution of consumption, it is possible to predict that :-

(1) an increase in per capita consumption indicates

(2) generally heavier drinking throughout the population, in turn indicating

(3) a greater proportion of the population in high-risk categories, and therefore

(4) a higher prevalence of alcohol related problems.

Most attempts to substantiate this model have used the evidence of national statistics of alcohol consumption and rates of various alcohol related problems such as deaths from cirrhosis and convictions for drunkenness. When increases in per capita consumption are recorded in a society, rates of alcohol related problems tend to increase as well. When the per capita consumptions of different countries are compared, countries with a higher consumption also tend to have higher rates of alcohol related problems (Popham et al. 1975).

However, considerable doubts remain as to the validity of these associations, because they relate per capita consumption to alcohol related problems without substantiating the intermediate steps of the Ledermann model. Comparisons of national statistics do not validate the propositions that increased per capita consumption indicates generally heavier drinking throughout the population or that this means a greater proportion of people are at greater risk of experiencing alcohol related problems. If these intermediate steps are not established, then it could be argued, for example, that countries who report higher consumption also report more alcohol related problems because their methods of keeping records on both consumption and problems are more efficient than those of countries who report both low consumption and low problems.

The lack of evidence to support the intermediate steps of the Ledermann model creates further doubts. Critics have argued that an increase in per capita consumption might not result from a general move toward heavier drinking, but may simply represent a reduction in the proportion of abstainers in the population, or may even be due to a slight trend towards more frequent drinking, rather than heavier drinking.

To derive more substantial data for this debate requires surveys of the drinking patterns of similar populations at different points in time. Unfortunately, this undertaking is very costly of both time and money compared to the ease of collecting statistics from national records.

Yet sample surveys are necessary not only to assess the intermediate steps of the model but also to ascertain whether national rates of alcohol related problems are valid indicators of the prevalence of the problems they purport to measure. For instance, an increase in convictions for drunkenness might not indicate an increase in episodes of unsociable drunken behaviour. The increase may have been created by some change of police policy or court procedure. Furthermore, a large proportion of the convictions are multiple convictions of the same individuals, particularly from the homeless group of drinkers. Similarly, increased admissions of patients diagnosed as alcoholics cannot be unproblematically accepted as evidence of increased prevalence. Increased alcoholic admissions might reflect a greater availability of facilities designed for such patients, a greater awareness of the problem amongst the public and amongst doctors, or an increase in the tendency of doctors to label patients as alcoholics.

The sample survey approach circumvents many of these difficulties by studying the consumption and behaviour of individuals. Although the survey approach also entails various difficulties, its strengths in the present context lie in exactly those areas where the usage of national statistics is weakest. These issues will not be discussed here. This paper attempts to use both national statistics and sample survey data to evaluate the use of per capita consumption as an indicator of the prevalence of alcohol related problems. The statistics used have been taken from government publications. The survey data refers to two general population studies made in Camberwell - a suburb in South London - conducted in 1965 and replicated in 1974.

METHODOLOGY

1. National Statistics

(a) Admissions of diagnosed alcoholics. The number of persons admitted to mental illness hospitals and special units in England and Wales with a primary diagnosis of alcohol psychosis or alcoholism is published in the Department of Health and Social Security's special report series (DH & SS 1974). Figures used in this paper are taken directly from the published figures. The figures do not take account of admissions to general hospitals.

(b) Convictions for drunkenness. The number of persons convicted for drunkenness offences is published in the Annual Abstract of Statistics (1974). The figures used in this paper are taken directly from that publication.

(c) Per capita consumption. Consumption of beers, wines and spirits is published in the Annual Abstract of Statistics (1974). To calculate per capita consumption it is necessary to convert the published figures of the various alcoholic beverages into a common measure of volume of absolute alcohol. As there is some variation in the amount of alcohol contained in beverages even of the same type any conversion factor is subject to some inaccuracies. The conversions from national statistics have been based on the following averages :

Spirits containing 39.9% alcohol on average
Wines containing 16% alcohol on average
Beers containing 4% alcohol on average

Per capita consumption is devised by dividing the number of litres of absolute alcohol consumed by the adult population. The figures used here are the figures published in the Annual Abstract of Statistics (1974) of the population over the age of 15.

These figures underestimate the total consumption in the population, especially because they do not document consumption of home-made alcoholic beverages.

2. General Population Surveys

(a) The samples. The two samples analysed in this paper were taken from the population of the same London suburb in 1965 and in 1974.

Each has been described in detail elsewhere and only basic information is provided here.

(i) The 1965 sample: The 1965 sample was interviewed by Edwards et al (1972) to assess the drinking habits and drinking problems of the local population. A stratified sampling procedure was used; respondents being randomly selected from households on six housing estates in the area. (It was thought that the characteristics of their populations would produce greater demographic variety). The effective response rate was 89% and the interviewed sample comprised 928 persons aged 18 and over. These 928 persons are referred to as the 1965 sample.

(ii) The 1974 sample: The 1974 study was conducted by the Maudsley Alcohol Pilot Project (Cartwright et al. 1975a) as part of an overall assessment of the prevalence of alcohol abuse in the local area. The sample was selected using a random sampling procedure based on the electoral register but employing the Marchant-Blythe self-weighting random sampling technique (Blyth & Marchant 1972) to include persons not registered as electors. The effective response rate was 80% and the sample comprised 286 respondents are referred to as the 1974 sample.

(b) Comparability of the samples: As demographic characteristics such as age, sex and occupational status affect both alcohol consumption and alcohol related problems, a comparison of the two samples was made in terms of these three factors to ensure that neither sample contained an over represented high risk group. As a result of this analysis which is described in an earlier paper (Cartwright et al. 1975b), it was concluded that the samples were sufficiently alike in their age, sex and occupational status distributions to allow meaningful comparisons to be made between them. However, where appropriate, tests of significance will be computed with these factors controlled.

(c) The measurement of prevalence in the two samples: Each sample were asked a variety of questions about behavioural effects of their alcohol consumption during the 12 months prior to their interview and though the perspective of the two research teams was somewhat different, seven questions were asked of both samples. It is the response to these seven questions that forms the present measurement of the prevalence of alcohol related problems.

For the purpose of the analysis scores on these items have been summed and treated as a scale of alcohol related problems.

Each person who reported experiencing an item in the 12 months prior to interview was given a score of one for that item and then the scores for the seven items were totalled to give a score of alcohol related problems ranging between zero and seven. This scale has been called the index of alcohol related problems.

TABLE I : ITEMS COMPRISING THE INDEX OF ALCOHOL RELATED PROBLEMS

In the last 12 months have you ...
1. ever found that your hands were shaky in the morning after drinking?
2. ever had a drink first thing in the morning to steady your nerves?
3. ever found you could not remember the night before after drinking?
4. been criticised for your drinking?
5. on any occasion been unable to stop drinking?
6. been in arguments with family or friends after drinking?
7. been in fights with family or friends after drinking?

(d) Validity of the index of alcohol related problems: (i) The validity of adopting a single scale. If each of the items in the index was unrelated to any of the others, then creating an index from them would be meaningless because it would not be possible to predict a score on any one item from a score on another, and little, if any relationship would exist between scores on any single item and overall scores on the index. Hence, the ability to predict the score on the index from the individual items is a measure of the extent to which the items can be treated as representing a unified dimension. The validity of the index was investigated by item analysis. The index was dichotomised and the hypothesis that the total score was related to individual scores was tested with a X^2 test. Each item was significantly related to the total score with a chance occurrence approaching infinity and therefore the index could be considered to be measuring a single dimension.

(ii) The validity of the severity hypothesis: The second assumption made in constructing the index was that a higher score indicated a greater severity of problems. Tests of this assumption were made by an analysis of data collected as part of a small pilot study of patients referred to a hospital alcoholism clinic. Each of the patients had been asked the seven items on the index.

Since it is assumed that a large proportion of people referred to such clinics suffer from various alcohol related problems, their scores on the index of alcohol related problems could be used as a validation of the severity hypothesis.

Hence, there seems to be a prima facie case for suggesting that the number of items reported is an indication of severity.

TABLE II : SCORES ON INDEX OF ALCOHOL RELATED PROBLEMS OF 18 SUBJECTS ATTENDING OUT-PATIENT ALCOHOLISM CLINIC

Score	No. of patients
7	3
6	6
5	1
4	2
3	5
2	0
1	1

(e) The measurement of alcohol consumption: A similar approach to the measurement of the respondents' consumption of alcohol was used in each survey. First it was ascertained how long it was since the person last drank. Detailed data was only collected from those respondents who had consumed alcohol during the seven days prior to interview. In both surveys the respondents were asked what types of alcohol they had drunk on each of the seven days prior to interview before being asked how many drinks of each type they had consumed. Interviewers were instructed to ensure that as far as possible respondents did not omit any drinks, and that the size of the drinks were recorded as accurately as possible.

For the purpose of this analysis the information was coded so that drinks with approximately the same concentration of alcohol were each rated as one drink. Hence, half a pint of beer, a glass of wine, a small sherry or a single tot of whisky - each of which contains approximately one centilitre of absolute alcohol - were classified as one drink.

The population study can be divided into two general groups. Firstly, abstainers, who had not drunk in the year prior to interview. Secondly, the drinking population, who drank within the twelve months prior to interview.

The measure of consumption that this paper will be concerned with is the total consumption of the drinking population. The total consumption was calculated by adding the total number of drinks respondents consumed during the seven days prior to interview.

In summary then, the analysis will be concerned with three indicators of the prevalence of alcohol related problems. These are hospital admissions of diagnosed alcoholics,convictions for drunkenness offences, and scores on the index of alcohol related problems. The analysis also employs two measures of alcohol consumption - the per capita consumption of the United Kingdom population and the consumption of the survey respondents.

THE ANALYSIS

The first condition of the model was that the per capita consumption of alcohol in the area under study should have increased. The per capita consumption of alcohol rose in the United Kingdom from 6.1 litres per person over the age of 15 in 1965 to 9.1 litres in 1973.

As the following analysis is based upon survey data it is necessary that the per capita consumption of the survey samples should have also risen. In the 1965 sample the average number of drinks consumed in the seven days prior to interview was 5.3 whilst in the 1974 sample it was 8.

The average yearly increase in consumption indicated by national statistics was 4.5% while that for the survey data was 5.6%. Hence the two measures show similar trends and the first condition is fulfilled.

The second condition necessary to verify the model was that the change in per capita consumption should be the result of a general trend towards heavier consumption throughout the population. By their nature, national statistics cannot be used to examine this condition. The survey data are amenable to this form of analysis.

The first question is how far has the change in per capita consumption been the result of a reduction in the proportion of abstainers in the population. Table III presents the time period elapsing between the date of interview and the last consumption of alcohol beverage reported by respondents.

TABLE III : TIME OF LAST DRINK

	Sample			
	1965		1974	
	N	%	N	%
During the last 7 days	548	59.1	175	61.2
8 to 28 days ago	124	13.4	35	12.2
28 days to 12 months ago	153	16.5	41	14.3
More than 12 months ago	103	11.1	34	11.9
Not known	0		1	0.4

$X^2 = 1.131$ 3df N.S.

There was no statistically significant difference between the two samples and the proportion of abstainers in the populations were similar. The survey data refute the contention that per capita consumption has been rising because there were fewer abstainers in the population.

Having eliminated the role of changing patterns of abstention in the rise in per capita consumption, it is now possible to consider how far changes have been the result of a general move towards heavier drinking.

Table IV displays the number of respondents in each of seven consumption categories. There is a general tendency for respondents in the 1974 sample to report consuming more drinks than respondents in the 1965 sample.

It could be argued that this tendency towards heavier consumption was the result of an increase in light frequent drinking. An earlier analysis of this data has shown this not to be the case. The number of days on which members of the 1965 sample reported drinking was 1.9 while the number of days reported by the 1974 sample was 2. The increase in total consumption was found to be due to respondents drinking more heavily on those occasions when they drank. In 1965 respondents consumed on average 1.8 drinks on each drinking occasion, but in 1974 they reported consuming an average 2.7 drinks - an increase of 48% (Cartwright et al. 1975b).

TABLE IV : PERCENTAGE OF DRINKING POPULATION SAMPLES CONSUMING GIVEN NUMBERS OF DRINKS

	1965		1974	
	N	%	N	%
0 - 9.9	393	71.7	112	64.0
10 - 19.9	93	17.0	28	16.0
20 - 29.9	32	5.8	15	8.6
30 - 39.9	13	2.4	4	2.3
40 - 49.9	6	1.1	8	4.6
50 - 59.9	4	0.7	5	2.9
60 +	7	1.3	3	1.7

* X^2 = 19.9 6 df Sig. 0.003

* F = 10.438 Sig. 0.002

Controlled for age, sex and occupational status

Hence the survey data support the contention that a change in per capita consumption is an indication of a general trend towards heavier drinking, and this fulfills the second condition.

The third condition postulated was that more people would be found in high risk categories in populations with higher per capita consumption. Implicitly this means that as an individual consumes more alcohol they will be more likely to experience alcohol related problems. To examine this condition the two drinking samples have been combined and their scores on the index of alcohol related problems plotted within the seven consumption categories. This relationship is shown in figure 2.

The relationship is linear and the correlation of 0.44 is highly significant, hence this data support the contention that the more alcohol a person consumes the more likely they are to be in a high risk category.

Indeed, the probability that a person in the highest consumption category will report at least three items on the index of alcohol related problems is 36 times greater than the probability for a person person in the lowest consumption category.

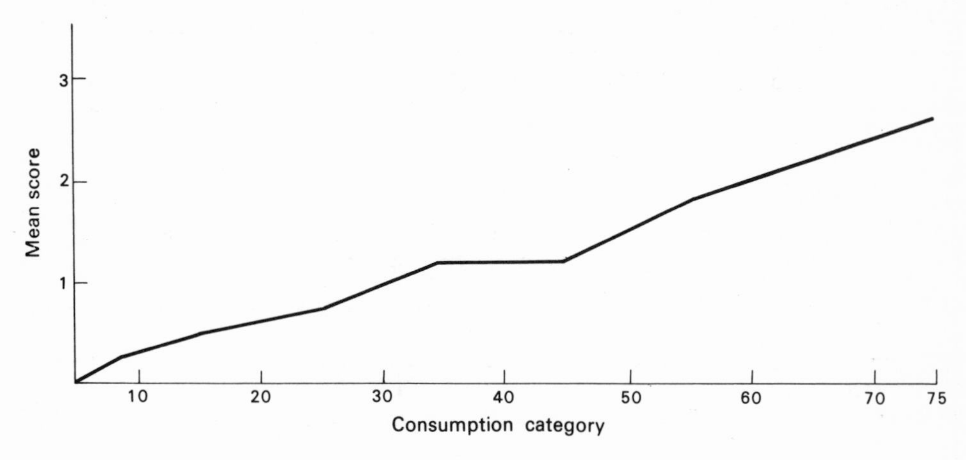

FIGURE 2. Mean Scores of Drinking Population Samples (Combined) on the Index of Alcohol Related Problems

In Table IV it was demonstrated that respondents in the 1974 sample were more likely to be in high consumption categories than respondents in the 1965 sample. As high consumption categories have been shown to be the equivalent of high risk categories, there were more respondents in high risk categories in the 1974 sample, and this fulfils the third condition.

The fourth condition postulated was that with a rise in consumption, the prevalence of alcohol related problems should increase.

Table V shows the scores of the two samples on the index of alcohol related problems. The mean number of items reported by the 1965 sample was 0.24, while the 1974 sample reported a mean 0.43 - almost double. There was a strong trend for the 1974 respondents as a group to report more items on the index.

The analysis of variance indicates that the differences between the scores of the two samples on the index of alcohol related problems was statistically significant. It has already been argued that the higher scores on the index of alcohol related problems of respondents in the 1974 sample were due to their general tendency to consume more alcohol. If this is indeed the case then introducing the co-variate of total consumption into the model should eliminate the difference between the samples. This is precisely what happened; the value of F being reduced from a significant level of 7.2 to a non-significant level of 2.2.

TABLE V : SCORES OF THE DRINKING POPULATION SAMPLES ON THE INDEX OF ALCOHOL RELATED PROBLEMS

Score	Sample			
	1965		1974	
	N	%	N	%
0	726	88.0	194	77.3
1	55	6.7	32	12.7
2	20	2.4	7	2.8
3	8	1.0	12	4.8
4	11	1.3	4	1.6
5	1	0.1	1	0.4
6	2	0.2	1	0.4
7	2	0.2	0	0.0

Difference between categories

X^2 = 28.19, df = 7, Sig. = 0.002

Difference between sample means

*F = 7.204, Sig. = 0.007

Difference between sample means controlling for consumption

*F = 2.236, Sig. = 0.13

x Controlled for age, sex and occupational status.

CONCLUSION

Four conditions were postulated to test the validity of the notion that per capita consumption was an indicator of the prevalence of alcohol related problems. The conditions were that :-

(1) a study should be carried out in an area where the per capita consumption of alcohol was rising.

(2) increases in this per capita consumption should be related to a general trend towards heavier drinking.

(3) general heavier drinking should be associated with a higher risk of alcohol related problems.

(4) the prevalence of these alcohol related problems should have increased.

Within the limitations of the data, these four conditions have been fulfilled, and therefore it is reasonable to assume that per capita consumption is a valid indicator of the prevalence of alcohol related problems.

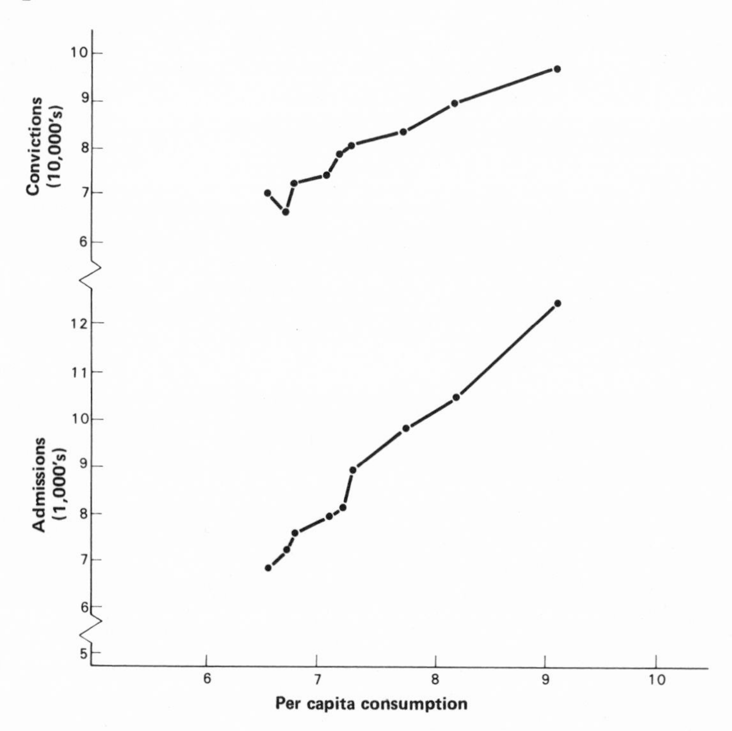

FIGURE 3. Alcoholism Admissions and Drunkenness Convictions by Per Capita Consumption

Having validated the model it is relevant to return to national statistics which relate per capita consumption to the prevalence of alcohol related problems. Figure 3 plots hospital admissions of diagnosed alcoholics, and convictions for drunkenness against per capita consumption. In each case, the correlation is greater than 0.95, which suggests that consumption explains 90% of the variance in prevalence. As the analysis of the model has confirmed, this association can be taken to be valid, since per capita consumption is a valid indicator of the prevalence of alcohol related problems.

REFERENCES

ANNUAL ABSTRACT OF STATISTICS (1974). H.M.S.O. London.

BLYTH, W.G. and MARCHANT, L.J. (1972). "A Self-Weighting Random Sampling Technique". Journal of the Market Research Society, 15, (3), 157-162.

CARTWRIGHT, A.K.J., SHAW, S.J. and SPRATLEY, T.A. (1975a). "Designing a Comprehensive Community Response to Problems of Alcohol Abuse". Report to the Department of Health and Social Security, Maudsley Alcohol Pilot Project.

CARTWRIGHT, A.K.J., SHAW, S.J. and SPRATLEY, T.A. (1975b). "Changing Patterns of Drinking in a London Suburb". Anglo-French Symposium on Alcoholism, Paris.

DEPARTMENT OF HEALTH & SOCIAL SECURITY (1974). "Background to Policy". Department of Health and Social Security, Paper for Meeting of Consultants.

EDWARDS, G., CHANDLER, J. and HENSMAN, C. (1972). "Drinking in a London Suburb I: Correlates of Normal Drinking". Quarterly Journal Studies on Alcohol, 6, 69-93.

LEDERMANN, S. (1964). "Alcool, Alcoolisme, Alcoolisation: donnés scientifiques de caractère physiologique, economique et social". Institut National d'Etades Demographiques, Travaux et Documents, 29, Presses Universitaires, Paris.

POPHAM, R.E., SCHMIDT, W. and de LINT, J. (1975). "The Prevention of Alcoholism: Epidemiological Studies of the Effects of Government Control Measures". British Journal of Addiction, 70, 125-144.

ON THE NEED TO RECONCILE THE AETIOLOGIES OF DRUG ABUSE

CINDY FAZEY

DEPARTMENT OF SOCIAL STUDIES

PRESTON POLYTECHNIC, PRESTON

To begin by saying that one makes no apology for the views expressed here, does in itself rather sound like an apology. But the aim is to look at the overall trends in aetiological research in drug dependence, including of course alcohol and tobacco, perhaps in a somewhat polemical fashion - and to examine some of the problems and issues raised. The author is aware that there are some research projects to which the following remarks do not apply.

The main theme of this paper is to suggest that while on the one hand there is an abundance of research "findings", on the other hand little progress is being made in understanding. This is due to several factors. First, there is a lack of genuinely interdisciplinary research; second, there is a profusion of, and trend towards epidemiological projects; thirdly much research is beset by methodological inadequacies and limitations and fourthly, the organisation of much research leaves a great deal to be desired.

During the course of this paper, it is hoped to make tentative suggestions as to how some of these shortcomings may be alleviated. Primarily this can be done by overcoming the disciplinary isolation, subject-field isolation (that is the study of alcohol, or cigarette smoking, or heroin use as separate and distinct entities) and also by changing the scale of research. Several workers have recently acknowledged the need for a multidisciplinary approach. But they have done so from their own disciplinary fortresses.

Academic tokenism has been the result. There has been an understandable tendency to carry on in the old ways from discipline-based conceptual frameworks at the same time acknowledging the contribution that other disciplines may, could, would or should make to a fuller understanding. A rare exception is in a paper by Sytinsky (1973). Alternatively, a shopping list of other variables usually felt to be other disciplines' preserves is cited which it is said will also affect the situation. But is anything done about it? Partly this seems to be a function of the scale of research but partly it seems that there is a genuine lack of effort to appreciate what other disciplines are saying and doing. This failure is evident not only within the traditionally circumscribed subject-fields which have characterised the recent developments in drug use/abuse study, but particularly across these fields. For example, studies in the possible biological differential susceptibility of people to alcohol - in an individual or national basis - are often only cited en passant when personality variables are being studied. But both may be given fleeting reference by others looking at the social situations in which heavy/problem drinking and/or alcoholism occur.

But more than this, by keeping separate the fields of problem drinking and alcoholism from tobacco research, from other forms of drug dependence research, many interesting theoretical and methodological advances made by studies in one area which could prove fruitful to others, have gone unrecognised. For example, the work by Tonkins (1966), McKennel and Thomas (1967) and McKennel (1973) in the field of the aetiology of smoking might prove very useful applied to other forms of drug use. Further, this subject-field restriction ignores the reality that many people are multi-drug users (Freed 1973). Why should we pick on one drug in these cases and ignore the others? By the same token, to overspill as it were, the medical model into all forms of drug abuse, with contagion theory, is to expand the concept of contagion to a level of generality which is surely nonsense (Bejerot 1970).

The point about contagion theory is that people catch a disease by contact willy nilly of whether they wish to. It may be considered that television is a social ill but surely not a social disease, and yet people see T.V. before they buy one and their number has increased almost logarithmically in most countries in recent years. This use of "contagion" denies any free will in the actor at all. Certainly drug abuse, particularly in relation to marijuana and amphetamines can be seen to have spread through social contact, but not by contact alone. The social geographers' concept of diffusion might provide a more useful analogy here. But another result of field specialisation has been a concentration on the more extreme forms of drug behaviour, e.g. alcoholism and heroin addiction, so that it is very difficult to explain why heavy drinkers are not alcoholics and occasional users of heroin do not become addicted. It has been assumed that physical dependence upon some drugs is both quantitatively and qualitatively different, but how far is this assumption justified ? Some people, for example, are physically dependent on drugs to sleep at night but conceptually are put in a different aetiological pigeon-hole from other forms of drug dependence.

There is not only a need to integrate vertically within a subject field, that is to look at different levels of explanation, but also across subject fields, horizontally, that is to look at the same level of explanation across different forms of drug use. When this is done not only can competing explanations be seen as such, but many apparently differing views reconciled. An attempt to integrate both psychological and sociological variables, that is to reconcile different levels of explanation, in the field of alcoholism is seen in the work of Richard Jessor (1968) (1975) and Peter Würthrich (1974). A demonstration of how one concept, Jellinek's "loss of control", can be seen from another perspective is found in a recent paper by Bacon (1973). Further, data supporting the "dependency-conflict" hypothesis in a study by Bacon (1974) could equally be used to support the "need for personalised power" hypothesis of McClellan (1972) and of Wanner (1972). Also, reinforcement theories, (Wikler, 1965, 1973) social learning theories (Kingham 1958) and interactionist theories seem to be describing the same process, but with differing emphases on the self, the other or the process, whilst using different terminology to describe essentially the same concepts.

As some recent papers have pointed out, different disciplines have used different models for the study of drug use, so that not only is the aetiology seen from different perspectives, but the ramifications of this are that different treatment programmes are recommended and different social action advised. There have been wholesale shifts in the definition of problems of drug use (even in defining certain patterns of drug use as problems) and with them go the assumptions underlying these definitions (Siegler and Osmond 1968, Siegler et al. 1968, Steiner 1969). Defining a priori certain patterns of behaviour as ill or criminal simplifies and reifies the behaviour concerned. In other words, the model becomes the reality, often not least because these judgements often contain implicit moral judgements.

Perhaps in response to demands by governments and government agencies there appears to be a trend towards more headcounting - a headlong rush into the field to find out who is doing what by age, sex and social class. So we have a proliferation of methodologically neat studies - often circumscribed admittedly by the finance available, which are carefully defined and which produce, after a short time, "facts" and "figures" to justify the exercise. But the result has been a dearth of genuinely longitudinal studies excepting McCord and McCord (1960), so that even where the finance exists, not one, but a plethora of more easily managed smaller projects, with emphasis on quantification, result. There is no justification, from an examination of the development of the natural sciences, to assume that knowledge develops in a building brick fashion, and that the contribution will be to collect as many "facts" as possible and to await the day when some theoretician will wave a magic wand and produce an edifice of aetiological theory.

Without trying to minimise the complex and difficult methodological problems which beset research in this field, it does seem that some researchers are not even cognizant with them, let alone trying to overcome them. (An excellent review of some of the methodological problems associated with the study of illicit drug use is provided by Sadava (1975)). Some of the major methodological problems which beset research are those of :

1. Sampling: Generalisations from pitifully small non-random samples to general population are unacceptable. Nor does any increase in non-random sample size increase accuracy. (For an evaluation of results due to sampling errors see Gendreau and Gendreau (1970), Peters and Ferris (1967) and Tinklenberg (1975)).

2. Lack of genuine hypotheses: Lack of specification of conditions under which a theory can be accepted or disproved.

3. Proof: A touching faith in the logic of induction does not overcome the fact that conclusive proof is not possible, only conclusive refutation. Apparent support for a theory means no more than that - apparent support.

This is not to deny the heuristic nature of much research, but that many ideas are put forward and evidence collected to support them, and presented in such a way that they cannot be tested - for there are not conditions under which they can be refuted. Theories of anxiety and tension reduction often come into this category (Cappell, 1975).

4. Cause: Apart from the gratuitous assumption that correlation means causal relationship, there is the problem of ascertaining and often confusing cause and effect, that is the assumption of antecedent variables. (Often the result of, for theoretical reasons, minimising or ignoring the capacity for change through interaction with the environment in the widest sense). For example, models of anxiety reduction and tension reduction have often been put forward to account for alcohol use (Horton 1943, Field 1962) but it is surely possible for anxiety and tension to be the result of dependency on a drug, and become evident on the potential withdrawal or loss of supply.

5. Control groups: The lack of sophisticated, or in many cases, any, control groups, renders many interesting and promising lines of enquiry of limited use. In turn, this is seen in the lack of

6. Replication studies: Their deficit is partly a function of the organisation of research but replication studies and a triangulation of research - that is measuring the same variable by different methods - would at least enable the theory field to shed some of its mistaken ideas. Even where some replication has been attempted, the independent variables e.g. ethnic group, have been selected on different criteria (Heath 1975).

Many of the foregoing problems, it seems, arise from the organisation of research. People or rather usually governments want quick answers, and it is the contention here that we have, as far as aetiology is concerned, gone beyond the stage of the small studies. However the problem is perhaps more fundamental than this for it derives from the finding of research and the relationship between research and career patterns. The level of sophistication of the natural sciences may not have been reached but we have got beyond the stage of the alchemists. Case studies, for example, are an important and valid approach in medicine because the symptoms and progress of a specific disease are almost universal. In another way, case studies are useful. Before dashing into the field there is a need to have some idea of what might constitute relevant or significant variables. But in the field of drug abuse we are surely past this stage.

It is argued here that if strides are to be made in understanding the aetiology of drug use, then the scale and orientation of research must be enlarged. Not only should longitudinal studies be undertaken but the phenomenon which it is sought to explain must be studied at all levels of explanation. Instead of studying one variable or small group of variables from the point of view of one discipline, it is necessary to study theoretically relevant variables from different disciplines at the same time and over the same population. Indeed, it can be seen that many disciplines, using their own language are perhaps talking about essentially the same thing. For example the ideal that addicts "mature out of narcotic addiction" has never been adequately tested because the statistics on which it was based were shown to be at fault. However Jessor (1968) related onset of adolescent drinking to "transition prone" problems.

The communication (Gorad 1971), transactional (Steiner 1969), systems theory (Ward and Faillace 1970) and interactionist (Rubington 1967, Carey 1968) approaches are essentially concerned with the same thing, that is the dynamic relationship between the drug taker and his environment, and in particular his family. These analyses need not be confined to the drug with which they were originally concerned, emphasising the process of drug involvement. They can also be related, often retrospectively, to the same group of interactions, emphasising personality factors.

Short term convenience has stultified any such research development. Differential susceptibility to the drug, the relationship between physiology and personality, personality and modes of interaction, the way in which the individual perceives his environment, as well as the micro- and macro-social forces which structure the pattern of relationships are among variables which need to be taken into account. To isolate one level or aspect of explanation seems to merely perpetuate misunderstanding. Nor is it enough to list the variables which might be relevant. Shopping lists are not explanations. An endeavour must be made to work towards an understanding of drug abuse/use, through the development of a theoretical framework which recognise not only different levels of analysis but also the differing perceptions of the nature of reality within which this pattern of behaviour comes to be developed. A framework should be structured that encompasses not only the individuals' construction of events, but their belief of the effects of the drug, for example, the use of opiates (Ding 1972, Leong 1974) and marijuana (El Jabbory 1972) for aphrodisiac purposes. Patently also any assumption about the uniformity of drug effects must be abandoned in view of research which suggests that the effects of a drug are not only individually variable, but that the display of symptoms are socially and psychologically conditioned (MacAndrew et al. 1970, Teasdale 1973).

Ironically perhaps, the plea of this researcher is to stop research, to stop that is unless we can produce more sustained, longer term, genuinely interdisciplinary research which is not pigeon-holed by type of drug. Some time ago the "rush to combine" the study of alcohol and other drugs was criticised. Quite rightly the naive lumping together of alcohol and heroin addiction was criticised but it is argued that at a more sophisticated level this is essential. Simply a look at the choice of drug in various cultures (e.g. by Carstairs 1954, Westermeyer 1971, Singer 1971), reveals that we need to look at drug use not as a pattern of behaviour, but in terms of patterns of behaviour which may not always be directly related to the physiological effect of the drug in question but to legal, social and religious definitions of its effects. Different drugs may be used or abused in different cultures, or within the same culture, for very similar reasons, and conversely the same drug used for very different reasons. The search for a reason why people take a drug involves a simplification and reification of behaviour to suit our particular disciplines.

It has been the contention of this paper that unless the scale of research increases, and unless we break out of our disciplinary fortresses and circumscribed subject-fields, we may be in danger of becoming not a "fachmeister" but a "fachidioten".

REFERENCES

BACON, M.K. (1974). "The Dependency-Conflict Hypothesis and the Frequency of Drunkenness: Further Evidence from a Cross-Cultural Study". Quarterly Journal of Studies on Alcohol, 35, 3, 863-876.

BACON, S. (1973). "The process of Addiction to Alcohol: Social Aspects" Quarterly Journal of Studies on Alcohol, 30, 4, 920-93.

BEJEROT, N. (1970). Addiction and Society. Springfield Illinois, Charles C. Thomas.

CAPPELL, H. (1975) "An Evaluation of Tension Models of Alcohol Consumption". Gibbens, Robert J. et al. (eds) Research Advances in Alcohol and Drug Problems 2, 177-209, New York, Wiley.

CAREY, T. (1968) The College Drug Scene, Englewood Cliffs, N.J., Prentice-Hall

CARSTAIRS, G.M. (1954) "Cultural factors in the Choice of Intoxicant". Quarterly Journal of Studies on Alcohol, 15, 220-237.

DING, L.K. (1972) "The Role of Sex in Narcotic Addiction in Hong Kong". Asian Journal of Medicine, 3, 119-121.

EL JABBORY, T.K. (1972) Research into the Phenomenon of Drug Taking in Iraq, National Centre for Social and Criminal Research, Ministry of Employment and Social Affairs, Iraq.

FIELD, P.B. (1962) "A New Cross-Cultural Study of Drunkenness". Pittman, D.J. and Snyder, C.R. (eds), Society, Culture and Drinking Patterns, 48-74, Wiley, London.

FREED, E.X. (1973) "Drug Abuse by Alcoholics: A Review". The International Journal of the Addictions, 8, 3, 451-473.

GENDREAU, P. and GENDREAU, L.P. (1970). "The 'addiction-prone' personality: A Study of Canadian Heroin Addicts". Canadian Journal of Behavioural Science, 2, 1, 18-25.

GORAD, S.L. et al. (1971). "A Communications Approach to Alcoholism". Quarterly Journal of Studies on Alcohol, 32, 651-668.

HEATH, D.B. (1975). "A critical Review of Ethnographic Studies of Alcohol Use: Gibbens, R.J. et al (eds.). Research Advances in Alcohol and Drug Problems Volume 2. 1-92, Wiley, New York.

HORTON, D. (1943). "The Functions of Alcohol in Primitive Societies: A Cross-Cultural Study". Quarterly Journal of Studies on Alcohol, 4, 199-320.

JESSOR, R. (1968). Society, Personality and Deviant Behaviour: A Study of a Tri-ethnic Community Holt Rinehart and Winstone, New York.

JESSOR, R. (1975). "Adolescent Development and the Onset of Drinking: A Longitudinal Study". Journal of Studies on Alcohol, 36, 1, 27-51.

KINGHAM, R.J. (1958). "Alcoholism and the Reinforcement Theory of Learning". Quarterly Journal of Studies on Alcohol, 19, 320-330.

LEONG, J.H.K. (1974). "Cross-cultural Influences on Ideas about Drugs". Bulletin on Narcotics, 26, 4, 1-8.

MacANDREW, C. and EDGERTON, R.B. (1970). Drunken Comportment: A Social Explanation, Nelson, London.

McCLELLAND, D.C. (1972). Summary: The Drinking Man: Alcohol and Human Motivation, 332-336, Free Press, New York.

McCORD, W. and McCORD J. (1960). Origins of Alcoholism Tavistock, London.

McKENNELL, A.C. and THOMAS, R.K. (1967). Adults and Adolescents Smoking Habits and Attitudes Ministry of Health, H.M.S.O. London.

McKENNELL, A.C. (1973). A Comparison of Two Smoking Typologies. Tobacco Research Council, London.

PETERS, J.M. and FERRIS, B.G. (1967). "Morphological Constitution and Smoking". Archives of Environmental Health, 14, 678-681.

RUBINGTON E. (1967). "Drug Addiction as a Deviant Career". The International Journal of the Addictions, 2, 1, 3-20.

SADAVA, S.W. (1975). "Research Approaches in Illicit Drug Use: A Critical Review". Genetic Psychology Monographs, 91, 3-59.

SIEGLER, M. and OSMOND, H. (1968). "Models of Drug Addiction". International Journal of the Addictions, 3, 1, 3-24.

SIEGLER, M. et al (1968). "Models of Alcoholism". Quarterly Journal of Studies on Alcohol, 29, 3, 571-591.

SINGER, K. (1974). "The Choice of Intoxicant among the Chinese". British Journal of Addiction, 69, 3, 257-268.

STEINER, C.M. (1969). "The Alcohol Game". Quarterly Journal of Studies on Alcohol, 30, 4, 920-938.

SYTINSKY, I.A. (1973), "A Schema of the Etiology of Alcoholism as a Pathological Motivation: A Working Hypothesis involving the Interplay of Sociological, Psychological and Physiobiological Factors on Molecular, Cellular and Organosystemic Levels". Quarterly Journal of Studies on Alcohol, 34, 4.

TEASDALE, J.D. (1973). "Conditioned Abstinence in Narcotic Addicts". The International Journal of the Additions, 8, 2, 273-292.

TINKLENBERG, J.R. (1975). "Assessing the Effects of Drug Use on Antisocial Behaviour". Annals of the American Academy of Political and Social Science, 417, 66-75.

TONKINS, S.S. (1966). "Psychological Model for Smoking Behaviour". American Journal of Public Health, 56, 17-20.

WANNER, E. (1972). "Power and Inhibition: A Revision of the Magical Potency Theory". In: McClelland, D.C. et al (eds). The Drinking Man: Alcohol and Human Motivation, 73-98, New York Free Press.

WARD, R.F. and FAILLANCE, L.A. (1970). "The Alcoholic and His Helpers. A Systematic View". Quarterly Journal of Studies on Alcohol, 31, 3, 684-691.

WESTERMEYER, J. (1971). "Use of Alcohol and Opium by the Meo of Laos". American Journal of Psychiatry, 127, 8, 1019-1023.

WIKLER, A. (1965). "Conditioning factors in Opiate Addiction and Relapse". In: Wilner, D.M. and Kassebaum, G.G. (eds). Narcotics, 85-100, McGraw Hill, New York.

WIKLER, A. (1973), "Dynamics of Drug Dependence. Implications of a Conditioning Theory for Research and Treatment". Fisher, S. and Freedman, A.M. (eds). Opiate Addiction: Origins and Treatment, 7-21, H.V. Winston, Washington D.C.

WURTHRICH, P. (1974). Zur Sociogenises of Chronischen Alkohismus, S. Karger, Basel.

DRINKING PATTERNS OF YOUNG PEOPLE

ANN HAWKER

MEDICAL COUNCIL ON ALCOHOLISM

LONDON

For the last year the drinking behaviour of young people has made headlines at least once a week in the daily press in this country and there is no doubt that there is a growing universal concern about teenage drinking which is by no means confined to only a few individuals or even a few individual areas.

If one looks briefly at the more recent research from the U.S.A., Scandinavia and what little is available in Great Britain, reports from a variety of sources indicate that children are generally drinking or beginning to drink at an early age. In this country this is a relatively new phenomenon but in America this is not so and consequently there is much in the American literature that shows the patterns of drinking of school children and university students. The first surveys on young people were conducted on students attending university and were the first major studies of their kind. These included work not only from the U.S.A. but Scandinavia and the U.K. and the populations studied were groups of young people who on the whole were legally entitled to drink. The majority of reports indicate that young persons are starting to drink at an early age and that by the age of 18 most youngsters had already been drinking and purchasing alcohol for many years.

In America, where social workers regard alcoholism as a major problem, results from a variety of studies show that 63% of boys between the ages of 11 and 13 and 53% of girls in the same age group had at least tried alcohol and there was sufficient evidence to categorically say that one in seven of 17 year olds were getting drunk once a week. These young people do not appear to be an isolated group who have academic or emotional problems but a complete cross-section of young people attending school. A recent survey in New York with a sample of 10,000 high school students aged 16-19 was carried out by two psychologists who estimated that one in eight had either a drinking problem or were already showing symptoms of dependency on alcohol. Some of these students claimed that by the age of 12 they were drinking on a weekly basis and nearly a quarter of them said they liked drinking because they liked getting drunk. The quantity and frequency of their drinking was directly related to their parents' drinking behaviour and the availability of alcohol at home. One point worth noting is that in many states in America, beer is cheaper than coca-cola. Dr. Morris Chafetz has recently stated that "the situation in America will be catastrophic if our young people do not adopt more responsible attitudes towards the taking of alcohol than today's adults".

Denmark has recently launched a public enquiry under the auspices of the Ministry of Education asking a large group of 15 year olds to give their views about drinking and alcoholism; the results will be used with a view to applying preventative measures. A similar study has been completed in Sweden.

In this country researchers at the Addiction Research Unit of the Institute of Psychiatry have examined the attitudes and drinking behaviour of students attending two colleges of the University of London, Postoyan & Orford 1970) while Davies & Stacey (1972) published their report "Teenagers and Alcohol" - a developmental study study carried out in Glasgow which was sponsored by the Scottish Health Education Unit. The latter study was designed to elicit three classes of information :

1. Commencement and maintenance of youthful drinking behaviour and their relationship to selected demographic variables.

2. Attitudes of young people towards drink and drinking and towards a variety of other related activities.

3. Self images of young people and their stereotypes of other young people - notably of heavy drinkers and abstainers.

The results showed that 85% of the girls and 92% of the boys had tasted alcohol by the age of 14. This first drink had usually been offered by a parent or relative on some special occasion. With increasing age larger quantities of drink were consumed at a friend's house rather than the parental home. The results showed that many of the heavy drinkers in the sample were introduced to alcohol later than the more moderate drinkers - this suggested that they learned to use alcohol alone rather than in the parental home. These included boys who, at first started to drink in the open air -

parks etc., at the age of 14 and then progressed to drinking in public houses by the age of 17 or just under.

Another report produced at the same time, also from Glasgow was "Children and Alcohol" (Jahoda & Cramond 1972). The sample of 240 children were divided into three age groups of six, eight and ten years. Six children out of every ten had tasted alcohol. Many were able to identify alcoholic drinks by their smell alone. They recognized the behavioural manifestations of drunkeness and by the time they had reached the age of eight they had some degree of understanding of the role that alcohol plays in society as they knew it. In the older groups some of the children expressed negative attitudes about their own future individual use of alcohol, but as the results of the parallel study on 14-17 year olds show that the large majority of both boys and girls were drinking, a marked change in attitudes must occur in the intervening years.

If we examine the major findings of the work already briefly described, the various researchers have reached conclusions that have some similarity. Without doubt young people appear to be experimenting with alcohol at an earlier age than was previously the case; although the reasons are not clear some authorities speculate that earlier social maturation of today's young people leads to earlier experimentation in a variety of so-called adult behaviour.

Young people drink for a variety of reasons - the most important would appear to be parental and peer influences. As parents serve as role models for adult behaviour their attitudes towards drinking and their own drinking behaviour must play a major role in determining their children's approach to alcohol during adolescence. Davies and Stacey found that parents with extreme prohibitive attitudes towards drinking had children who learned to drink outside the home environment and tended from the outset of their drinking to develop devious attitudes towards it. Peer influence is demonstrably a factor in much teenage drinking. Both "sociability" and the desire to do something "grown up" have been emphasised as reasons for drinking in the majority of the surveys.

Apart from the Scottish reports there is little factual evidence about the real extent of teenage drinking in this country. The various indices that are used to measure prevalence of problem drinking and alcoholism in an adult population are difficult to apply to young people with the exception of arrests for drunkenness. Certainly these figures have shown an increase for the last three years for those under 18 but one has to bear in mind that the average 15 year old who drinks will normally only be approached by the police if he or she exhibits either aggressive or violent behaviour. The upsurge of violence and hooliganism at football matches is usually a result of drinking and Professor Sparkes, a former Dean of one of the new universities, when reporting on student violence and rebellion was convinced that this only took place after students had been drinking.

The Medical Council on Alcoholism, because of its concern about the patterns of drinking amongst young people, has sponsored the production of two films with accompanying notes and for the last two years these have been shown to a large number of young people attending school. Many local education authorities now recognise the need for including education about alcohol and alcoholism in their education programmes and interest in the educational aspects has increased amongst youth leaders who are increasingly under pressure because of irresponsible drinking amongst their members.

Apart from involvement in providing information and audio-visual aids, the Medical Council on Alcoholism has sponsored two research projects dealing specifically with the drinking behaviour of young people. The first is a major cross-cultural study by Joyce O'Connor which has now been completed. She has examined the cultural influences on the drinking behaviour of a group of Irish, Anglo-Irish and English young people between the ages of 18 and 21; all the respondents and their parents were interviewed at length. The full interpretation and analysis of this work will give detailed information on drinking behaviour and attitudes towards drink in two different societies and the subsequent influence on the drinking patterns and attitudes of their offspring.

The second project for which the author has been awarded a two year Research Fellowship by the Medical Council on Alcoholism, is a study designed to provide some factual information about the drinking habits of children who are still attending school. Local education authorities in various parts of the country have been approached and asked to co-operate by providing schools who would be willing to participate. The final sample will be in the region of 9,000-10,000 children mainly between the ages of 13 and 16 with a small sample of 16-18 year olds. Comprehensive and grammar schools have been used in large industrial and small market towns in different areas, with a proportion of rural schools. The purpose of this research is to examine the actual drinking behaviour of young people - to discover whether there are any differences in the patterns of drinking that are significant in samples from urban and rural areas and to see to what extent, if any, social damage is a direct result of irresponsible drinking. It is known for instance that Alcoholics Anonymous reports a larger number of young alcoholics than ever before and agencies responsible for counselling, referral and treatment corroborate this. It is not suggested that children at school are all drinking in an addictive manner, or that children who report that they get drunk on occasions are necessarily potential alcoholics, but when one reaches the present position where there is widespread concern and often alarm about the drinking behaviour of young people, it becomes of paramount importance to establish factual evidence about the incidence and when necessary refute some of the sensational publicity that surrounds the whole area.

In order to undertake such a study it has of course been necessary to complete a pilot project. The purpose of any pilot study, particularly where any form of questionnaire is to be used, is not only to test out the methodology but to evaluate the questionnaire and to make some assessment of the validity of the responses. One of the problems in designing this type of questionnaire (which in the present case because of the size of the sample had to be self-administered), is to construct the questions in such a way that they are acceptable to children with a varying age range as well as with a variety of academic capabilities. The final version of the questionnaire includes questions on demographic data, questions about spare-time activities, questions about smoking, and questions relating to a fairly detailed account of drinking behaviour. The instructions stress that anonymity will be maintained, not only the respondent, but also the name of the school will remain confidential.

It will be appreciated that as the main study has not been completed it is impossible to give results, but the report will be published early next year.

Returning to the pilot project which was carried out in June 1975, it might be interesting to look briefly at some of the findings. The author was fortuntate in gaining the co-operation of a headmaster and a senior teacher in charge of a social studies programme who agreed to the pilot study being carried out in their school. This was a large, fairly new comprehensive school in Middlesex. The majority of children came from fairly affluent backgrounds. Most of them lived on council estates with few recreational facilities and the school described them as coming from culturally deprived homes. All the parents of the children taking part in the survey were informed by the Headmaster that the school had agreed to the questionnaires being completed during school hours and if parents had any queries or objections they should contact the school. They were also informed that questionnaires were available for inspection. No parents objected or raised any queries.

The total number of completed questionnaires was 326. Four were spoilt so the final sample was 160 girls and 160 boys with an age range of 13-16.

The section on smoking included a question on education about smoking and a number of both girls and boys said that after seeing an educational film on the effects of smoking they had either cut down or stopped smoking altogether. Far more parents disapproved of smoking than drinking; the number of children who smoked in excess of ten cigarettes daily was comparatively small, boys smoking more than girls.

Previous studies have reported that the majority of teenagers have tasted alcohol by the age of 14; the results in this study were similar. 20% of the girls had their first alcoholic drink at the age of ten or under, compared to 33% of the boys and by the age of 17+ 90% of the girls were drinking compared to 98% of the boys. Ten girls and seven boys did not start drinking until the age of 14.

The first drinking experience for both girls and boys alike - 55% of the sample took place at home. 12% of the boys had their first drink in a pub compared to 4% of the girls; 15% of the boys and 18% of the girls drank for the first time at a party.

TABLE I: PLACE OF FIRST DRINK

	(N = 157) BOYS	(N = 145) GIRLS
Home	55%	55%
Pub	12%	4%
Party	15%	18%
Disco/Friend's House/Open Air	18%	23%

In respect to frequency of drinking 30% of the girls and 8% of the boys drank at Christmas or special celebrations only. 20% of the girls and 30% of the boys in all age groups drank once a week, three 15 year old girls drank every day and four 15 year old boys, five 16 year olds and two 17+ also drank daily.

TABLE II : FREQUENCY OF DRINKING

	(N = 157) BOYS	(N = 145) GIRLS
Christmas and special celebrations only	8%	30%
Once or twice a month	30%	32%
Once or twice a week	48%	30%
Every day	14%	8%

Questions relating to the effects of a hangover included physical effects, amnesic periods, and inability to attend school; the results were combined "once only" or "more than once". 26% of the girls (30 aged 15 and under) "once only"; 41% of the girls (40 aged 15 and under) more than once"; 25% of the boys (22 aged 15 and under) "once only", and 49% of the boys (40 aged 15 and under) "more than once".

TABLE III : EXPERIENCE OF A HANGOVER DURING THE LAST YEAR

	(N = 157) BOYS	(N = 145) GIRLS
Once only	25%	26%
More than once	49%	41%
No experience	26%	33%

Three girls reported trouble with the police as a direct result of drinking and so did 14 boys - ten under the age of 15. The question did not specify details so it cannot be assumed that these were drunkenness offences charges.

With regard to the results of having a lot to drink 10% of the girls and 7% of the boys answered "afraid to go home", 5% of the girls and 6% of the boys "aggressive at sports match", 4% of the girls and 10% of the boys "started a fight or an argument with a friend", 7% of the boys "feel like smashing things", two girls and five boys had taken "something belonging to someone else".

TABLE IV : RESULTS OF DRINKING TOO MUCH

	(N = 157) BOYS	(N = 145) GIRLS
Afraid to go home	7%	10%
Aggressive behaviour at a sports match	6%	5%
Started a fight or argument with a friend	10%	4%
Felt like smashing things	7%	-
Taken something belonging to someone else	4%	2%
None of the above	66%	79%

Information about pocket money and any earnings from part-time or holiday jobs showed that the majority were receiving in excess of £2 a week.

The majority of parents approved of their children drinking (61% of the parents of the girls and 70% of the parents of the boys). 13% of the girls said their parents did not know they drank, compared to 7% of the boys.

The respondents were asked about parents' attitudes towards their children drinking a lot or too much: 26% of the girls' and 30% of the boys' "parents strongly disapprove", 35% of the girls and 30% of the boys did not know what their parents' reactions would be. Very few thought that their parents would be either sympathetic or discuss the matter and offer advice.

Attitudes towards people who drink too much were elicited; 31% of the girls and boys saw them as "people who need help", 19% of the girls and 16% of the boys as "weak willed", 41% of the girls and 37% of the boys as "stupid people" and 9% of the girls and 16% of the boys as "amusing people".

TABLE V : ATTITUDES TOWARDS PEOPLE WHO REGULARLY DRINK TOO MUCH

	(N = 160) BOYS	(N = 160) GIRLS
People who need help	16%	15%
Weak willed	16%	19%
Stupid people	37%	41%
Amusing people	16%	9%
Don't know	15%	16%

Although 24% of the girls and 43% of the boys had on one occasion or more been refused the sale of alcohol because of being under age nearly half of them thought that the existing legislation was correct - 48% of the girls and 51% of the boys. 29% of the girls and 25% of the boys thought that the "age limit should be lowered" and 23% of the girls and 24% of the boys throught that children should be "allowed to drink when they want to". One girl qualified this by saying only if the parents agreed.

TABLE VI : ATTITUDES TOWARDS LEGISLATION

	(N = 160) BOYS	(N = 160) GIRLS
Present age limit correct	51%	48%
Age limit should be raised	-	-
Age limit should be lowered	25%	29%
Young people should be allowed to drink when they want to	24%	23%

A great deal of criticism about increasing the availability of alcohol by allowing supermarkets and shops to hold licences has been made recently but this study shows that the majority of teenagers bought their alcohol in pubs or off-licences and only a small minority elsewhere.

It was decided, as the project was a pilot study, not to include analysis of the variables so it is not possible to say which, if any, of the demographic factors influence this group of teenagers' drinking behaviour but the results do show that in this particular sample many of them are introduced to alcohol at a young age, often at home, and a proportion of them were drinking quite regularly. A minority appear to have suffered some social damage as a direct result of drinking, for example trouble with police, absenteeism from school and behavioural problems with friends and at social functions.

Another interesting fact that emerged was that quite a large number of children thought that the increasing publicity given to young peoples' drinking behaviour contributed to the increase in teenage drinking. A possible deterrent advocated by many was the banning of advertisements that promoted alcoholic drinks.

To try to estimate the actual amount of alcohol consumed either on any particular occasion, or an estimated average, is, an impossibility. Any of the proven quantity frequency indexes are not really applicable to children who probably drink from a variety of glasses, have no real idea of the accepted standard measures and are more inclined to over-estimate the actual amount they drink. It is more realistic to establish where they drink, who they drink with and how often they drink, and of course their reasons for drinking.

As indicated, the purpose of this research is to supply factual evidence of how and to what extent children are using alcohol. The author is not an expert in health education and does not make recommendations on the basis of the results. It is hoped that the completed report will be of use to those who are in a position to implement education on alcohol and alcoholism, if they see education as a logical step towards the prevention of the increasing problem of excessive drinking.

REFERENCES

DAVIES, J. and STACEY, B. (1972). "Teenagers and Alcohol : A Developmental Study in Glasgow" H.M.S.O., London.

JAHODA, G. and CRAMOND, J. (1972). "Children and Alcohol : A Developmental Study in Glasgow". H.M.S.O., London.

POSTOYAN, S. and ORFORD, J. (1970). "Drinking Behaviour and its Determinants amongst University Students in London". Proceedings of the Third International Conference on Alcoholism and Drug Dependence, Cardiff.

ALCOHOLISM AND PSYCHOLOGY - SOME RECENT TRENDS AND METHODS

G. LOWE

DEPARTMENT OF PSYCHOLOGY

UNIVERSITY OF HULL

With alcoholism increasing rapidly, effective measures are needed to control it. Psychologists approach it as a behavioural disorder rather than a disease. Current research stresses the role of individual differences in the aetiology and treatment of alcoholism.

There are many varieties of problem drinker, but the full-blown alcoholic has a fairly characteristic life-style. Imagine a 39 year old male who has been in hospital at least 20 times for alcoholism and who drinks up to two quarts of gin per day. For ten years he has had this drinking problem. Very few of us would fail to recognise this as a case of chronic alcoholism. Moreoever, it is likely that the only attention such a person has received would be from the police, hospital personnel, or (possibly) next of kin.

It is estimated that 75% of the adult population drink alcoholic beverages at least occasionally. Amongst these drinkers approximately 6% consume 'hazardous' amounts (more than 100 ml - about 3½ ounces or five 'social' drinks) per day. And although only a much smaller percentage become addicted to alcohol, this nevertheless amounts to an estimated 350,000 - 400,000 in Britain.

In recent years the disease concept of alcoholism has gained widespread acceptance among both professionals and non-professionals. However, many do not accept this thesis because they see alcoholism differing greatly from other diseases like heart ailments and pneumonia. The disease theory, which holds that addiction is the result of a chemical imbalance, has not been proved. A more serviceable concept may be to term alcoholism as a behavioural disorder.

As with other such disorders, three different categories of factors or causes may operate in the development of alcoholism - physiological, psychological and social. Max Glatt, in discussing addiction to alcohol and other drugs, stresses the interacting triad of the 'host' (i.e. the individual), the environment, and the 'agent' (i.e. the pharmacological nature of the drug concerned).

Since only a small proportion of the millions of drinkers become alcoholics, the possibility must be considered that the physiology of some persons makes them particularly susceptible. Much effort has been exerted to find chemicals in specific beverages which might be responsible for alcohol addiction, or physiological, nutritional, metabolic, or genetic defects which could explain excessive drinking. To date, these attempts have not succeeded. So far, it has been impossible to produce clear-cut alcohol addiction by any practical means in experimental animals.

The clearest case for the role of physiological factors in the development of alcoholism would be provided by the existence of hereditary tendencies. It is true that the children of alcoholics often end up the same way. But alcoholism also occurs in those whose parents are abstainers. Moreover, children of alcoholics can be protected if they are reared away from their parents. This strengthens the belief that alcoholism is related more to environmental than to genetic factors. Nevertheless, Gerald McLearn of the Institute of Behavioural Genetics, University of Colorado, argues strongly for a model of alcoholism that includes genetic parameters. His results from a selective breeding programme in rats and mice demonstrated that avidity for alcohol and differential sensitivity to its effects have heritable bases. The problem remains however, of extrapolation from animals to man.

It is believed by some people that alcoholics are psychologically 'different' that they possess a number of traits which in common make up the 'alcoholic personality'. There is, however, no agreement on the identity of these traits, nor on whether they may be the causes or the results of excessive drinking. Moreover, many of these same qualities and experiences have been observed in men and women who are not alcoholics, but who may be suffering from bizarre phobias or various mental ailments from mild neuroses to severe psychoses, or who may even be leading reasonably normal lives.

However, certain factors common to many alcoholics may play a part in the disorder. One of the most popular theories is that the individual drinks to remove anxiety. Intuitively this seems quite plausible, which may explain the considerable attention it has received. Several experiments with animals have clearly illustrated the efficacy of alcohol in reducing fear. Some human studies have used psychophysiological indicators - basal skin conductance (BSC) which reflects chronic tension level, and galvanic skin response (GSR) which reflects reactivity to emotionally arousing stimuli. The results confirm the widespread assumption that very moderate amounts of alcohol may reduce emotional tension. For some highly anxious individuals, tension reduction is probably a vital factor in producing drinking behaviour. And it is of interest to note that Lynn (1975) relies heavily on alcoholism as one index of anxiety. For most alcoholics, however, anxiety reduction is probably just one of several reinforcers of drinking and may not have provided the initial motivation.

Another notion is that drinking behaviour is somehow related to the psychological state of dependence. Although there is a tendency for the alcoholic to assume a dependent role in marriage with a dominant wife or husband, this could well be a consequence rather than a precursor of the alcoholic's state. Further theories have been concerned with conflict over sex-role, low frustration tolerance, and feelings of guilt - to mention just a few, but there is no conclusive evidence of an alcohol-prone personality.

Obviously the treatment of alcoholism cannot be undertaken except on a hit-or-miss basis until the problem of causation is solved. The evidence suggests that alcoholism is a complex product of possibly physiological and certainly psychological and sociological factors. This complexity bears upon the problem of treatment, for characteristically some things work in some cases, but not in others. The best results seem to be brought about by a combination of treatments - physiological, psychological and social.

Physiological treatment is primarily directed toward detoxification, mitigation of withdrawal symptoms, and the treatment of the deleterious physiological effects of prolonged alcohol use. Antipsychotic and antidepressant drugs are typically of little, if any, value in the treatment of chronic alcoholism, and tranquillizers may simply lead to an exchange of one form of dependence to another. Nevertheless, a survey of psychiatrists in the U.S.A. found alcoholism to be the third most frequent disorder for which psychotropic drugs were prescribed, with schizophrenia and depression ranking first and second respectively. It is surprising to find such widespread use of psychotropic drugs in a condition for which the efficacy of drug treatment remains largely unproven.

Another basically physiological approach which has been more commonly used is deterrent therapy, using drugs which sensitize the body to alcohol, such as disulfiram (Antabuse). Various distressing symptoms occur whenever alcohol is subsequently ingested (or in the case of some recently developed drugs whenever some limit is surpassed). These drugs act as a 'chemical fence' around the alcoholic, forcing him to suppress his drinking behaviour and supplying time for other therapies to be attempted.

Recent psychological approaches consider alcoholism as constituting a complex behaviour pattern and thus theoretically amenable to various methods of behavioural control. In one respect the conception of alcoholism has not changed much. Its causes were, and still are, thought to lie primarily in the individual. However, the manner in which the psychological factors in alcoholism must be handled has been revolutionized. It is generally agreed to be useless to harangue, denounce, or admonish the alcoholic, or to exhort him to strengthen his will. If the alcoholic drinks in the first place out of a sense of inadequacy, weakness, inferiority or guilt, it is hardly plausible that he will be inclined to give up drinking by intensifying these feelings.

Psychotherapy is a general label for a wide variety of procedures which attack the problem of alcoholism at the level of the individual's conscious and emotional experience. Trained professionals seek to raise the level of the individual's insight into his own problems, while at the same time giving him whatever counselling and guidance will help bring his drinking under control. When psychotherapy is successful, it is presumably because of a transformation of attitude and emotion on the part of the patient that renders drinking behaviour irrelevant.

The specific form that psychotherapy takes varies with the resources of the patient, the training and theoretical persuasion of the therapist, and the particular diagnosis. Group therapy (less expensive) is often the main method used in alcoholic units in mental hospitals.

The systematic use of behavioural modification and treatments based upon social learning formulations offers a promising new alternative to traditional methods. Within this framework alcohol abuse is seen as a learned behaviour pattern (socially acquired), and shaped and maintained by reinforcement contingencies. However, the shaping and control of behaviour is not necessarily the same as learning new patterns. Thus a comprehensive behavioural model requires a two-fold approach to treatment - firstly, techniques which suppress excessive drinking by decreasing the immediate reinforcing properties of alcohol, and secondly, techniques specifically designed to encourage non-drinking behaviour or behaviour incompatible with alcohol abuse.

Aversion therapy is based primarily on the Pavlovian or classical conditioning paradigm, and in the case of alcoholism, unpleasant stimulation is associated with the whole sequence from urges to actual drinking behaviour. A conditioned anxiety response is assumed to develop during this process, and any anxiety reduction following the appropriate avoidance behaviour serves as a reinforcer. The most common aversive stimuli used with alcoholics are chemical, electrical and verbal.

Chemical aversion techniques have typically used nausea-producing agents such as emetine or apomorphine, administered with an alcoholic beverage, and are intended to develop a conditioned reflex aversion to alcohol in any form. Nausea and vomiting occur shortly after drinking, and this procedure is repeated frequently during the course of treatment. The process is very different from the direct pharmacological action of Antabuse upon alcohol metabolism. Although this type of treatment is quick and provides the patient with the feeling that something positive is being done to help him, it is a drastic, painful procedure which pays little heed to any psychological problems the alcoholic may have. There are also difficulties in controlling the necessary temporal contiguity of alcohol intake and onset of nausea. Several investigators have observed that even when aversion is successfully learned to one type of alcoholic beverage (such as whisky), patients often drink beer, wine or rum while retaining their aversion to whisky. Furthermore, as with other forms of aversion treatment, regular 'booster' sessions are often required.

A more drastic aversive stimulus used in chemical aversion therapy has been scoline, a curarizing drug that momentarily produces total paralysis, including inability to breathe. This experience is associated with the sight and smell of alcohol. Very few anaesthetists these days would be willing to administer this procedure; the experience is (deliberately) terrifying and traumatic. And the results are in any case disappointing.

Other aversion procedures have been attempted using electric shock as the unconditioned stimulus. Here there is more precise control of intensity and temporal contiguity, but the approach has not been demonstrated to be any more effective than chemical aversion. In those cases where electrical aversion therapy is effective, there is evidence of increased psychophysiological responses to imaginal alcoholic stimuli. But such conditioned anxiety is generally observed in most successfully treated patients, whatever type of treatment is involved.

One of the newer developments in the treatment of alcoholism by aversion therapy is the use of noxious verbal images as the aversive stimulus. This method, called 'covert sensitization', involves intitial relaxation training followed by vivid descriptions of the gradual onset of violent nausea and vomiting. These are associated with all aspects of the sequence of behaviour leading to drinking. Emphasis is placed on self control and patients are advised to practise such sequences regularly.

Apart from indications that female therapists tend to be more effective in covert sensitization than males, the results of controlled studies with this technique are also unimpressive, especially in terms of long term abstinence/improvement. The fact that verbal aversion may be applicable to more people than other aversion procedures (since it is less traumatic, has no medical contra-indications, and does not require elaborate equipment) is of no help if the treatment is not significantly effective.

It seems likely, however, that specific aversion techniques may be appropriate for particular types of patient. Moreover, as Dr Gloria Litman of the Addiction Research Unit at the Institute of Psychiatry in London points out, aversion therapies are only initial procedures which may be useful in temporarily suppressing the deviant behaviour. Proponents of the behaviour modification approach would argue that the therapeutic goal should be the training of appropriate and positive alternative behaviours to excessive drinking behaviour.

Relaxation training and systematic desensitization have been used to provide the alcoholic with alternative ways of reducing anxiety to stressful situations. The patient is required to go systematically through a hierarchy of imagined anxiety provoking situations, which typically induced excessive drinking, until the associated anxiety is gradually dissipated. The state of relaxation (which may be induced by hypnosis, hypnotic drugs and other techniques) is important since anxiety and relaxation are incompatible responses. A number of successful case studies have been reported with patients who drink primarily because of social anxiety. It is worth mentioning, however, the case of one patient who had 23 desensitizing treatments, but required a further 27 sessions for his dependence on the therapist.

Recent applications of the operant approach in hospital settings have involved contingency management, using token economy systems and enriched vs impoverished ward environments. The former provides tight control of a whole range of social and environmental rewards and punishments; the latter can be used to alter drinking behaviour specifically but in a much more direct manner. A recent highly experimental attempt to regulate the alcoholic intake of patients in a closed ward setting by a decision making procedure looks very promising. The idea is that the patients themselves decide whether or not to drink at fixed (hourly) intervals. Patients are trained to stop and think about each decision, rather than act impulsively.

Whilst operant and other treatment programmes may be set up relatively easily enough in the hospital setting, 'in vivo' treatment (outside hospital) is obviously a more difficult proposition. Indeed, social reinforcement contingencies in the environment may actually maintain alcohol abuse. Attention from fellow drinkers and police, or possibly negative attention from an otherwise non-attentive wife are examples of such consequences of drinking. Attempts have been made, however, to rearrange social reinforcement in an alcoholic's environment by making peer companionship and spouse attention contingent upon non-alcoholic drinking behaviour, e.g. getting his friends to leave him immediately whenever he orders an alcoholic drink and his wife to be pleasant to him whenever he is sober.

Other workers have used operant techniques to investigate, experimentally, drinking patterns in chronic alcoholics. In one study a 'Skinner box' was set up and patients could 'work' (by pressing a lever) for alcohol or money or tokens. Immediate reinforcement resulted in spaced, moderate drinking, whereas delayed reinforcement (tokens earned could not be used until next day) led to massed, 'binge' drinking. These studies suggest that alcoholic drinking behaviour might be altered through rearranging schedules of reinforcement in the environment; the drinking patterns of alcoholics (and other drinkers) in France, where the licensing hours, etc. are different, would seem to bear this out.

The notion that chronic alcoholics can learn to drink in moderation is a new and highly controversial one. Total abstinence has been the emphatic goal of traditional alcohlism treatment. However, some proponents of behaviour modification are attempting treatment based on supervised controlled drinking practice (Hodgson 1975). Indeed, recent clinical evidence suggests that some chronic alcoholics can return to and maintain social drinking patterns. Such methods might be especially appropriate in the case of 'loss-of-control' alcoholics.

One point which may be relevant to the notion of controlled drinking is the likely incidence of alcoholism in the offspring of alcoholic parents. Since the children of both alcoholic and strictly abstemious parents (i.e. those with extreme views) are more likely to become alcoholic than children of more moderate parents, it would be eventually worthwhile to determine the extent of this risk in the children of 'controlled' as opposed to 'abstinent' formerly alcoholic parents.

An ingenious way to produce moderate drinking patterns involves training patients to discriminate their own blood alcohol levels. Through periodic feedback while drinking, patients are trained initially to discriminate the behavioural effects which typically accompany various blood alcohol levels, and asked to estimate their own levels. Electric shock is then made contingent upon concentrations exceeding some criterion level (typically 0.065%). Some success has been achieved with some patients, but the results can only be regarded as tentative at this stage. The discrimination of blood alcohol levels is not a universal phenomenon. Doug Cameron at Crichton Royal Hospital in Scotland recently attempted such discrimination training with normal social drinkers and found they generally could not estimate their own levels very well. Since alcoholics are more experienced drinkers, they are probably more 'tuned in' to the bodily and emotional changes due to alcohol. Moreover, some alcoholics will be more 'tuned in' than others.

Attempts to group alcoholics together (or, indeed with other drug addicts) are not likely to be helpful, in view of these and many other individual differences. Regardless of type of treatment, patient characteristics are highly related to outcome: notably, job stability, co-operation in treatment regime, and living with friends or relatives. There is currently a good deal of emphasis on a comprehensive treatment approach, where psychological, social and pharmacological techniques are combined to produce an individually tailored treatment for each patient. On a purely academic basis, it may be of interest to arrange specific treatment techniques so that proper controlled evaluation of them can be made. Unfortunately such an approach would not seem to be in the best interests of the patients. However, in the few intensive studies where such comparisons have been possible, it has generally been found that psychotherapy is best for alcoholics who are socially and psychologically stable, with drug therapy better for socially intact but psychologically below par individuals. Socially unstable alcoholics seem to do best in inpatient rehabilitation programmes. With individually tailored treatments, especially where better coping mechanisms are developed through social learning, the prognosis should be quite good - according to Max Glatt, roughly two in three alcoholics can be expected to improve greatly under such a therapeutic regime.

Nevertheless, many questions evidently remain unanswered. Is the quasi-mathematical connection between the prevalence of alcoholism and per capita consumption absolutely unalterable ? What specifically determines a person's level of consumption at any point in time ? What leads him to decrease or increase his consumption ? Why is it that only a tiny proportion of drinkers progress to hazardous levels of consumption ? And, (as was noted in New Behaviour recently), why are the numbers of younger adolescent problem drinkers increasing ? New solutions to these problems now seem most likely to emerge from current research in the behavioural, rather than the medical, sciences.

REFERENCES

BOURNE, P.G. & FOX, R. (1973) Alcoholism: Progress in Research and Treatment Academic Press: New York.

GLATT, M.M. (1974) A Guide to Addiction and Its Treatment MTP: Lancaster.

HODGSON, R. (1975) "Alcohol" New Behaviour 1, 229

LYNN R. (1975) "Anxious Nations" New Behaviour 1, 264-6

MILLER P.M. & BARLOW D.H. (1973) 'Behavioural approaches to the treatment of alcoholism' Journal of Nervous and Mental Diseases 157, 10-20

A PSYCHIATRIC VIEW OF SUBSTANCE DEPENDENCY

J. S. MADDEN

REGIONAL ADDICTION UNIT

CHESTER

Admissions to psychiatric hospitals and units in England and Wales of patients with a primary or secondary diagnosis of alcoholism or alcoholic psychosis increased from 8,708 in 1970 to 12,258 in 1973. During 1974 drunkenness convictions for England and Wales rose to 97,857 - a level not reached since World War I, then government measures were taken to reduce the high amount of national alcohol consumption of that era, with its attendant problems. Over the four year period ending in 1974 the annual consumption of alcoholic beverages in the United Kingdom has increased in terms of absolute alcohol by 37% to 70.6 million gallons. There are issued in England and Wales approximately 18 million prescriptions a year for tranquillizers, costing around £10 million; these drugs are taken by about one woman in five and one male in ten. Since the early 1960's there has been a world wide increase in the abuse of drugs. A United Nations commission on narcotics has reported that despite more effective law enforcement the incidence of drug dependency continues to rise in some countries (The Times, 1976).

In the U.K. drug abuse has generally stabilised or somewhat reduced of late, as reflected perhaps in the fall during 1974 of 17% in the yearly total of persons convicted of offences against the Drugs Act; yet in the same year there were 887 new notifications of narcotic users, while on the final day of 1974 there were 9% more individuals dependent on narcotics, and known to be taking such drugs, than on the same date in the previous year. On a modest estimate some 38,000 men and 4,000 women in England and Wales die prematurely each year as a result of smoking; the mortality rate from smoking is over five times greater than the death rate from road accidents, (Russell 1971).

Confronted with these figures enquiries must be directed to their causes, with a view to prevention. Those who provide care to the victims of alcohol, drug or tobacco dependency feel as though they are in the position of patching up people who have tumbled over a cliff, while all the while new casualties are falling; it would be preferable to fence off the danger point. Apart from humanitarian considerations there would then be more economic allocation of the proportion of national resources that is allotted to the caring services.

There is some evidence for a hereditary factor in alcoholism derived from family studies, particularly from investigations of twins and children of alcoholics reared apart from their parents, (Goodwin et al 1973, Shuckit et al 1972). On Merseyside anomalies among alcoholics have been noticed in characteristics that are partly controlled by genes, namely colour vision, (Swinson 1972), phenylthiocarbamate tasting, (Swinson 1973) and blood group secretor status, (Swinson & Madden 1973), but the anomalies may not have been inherited but secondarily acquired from alcoholism or from smoking. There is one overriding consideration that leads to the rebutall of claims for a strong hereditary component in substance dependency, whether the substance is alcohol, drug or tobacco: the marked fluctuations that occur within a few years in the incidences of alcohol, drug and tobacco problems cannot be attributed to alterations in the genetic endowment of the population. The fluctuations depend on national consumption, which in turn varies with attitudes towards usage and with availability of the substance consumed; these socio-economic factors are the main determinants of substances use and abuse and not hereditary or acquired features within the physiques and temperaments of individuals.

Availability varies with price; the Toronto researchers (de Lint 1973) have produced data to show that in many countries there has been in recent years a fall in the cost of alcoholic beverages relative to personal disposable income, with a consequent rise in per-caput alcohol consumption and therefore in the number of persons with alcohol problems. Fiscal policy has in the past been used to determine price in order to lower consumption, and affords a simple device that does not involve prohibition. In this country the Erroll Committee on Liquor Licensing Report of the Departmental Committee on Liquor Licensing (1972) recommended extension of the hours that permit the sales of alcoholic beverages. Common sense shows that additional hours would only be worked for profit, and that the profit would come from increased sales of alcohol. Fortunately official quarters are less than enthusiastic about the Erroll proposals. In respect to availability of drugs tribute must be paid to the work of police drug squads and of Customs and Excise officers, to the flexibility of the Dangerous Drugs Act, 1971, and to the efforts of doctors to reduce prescriptions of amphetamines and barbiturates. A further useful measure must lie in the inclusion of barbiturates in the Schedule of this Drugs Act. Acknowledgement should be made to the increased responsibility shown in recent years by advertisements for alcohol, cigarettes and analgesic preparations.

Personality factors help to explain why some people develop substance dependency and others within the same community do not. A study of patients treated in hospital for neurosis found that at follow-up 12 years later 10% portrayed evidence of alcohol or drug abuse, (Sims 1975). Both alcoholics and youngsters with drug problems have a high incidence of early parental deprivation and of parental alcoholism, while young drug abusers commonly show a background of parental mental illness (Rosenberg 1971). A considerable proportion of the predisposition to alcohol and drug abuse of persons from such backgrounds must arise from patterns of emotional instability acquired in childhood.

During 1973 there were admitted to mental hospitals because of drug dependence 328 patients aged 40 and over, of whom 272 were dependent on barbiturates. But at the present time the majority of persons with drug problems are from a younger age group. Youthful drug abusers frequently have a history of illegality preceeding their involvement with drugs; (Gordon 1973) that is, their drug taking is just one aspect of social deviancy. Their antiauthoritarian attitudes and hostility feelings, (Gossop & Roy 1976) can make therapy difficult, especially with those who have active, outgoing personality traits, although there are other drug takers with more inadequate temperaments who have drifted into drug abuse and are too passive to disrupt a treatment programme.

Alcoholics too have a wide range of personalities. Abstinence, for example, is not always correlated in alcoholism with satisfactory adjustment in other psychosocial spheres, (Pattison et al 1969) so that some alcoholics require long-term psychiatric guidance although they have stopped drinking. On the one hand, judging by their capabilities and performances when abstinent there are alcoholics who have virtually normal personalities; but for alcohol the latter would never need to consult a psychiatrist.

Substance dependency can result in illnesses but can the condition itself, whether superimposed on a normal or abnormal personality, be defined as an illness ? The disease concept of dependency requires the sickness to consister of a behaviour - the repeated consumption of a substance. Tobacco dependency, which occurs in about half the population of this country, may not be considered an illness unless it is thought that this proportion of the population is sick. It could then be argued that it is those who do not smoke who are abnormal, and ill. Since the medical model of alcoholism and drug dependency will continue to be disputed, it is appropriate for a doctor to say that there is much in favour of the viewpoint that these behaviours should not be defined only as diseases.

There are two kinds of definition. One, the essentialist definition, considers that the definition of a condition establishes its essential nature. Aristotle believed that "a definition is a statement that describes the essence of a thing". In contrast the nominalist type of definition avoids assaying profundities about an essence, but merely attaches a pragmatic label to cover phenomena with common properties. The first approach is inductive, and so, despite its distinguished adherents, lacks the qualities of modern scientific reasoning, (Popper 1969, 1968, 1972). It is preferable to use the latter procedure, avoiding attempts at definition of alcohol, drug or tobacco dependency that would convey the meaning that these conditions are essentially diseases, or essentially not diseases. Alcoholism and drug dependency should be classed within several categories, including the disease category; in order of decreasing impressiveness and the disease features include lowered life expectancy, abnormal somatic and mental functioning, treatment by drugs, restricted self-control and consideration by doctors as illnesses. But there are many other aspects to dependency, for instance, psychodynamic, interactional, learning theory, sociological, biochemical, spiritual and legal facets. Substance dependency must be viewed as a member of each of the categories with which it has common features, (Madden 1976).

The conclusion that substance dependency should not be subsumed exclusively within the medical model is further justified by the consideration that not only medical and paramedical professions provide effective care. Much of the assistance involves counselling; listening and talking to people in trouble, although sometimes dignified by the term 'psychotherapy' is not a specifically medical skill. It is provided by social workers, probation officers, hostel staff, members of voluntary agencies such as alcoholism and drug councils and the Samaritans, as well as by self help groups of persons dependent on alcohol, drugs or tobacco. Indeed, because of the quantity and nature of the problems it is fortunate that there are people who are not medically trained but who have relevant skills and experience.

The number of alcoholics in England and Wales has been officially estimated as between 300,000 and 400,000; the equivalent of 835 full time consultants who practice in adult psychiatry cannot provide a service to all the alcoholics requiring help; psychiatrists do not treat tobacco dependency as they have no special techniques for the condition. Naturally every specialist in adult psychiatry will continue to treat numerous alcoholics and drug abusers, with the special units providing centres of expertise for treatment, training and research. Psychiatrists must urge expansion of outpatient and day patient facilities, ask for provision of hostels and of alcohol detoxification units, and press for organisation of the present scattered caring resources. Perhaps local co-ordination is best achieved at the level of the Area Health Authority, (DH & SS 1975), through the existing Joint Consultative Committee of the caring services and through a specially established multidisciplinary working party.

There are many research avenues to be followed. Among the most necessary and intriguing are elucidation of the biochemical aspects of dependency, with particular reference to the role of the biogenic amines and of enkephalin. In substance dependency as in other conditions that interest psychiatrists, an amalgam of psychodynamic and learning theories is desirable to further understanding and therapy.

Treatments and treatment goals require improved methods of evaluation: it is difficult to gauge from published studies the value of different therapies employed by different workers who have different criteria to assess the outcome in different kinds of patients. There are hopes of increasing the currently small proportion of alcoholics who become able to drink in moderation. The advantages of methadone therapy for narcotic abuse require weighing against the disadvantages. The therapeutic potential of drugs which are opioid antagonists or blockers awaits assessment. It is important to establish whether, as has been claimed, there is a foetal alcohol syndrome - a specific pattern of malformations found in babies born to alcoholic mothers, (Jones & Smith 1973). Finally, further studies are needed in view of reports of a high proportion of chromosomal anomalies in alcoholics, (de Torok 1972 , Lilley 1975, Madden 1976).

Several of the suggested research topics do not lie strictly within the province of psychiatry. Conditions which have multifactorial causes need multidimensional measures for their comprehension and management. This applies to substance dependency, the control of which affords a paradigm for co-operation between psychiatry and many other disciplines.

REFERENCES

DEPARTMENT OF HEALTH AND SOCIAL SECURITY (1975). Better Services for the Mentally Ill. London, H.M.S.O.

GOODWIN, D.W., SCHULSINGER, F., HERMANSEN, L., GUZE, S.B. and WINOKUR, G. (1973). "Alcohol problems in adoptees raised apart from alcoholic biological parents". Archives of General Psychiatry, 28, 238 - 243.

GORDON, A.M. (1973). "Patterns of delinquency in drug addiction". British Journal of Psychiatry, 122, 205 - 210.

GOSSOP, M.R. and ROY, A. (1976). "Hostility in drug dependent individuals : its relation to specific drugs and oral or intravenous use". British Journal of Psychiatry, 128, 188 - 193.

JONES, K.L. and SMITH, D.W. (1973). "Recognition of the fetal alcohol syndrome in early infancy". Lancet, ii, 999 - 1001.

LILLY, L.J. (1975). "Investigations in vitro and in vivo, of the effects of disulphiram (antabuse) on human lymphocyte chromosomes". Toxicology, 4, 331 - 340.

de LINT, J. (1973). "The epidemiology of alcoholism : the elusive nature of the problem, estimating the prevalence of excessive alcohol use and alcohol-related mortality, current trends and the issue of prevention". Alcoholism : A Medical Profile. Proceedings of the First International Medical Conference on Alcoholism, N.Kessel, A. Hawker and H. Chalke (eds). London, B. Edsall & Co.

MADDEN, J.S. (1976). "Chromosome abnormalities in Dupuytren's disease". Lancet, i, 207.

MADDEN, J.S. (1976). "On defining alcoholism". British Journal of Addiction, 71, 137 - 140.

PATTISON, E.M., COE, R. and RHODES, J.R. (1969). "Evaluation of alcoholism treatment; a comparison of three facilities". Archives of General Psychiatry, 20, 478 - 488.

POPPER, K.R. (1966). The Open Society and its Enemies, volume I, chapter 3, section VI and volume II, chapter 11, section II. Fifth edition (first edition 1945). London, Routledge & Kegan Paul.

POPPER, K.R. (1968). The Logic of Scientific Discovery, p. 431. Third edition (first edition 1959). London, Hutchinson & Co.

POPPER, K.R. (1972). Conjectures and Refutations : the Growth of Scientific Knowledge, chapter 3, Fourth edition (first edition 1963). London, Routledge & Kegan Paul.

REPORT OF THE DEPARTMENTAL COMMITTEE ON LIQUOR LICENSING (1972). London, H.M.S.O.

ROSENBERG, C. M. (1971). "The young addict and his family". British Journal of Psychiatry, 118, 469 - 470.

RUSSELL, M.A.H. (1971). "Smoking in Britain : strategy for future emancipation". British Journal of Addiction, 66, 157 - 166.

SCHUCKIT, M.A., GOODWIN, D.A. and WINOKUR, G. (1972). "A study of alcoholism in half siblings". American Journal of Psychiatry, 128, 1132 - 1136.

SIMS, A. (1975). "Dependence on alcohol and drugs following treatment for neurosis". British Journal of Addiction, 70, 33 - 40.

SWINSON, R.P. (1972). "Colour vision defects in alcoholism". British Journal of Physiological Optics, 27, 43 - 50.

SWINSON, R.P. (1973). "Phenylthiocarbamate taste sensitivity in alcoholism". British Journal of Addiction, 68, 33 - 36.

SWINSON, R.P. and MADDEN, J.S. (1973). "ABO blood groups and ABH substance secretion in alcoholics". Quarterly Journal of Studies on Alcohol, 34, 64 - 70.

THE TIMES, (25 February 1976).

de TOROK, D. (1972). "Chromosomal irregularities of alcoholics". Annals of the New York Academy of Sciences, 197, 90 - 100.

HOW IMPORTANT IS ALCOHOL IN "ALCOHOLISM" ?

RON J. McKECHNIE

THE ALCOHOL RESEARCH & TREATMENT GROUP

CRICHTON ROYAL, DUMFRIES

The title of this paper is a question which the author finds himself asking more and more frequently. "How important is alcohol in alcoholism ? As a psychologist interested in analysing factors which influence behaviour it is a question whose answer would direct the author's attention. If the answer was that alcohol is of primary importance then attention should focus on alcohol, the amounts drunk, and what effects it has and only secondarily in where, how and why it is drunk. On the other hand if the answer is that alcohol is not important attention might be focussed on those psychological or social factors of personality or learning which differentiate those who abuse alcohol from those who do not.

As a clinician and therapist (with a psychological background) the question is more specific. It asks, "where does alcohol fit in this man's problem" ? "In what circumstances does drinking behaviour occur?" "What is the function of drinking behaviour in this situation?" Priority may be given to specific circumstances and environments which induce drinking behaviour in the individual if the choice of alcohol is not seen as important. The question, "how important is alcohol in alcoholism?" is both crucial and timely for it appears that within the field there is a pressure towards elevating alcohol to a position of unwarranted importance. This tendency may be seen in the popularity of the Ledermann model of alcohol consumption (Ledermann, 1956) and some of its implications. Figure 1 shows the presumed distribution of annual consumption of alcohol in France according to the model. The curve can be drawn if the mean per capita consumption is known. The value of the model lies in the fact that the hatched area under each curve relates proportionately to the incidence of liver disease in each country (de Lint and Schmidt 1971). The model is a model and as such

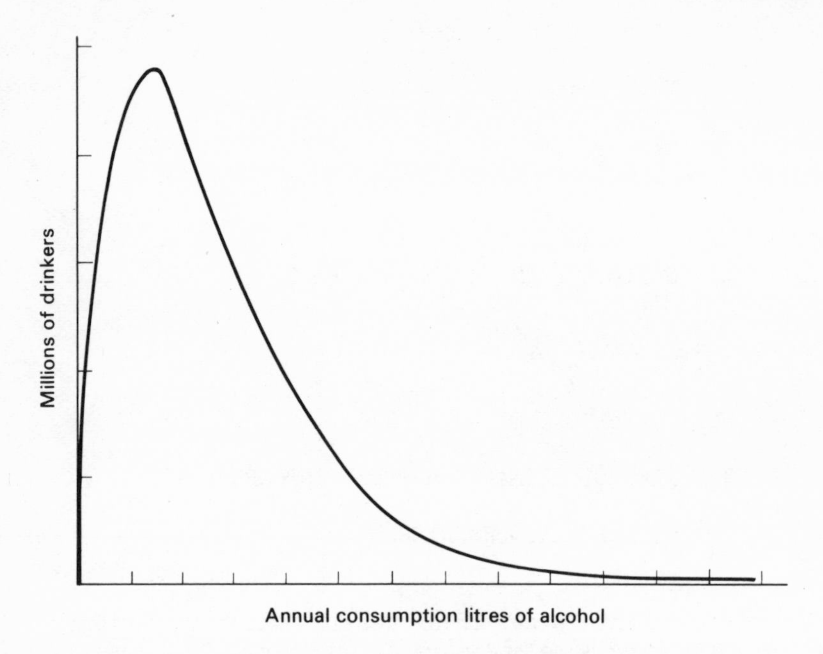

FIGURE 1. Distribution of Alcohol Consumption by 30 Million Adults in France if Consumption is Lederman Distributed

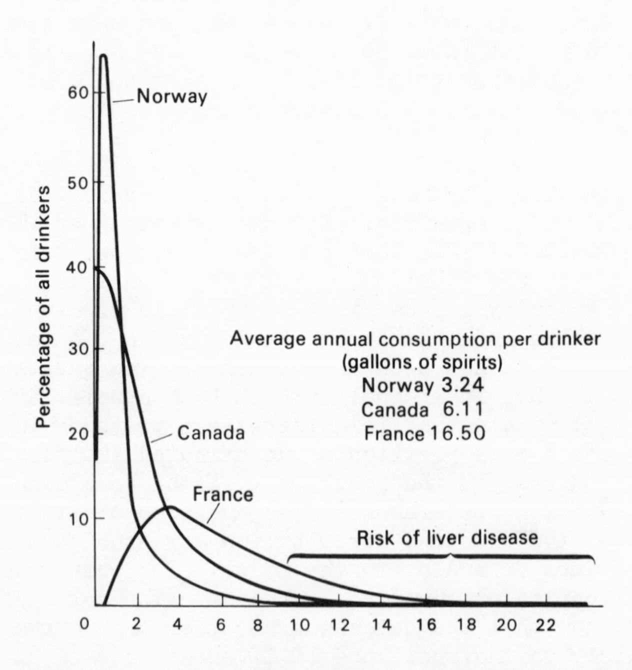

FIGURE 2. The Ledermann Consumption Model

represents an ideal rather than reality and has been accepted by some rather uncritically. Its acceptance is attractive because it simplifies. Some writers have used it to predict how much alcohol is safe and how much is not (Cartwright et al. 1975) which would make things very simple. Unfortunately it introduces simplicity where none exists. Miller and Agnew (1974) criticise the model on a number of points :

1. The model applies only to homogeneous populations. Factors such as sex, religious or cultural differences introduce heterogeneity. Therefore its use would have to be limited to small samples not whole nations.

2. There is no reason to assume that the proportion of heavy drinkers in the population is related to the mean consumption of the population.

3. The model may be useful in comparing populations but is inadequate to predict changes within a population as some have implied.

4. It refers to consumption of absolute alcohol and ignores drinking practices.

5. It implies a strong relationship between heavy drinking and alcoholism which is still controversial.

6. The validity of cirrhosis as an index of alcoholism is questionable as is its relationship to alcohol consumption. During prohibition in America the incidence of cirrhosis dropped (Terris 1967) but gross consumption of alcohol increased (Coffey 1976). Cerlek et al (1972) found evidence of cirrhosis from biopsy in half the 50 cases giving a history of heavy drinking, in a general hospital but only in 1 out of 50 cases admitted to a mental hospital and firmly diagnosed as alcoholic. Davidson and McLeod's (1971) report that, "A history of chronic alcoholism was formerly invariably present in cases of diffuse hepatic fibrosis (i.e. cirrhosis) occurring in Britain. While this association is still frequently encountered in the United States of America and South Africa it is now a less common feature of the disease in this country". Thus, the assumption that the proportion of cirrhosis deaths attributable to alcohol is constant internationally does not seem to hold.

7. The seventh criticism is that the model implies a cause and effect relationship. The relationship if it exists need not be causal and even if it were its direction would be uncertain.

8. The model also has the disadvantage of distracting attention away from other variables which might explain the

association between mean consumption and the percentage of heavy drinkers.

The requirement of the model that the population be homogeneous means that the populations to which it might apply should be small. Indeed it was derived from small samples (Ledermann 1956). It therefore seems inappropriate to attempt to describe whole nations. The whole of France is unlikely to be homogeneous, in fact if one considers alcohol consumption and cirrhosis mortality for each region (Sadoun et al. 1965) a different picture of the relationship emerges. A negative correlation of -0.65 (significant $p < 0.01$) exists between alcohol consumption and death from cirrhosis. That is the region with the highest consumption has the lowest death rate and vice versa. Notice from this example how important the question of scale can be.

The situation in this country and especially the comparison between Scotland and other parts of Britain should be looked at more specifically to see what implications the model might have. It is generally agreed that Scotland has a higher incidence of all drink related indices of harm than any other part of the British Isles. Figures 3 - 7 illustrate the difference between Scotland and England and Wales on some of those indices. Scottish figures are presented as a percentage of English figures, the English rate being shown as 100%. In Figure 3 convictions for drunkenness vary between 25 and 46% higher in Scotland. The overall increase in the period covered in the figures is 34% in England and Wales and 36% in Scotland (Annual Abstract of Statistics 1971).

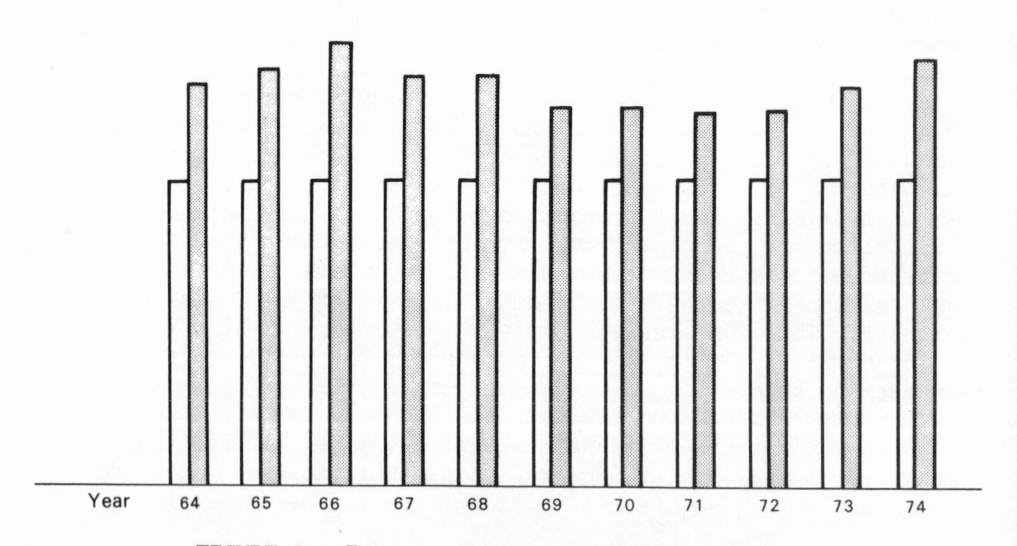

FIGURE 3. Persons Convicted for Drunkenness
England & Wales / Scotland

In Figure 4 other offences under the Intoxicating Liquor Laws are between 13 and 103% higher in Scotland. The increase in England and Wales is 30% whereas in Scotland it is 77% between 1964 - 74 (Annual Abstract of Statistics 1971).

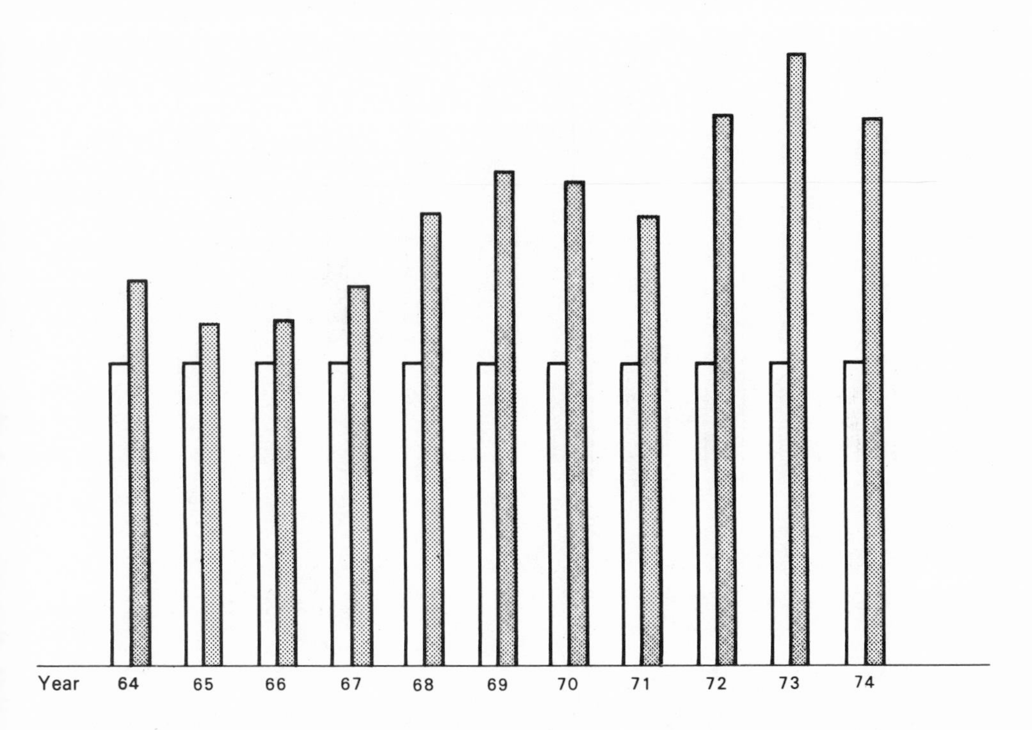

FIGURE 4. Persons Convicted for Other Offences Under Intoxicating Liquor Laws
England & Wales / Scotland

Figure 5 shows the Total Offences under the Intoxicating Liquor Laws (Drunkenness and Other Offences). They are between 25 and 45% higher in Scotland having increased overall by 30% in England and Wales and by 40% in Scotland (Annual Abstract of Statistics 1971).

Admissions to Psychiatric Hospitals for Alcoholic Psychosis and Alcoholism (Figure 6) are between four and five times higher in Scotland than in England. Between 1968 and 1972 they have increased by 48% in England and 38% in Scotland (Health and Personal Social Services Statistics for England 1975, Scottish Abstract of Statistics 1971).

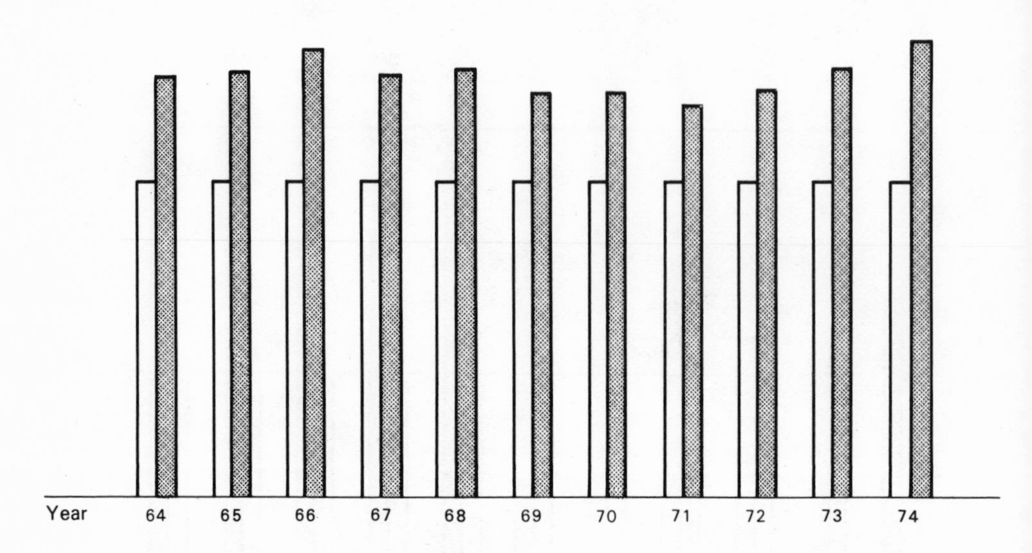

FIGURE 5. Total Offences Under Intoxication Liquor Laws England & Wales / Scotland

Deaths from cirrhosis (Figure 7) are between 31 and 63% higher in Scotland and have risen by 23% in England and Wales between 1968-73 against a rise of 3% in Scotland (Registrar General's Statistical Review of England and Wales 1974, Registrar General's Statistical Review of Scotland 1974).

The picture is rather gloomy and although each index used here is open to question together they add up to the conclusion that Scotland suffers much more from its use of alcohol than does England and Wales. These figures have led many to conclude that alcohol consumption in Scotland must be higher than in England and Wales. Working back through the Ledermann model that would also be its prediction. In the absence of field survey data on consumption the best available information on consumption comes from Family Expenditure Surveys (Scottish Abstract of Statistics 1971, Annual Abstract of Statistics 1971). Figure 8 reveals that family expenditure between 1965 and 1974 has been fairly similar to that of the rest of the United Kingdom. Although spending has ranged from being the same in 1972 to 25% higher in 1973 it has generally been less than 10% higher than the U.K. average.

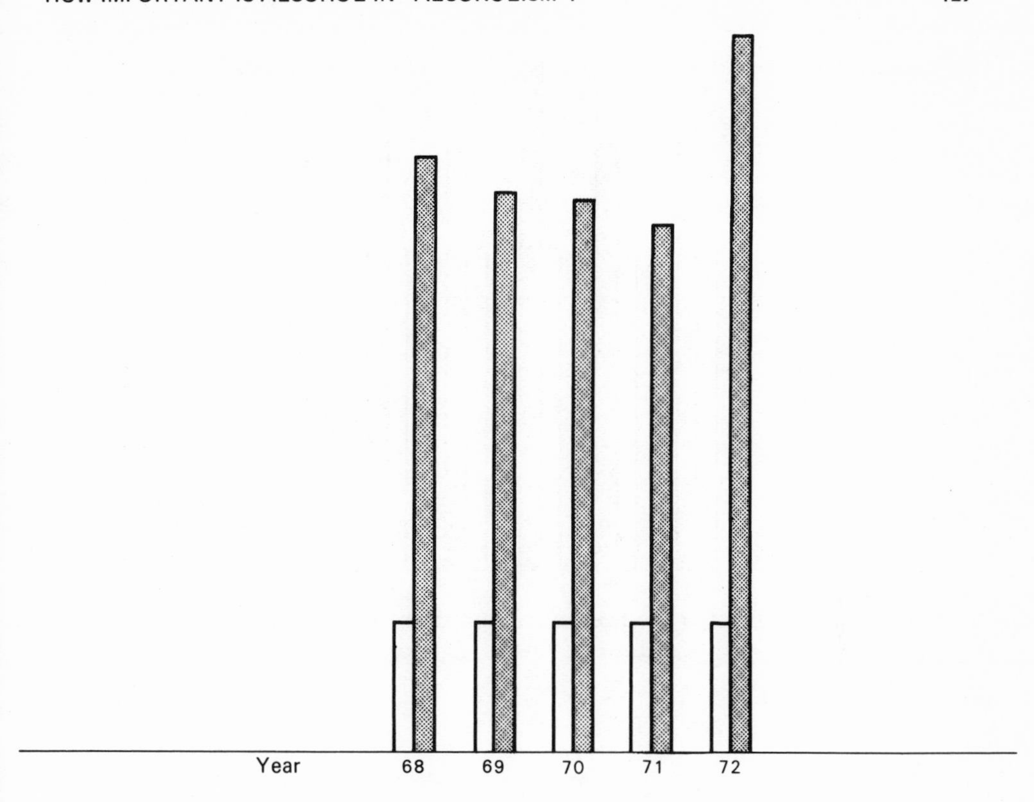

FIGURE 6. Admissions to Psychiatric Hospital for Alcoholic Psychosis and Alcoholism
England / Scotland

Data for 1961/63 (Creland and Murray 1969) show that the figure was 6% less than for the United Kingdom and recent figures (Scottish Daily Record 1975) for 1975 suggest that spending is again less, being 7% down on that for England and Wales. The overall picture is one of similarity in terms of absolute alcohol since the Scots prefer spirits (Wallace 1970) which are a little more expensive than other forms of beverage. These figures would suggest that Ledermann curves for Scotland and England and Wales would be very similar with similar predictions of death from cirrhosis which are seen not to be the case (Figure 7). Not only does this analysis severely question the Ledermann model, it leads back to the title of this paper. If consumption in Scotland is the same as in England yet Scotland's figures for drink-related harm are much higher we have to ask then "How important is alcohol in alcoholism?".

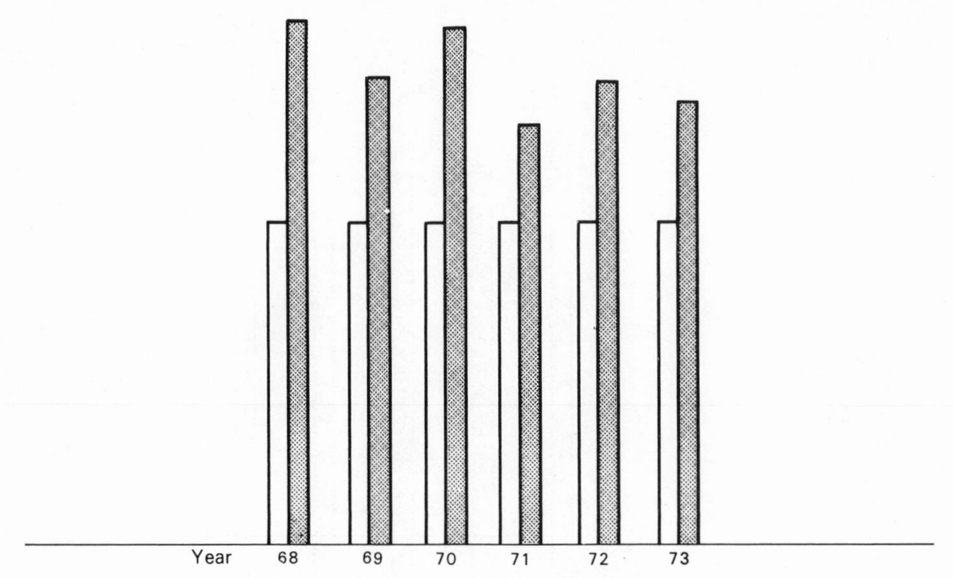

FIGURE 7. Deaths from Cirrhosis of Liver
England & Wales / Scotland

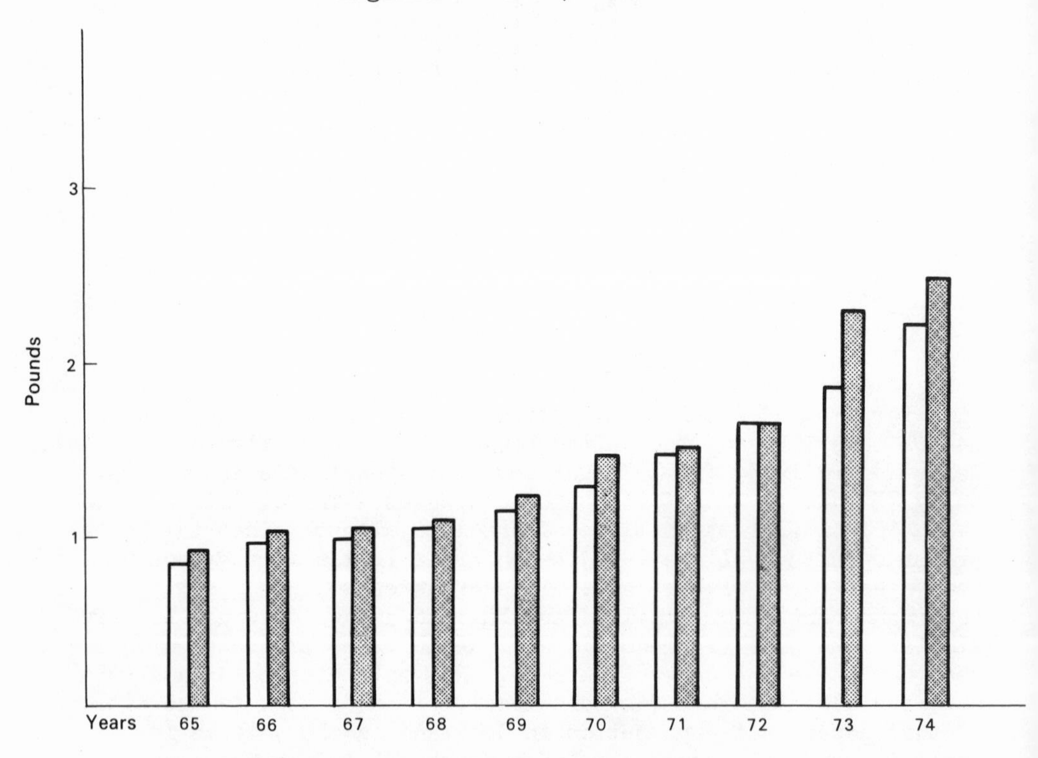

FIGURE 8. Family Expenditure Survey
U.K. / Scotland

Alcohol is a necessary ingredient of alcoholism or alcohol abuse but it is far from being a sufficient explanation. If it were a sufficient explanation the question should be asked why doesn't everyone who drinks become an excessive drinker or alcoholic ? From the above analysis gross consumption would not seem to be of any importance and one must look elsewhere in trying to account for cross-cultural and regional differences in rates of alcohol abuse. The literature contains frequent references to the fact that psychological, social, interpersonal and cultural factors are important in alcoholism. Yet one carries on as though they did not matter. It is repeatedly said that heavy drinking and drunkenness are not the same as alcoholism, yet this is ignored. Clients say that there are many people "out there" drinking more than them and not getting into the same difficulties. This is ignored on the assumption that it is not true. Perhaps our clients are saying "There's something different about me but it's not the drinking". The community behaves as though it was dazzled by alcohol. Like moths attracted by the light. There is a blinkered approach to the problem and it is not possible to see the surroundings. If alcohol itself is not important then the surroundings must be looked at - to look at drinking behaviour in its context. For example, how is the fact that between 1964 - 1974 in the United Kingdom consumption of spirits rose by 76%, beer by 30% and wine by 123% (Annual Abstract of Statistics 1975) to be evaluated.

Whether there is alarm, surprise or relief by those figures depends on perspective. Reaction depends on the context in which they are placed. What has happened to other drugs in the same period ? Between 1963 and 1974 prescriptions for tranquillizers rose by 186%, non-barbiturate hypnotics by 260% and anti-depressants by 180% (Office of Health Economics 1975). Against these figures the increases in alcohol consumption look paltry. Do these figures mean that the whole of society is changing in such a way or at such a rate that more and more people are having to rely on drugs of one shape or another. Given roughly equal levels of consumption why have death rates from cirrhosis risen almost eight times faster in England than Scotland ? Statistics by themselves don't mean much and can be interpreted in many ways. In the field of alcohol abuse statistics are always interpreted as having to do with alcohol. An increase in hospital admissions doesn't necessarily mean a worsening problem; it may mean more and better facilities. The figures say as much if not more about hospitals than they do about drinkers. Likewise convictions for drunken driving are as much about automobile abuse as they are about alcohol abuse and probably give most information about police activity. It is this constant focus on alcohol that we must move from if we are to begin to develop an adequate theory of drinking behaviour. What would such a theory take into consideration if it did not give much importance to alcohol ?

Considering the question of attitudes to drinking of alcohol, in France where there was a negative correlation between alcohol consumption and alcoholism as measured by cirrhosis, Sadoun, Lolli and Silverman (1965) point out that those regions with the lowest consumption in the North East and West displayed the strictest parental disapproval of drinking. One of the major differences between France and Italy was that drinking by children in France was an emotive topic being actively encouraged by some and discouraged by others unlike Italy where it was accepted as a normal part of growing up. It has been argued elsewhere (Davies and Stacey 1972, Globetti and Chamblin 1966, McKechnie 1976, McKechnie et al. in press) that parental or authoritarian disapproval is associated with heavy drinking. Parental disagreement (i.e. father and mother behaving differently or having different views) about drinking predisposes one towards alcoholism (Jackson and Conner 1953, McCord and McCord 1960).

Conflicting attitudes may not only be present within the individual but within the nation or the region under consideration. Does the presence of conflicting attitudes influence the incidence of alcohol related harm ? At an individual level, Gunn (1973) has demonstrated that the frequency and severity of hangovers relates more closely to attitude to drinking than they do to amounts consumed. The influence of religion might be important in this respect and it is well known that different religious groups have widely different rates of alcoholism. Bacon (1957) presents an interesting analysis of the attitudes of Jews and Mormons in America in relation to alcohol and alcoholism. The Jews allow and encourage the use of alcohol in religious and family ceremonies and have a low incidence of alcoholism. The Mormons on the other hand forbid the use of alcohol but there is a very high incidence of alcoholism amongst those who drink. These studies lead us to ask what role non-drinkers play in drinking problems ?

Scotland does seem to have slightly more abstainers than other parts of Britain (Wallace 1970) this being largely due to 32% of Scottish women who do not indulge. The highest proportion of abstainers is likely to be in the Highlands and Islands where there is a strong religious pressure against alcohol. In this region admission to psychiatric hospitals for alcoholism and alcoholic psychosis are three times higher than the Scottish average (Whittet 1970), that is almost 15 times as high as the English figure. Unfortunately there are many factors which are higher in Scotland than in England and higher in the Highlands and Islands than the rest of Scotland. For example, unemployment in the Highlands and Islands in 1968 was 7.7% against a Scottish average of 3.8% and a United Kingdom average of 2.4% (Highlands and Islands Development Board 1968). Although many alcoholics are unemployed when they present for treatment what part being unemployed might play in the development of drinking problems has not seriously been considered. On the other hand the role that certain occupations play in creating or maintaining drinking problems is not clear although it has been recognised that there are vast differences between occupations. Table I shows the Standard Mortality Rate for cirrhosis of the liver in England and Wales between 1949-53 and Table II shows more recent figures.

TABLE I : S.M.R. CIRRHOSIS OF LIVER (ENGLAND & WALES 1949-53)

Occupation	SMR
Publicans, owners and managers of hotels and inns	925
Makers of alcoholic drinks	250
Owners of wholesale businesses	243
Registered medical practitioners	240
Domestic Servants (Indoors)	233
Judges, Barristers, Solicitors	233
Proprietors of Retail Food Businesses	227
Teachers	44

McKEOWN & LOWE (1966)

TABLE II : STANDARD MORTALITY RATE FROM CIRRHOSIS

Occupation	SMR
Company Directors	2,200 %
Entertainers/Journalists	550 %
Hotel Keepers	450 %
Armed Forces	400 %
Doctors	350 %
General Population	100 %

Source: Registrar General "Occupational Mortality" H.M.S.O. 1970

It is difficult to accept that such variations are due to easy availability of alcohol and lack of supervision at work as some would suggest (Grant 1975) and more research requires to be done to tease out the complex interactions in this area.

The search for personality factors which distinguish alcoholics from others has been rather fruitless since whatever results emerge, if any, never establish any indication of whether they are primary or secondary factors, that is, whether they are the result of drinking problems or the cause of drinking problems. These approaches wish to have a single answer to a complex problem. The same is true of looking at the spouse of alcoholics. Only Rae and Drewery (1969) have avoided this to some extent. They identified a personality type in a proportion of wives of alcoholics which led to certain systematic distortions in the relationship between the alcoholic and his wife. They went on to focus on the interaction rather than the individual characteristics. Again, it is difficult to clarify whether this is cause or effect but the finding that a high proportion of women who divorce alcoholics remarry a man who is or becomes an alcoholic, might suggest there is a strong assertive mating effect taking place.

Drinking alcohol like many other behaviours in our society occurs in many environments and for many reasons. For most drinkers, drinking is a means rather than an end. Drink is taken for reasons or purposes, to cement friendship, to enhance enjoyment of a meal, to relax, to be uplifted, to heighten a mood, to assert oneself, to express anger, to be lifted out of oneself, to shut out the world. Alcohol may be offered with some intention and accepted with another. This is why a topographical description of any drinking episode which begins with the lifting or ordering of a drink is inadequate. It is inadequate because it says nothing about the function of the behaviour or its meaning for the individual. It looks upon drinking alcohol as an end rather than as a means. Drinking alcohol is not a random activity. It has purpose and meaning within the context it is used. This is as true for the alcoholic as it is for any drinker. The idea that the "real alcoholic" is the man who has no reason or purpose in drinking is both short-sighted and unhelpful. Chafetz (1970) goes some way towards this view in arguing for the status of symptoms for the drinking behaviour of the alcoholic. It is the alcoholic's way of protecting some even greater source of pain, like a man with a limp protecting his leg. The drinking has a restitutive function. Psychologists have been as guilty as any other discipline in looking at drinking behaviour as an end. They have tried to discover "the effects" of alcohol as though the consumption of alcohol were a biochemical event rather than primarily a social-psychological event. They have studied the reinforcing characteristics of alcohol and find that alcohol's reinforcing effects are especially potent in conflict situations (Masserman and Yum 1946, Conger 1956). They have spent little time or effort looking at the role of alcohol as a discriminative stimulus. That is as something which sets the scene for other behaviours.

Many alcoholics say that there are many things they can do having had a drink which they couldn't do otherwise. Such a statement is usually interpreted as referring to "dutch courage" or something similar. The behaviours usually reported in this context are being assertive, aggressive, affectionate, morbid or soft. It would appear that the alcoholic's relationship with others does not allow these behaviours to appear naturally with the "aid" of alcohol. The "aid" is not necessarily confidence or "dutch courage", it may be that such behaviours only occur in those situations which are defined as including alcohol. Or it may be that an escape route is left open from the consequences of the behaviour. Alcohol can be blamed for the behaviour and the individual freed from responsibility. Alcohol becomes the escape route since when you speak carefully to alcoholics it is clear that many of them know what behaviours are going to follow on their drinking. Thus, alcoholic drinking is no different from non-alcoholic drinking in the sense that it has meaning and purpose. It is not the chaotic shambles that many of them would have one believe. There is careful planning and organisation before and during drinking bouts or episodes probably more so than for non problem drinkers. The reasons, purposes and functions of such drinking may be difficult to unravel but it is here where lies the foundations from which a theory might be built. Steinglass gave such a description of a drinking gang (Steinglass and Wolin 1974), and of the relationship between two brothers who were problem drinkers (Steinglass et al. 1971). Their account of the organisation of roles and role reversal within those drinking systems is illuminating. The theory would have to be a general theory applicable to all. It should regard drinking behaviour as a response within a family of responses. It should examine the situations, events and relationships within which the response is provoked. Such a theory has to begin with detailed analysis of individual cases in their own context or environment to establish a general framework from which to proceed.

In this paper it is argued that drinking behaviour has to be considered against a wide background. In the time available, only a few of the social or cultural factors against which drinking for individual psychological or social purposes might take place have been presented. Many more could have been considered, e.g. sex, age, the association of drinking with masculinity, factors which produce changes in individual drinking patterns, the relationship between alcoholism and suicide to name but a few. A good theory will have to be able to consider many factors simultaneously. It should be capable of dealing with human interaction and relationships. If drinking behaviour is regarded as a means or a symptom, then a disturbed relationship with alcohol probably arises from a disturbed relationship to something or someone else, such as spouse, family, sex, emotions, religion, work or the conflicting attitudes and demands of the society in which we live.

We need urgently a theory of drinking based on human experience and not on laboratory animals who prefer water to alcohol given a free choice. A theory which will have relevance for therapy, guide research and point the nation towards a safer consumption of alcohol where all might enjoy its pleasures and few its discomforts.

This is a big challenge and requires many perspectives. It is a challenge which will not be met by one discipline but only by co-operation of the disciplines involved, sharing their insights, benefitting from each other's rigor and wisdom. A fresh outlook is needed and this can only come about when the question "How important is alcohol in alcoholism?" has been clearly answered and concluded that it is only of minor importance.

REFERENCES

ANNUAL ABSTRACT OF STATISTICS (1971). H.M.S.O. London.

BACON, S. (1957-1964). "Social Settings Conducive to Alcoholism". Journal of the American Medical Association, 177-181.

CARTWRIGHT, A.K.J., SHAW, S.J. and SPRATLEY, T.A. (1975). "Designing a Comprehensive Community Response to Problems of Alcohol Abuse". Maudsley Alcohol Pilot Project, London.

CERLEK, S., HUDOLIN, G., KNEZEVIC, M., FORENBACHER, S., LANG, M., and CERLEK, M. (1972). "Alcohol and Changes in Liver Parenchyma: Clinical and Experimental Observations".

CHAFETZ, M.E. (1970). "The Alcoholic Symptom and its Therapeutic Relevance". Quarterly Journal Studies on Alcohol, 31, 444-45.

COFFEY, T.M. (1976). "The Long Thirst: Prohibition in America 1920 - 1933". Hamilton.

CONGER, J.J. (1956). "Reinforcement Theory and the Dynamics of Alcoholism". Quarterly Journal Studies on Alcohol, 17, 296-305.

CRELAND, G.D. and MURRAY, G.T. (1969). "Scotland: a New Look". Scottish Television, London.

DAVIDSON, S. and McLEOD, J. (1971). Principles and Practice of Medicine, Churchill Livingston, Edinburgh.

DAVIES, J. and STACEY, B.G. (1972). "Teenagers and Alcohol", SS463 Vol. II. H.M.S.O. London.

GLOBETTI, G. and CHAMBLIN, F. (1966). Sociology - Anthropology Series, Mississippi State University, New Haven.

GRANT, M. (1975). Understanding Alcohol and Alcoholism in Scotland, Mimeographed.

GUNN, R.C. (1973). "Hangovers and Attitudes towards Drinking". Quarterly Journal Studies on Alcohol, 34. 194-198.

HEALTH AND PERSONAL SOCIAL SERVICES STATISTICS FOR ENGLAND (1975). H.M.S.O. London.

HIGHLANDS AND ISLANDS DEVELOPMENT BOARD (1968). Fourth Annual Report.

JACKSON, J. and CONNER, R. (1953). "Attitudes of Parents of Alcoholics, Moderate Drinkers and Non-Drinkers towards Drinking". Quarterly Journal Studies on Alcohol, 14. 590-613.

LEDERMANN, S. (1956). "Alcool, Alcoolisme, Alcoolisation; donnes Scientifiques de Caractere Physiologique, Economique et Social". Institut National d'Etudes Demographiques, Travaux et Documents, Cahier No. 29, Presses Universitaires, Paris.

de LINT, J. and SCHMIDT, W. (1971). "Consumption Averages and Alcoholism Prevalence. A Brief Review of Epidemiological Investigations". British Journal of Addiction, 66, 97-107.

McCORD, W. and McCORD, J. (1960). Origins of Alcoholism, Tavistock Publications, London.

McKECHNIE, R.J., CAMERON, D., CAMERON, I.A., and DREWERY, J. (In press) "Teenage Drinking in South-West Scotland". British Journal of Addiction.

McKECHNIE, R.J. (1976). "Parents-Children and Learning to Drink". Third International Conference on Alcoholism and Drug Dependence Liverpool.

McKEOWN, T. and LOWE, C.R. (1966). An Introduction to Social Medicine, Blackwells, Oxford.

MASSERMAN, J.H. and YUM, K.S. (1946). "An Analysis of the Influence of Alcohol on Experimental Neurosis in Cats". Psychosomatic Medicine, 8. 36-52.

MILLER, G.H. and AGNEW, M. (1974). "The Ledermann Model of Alcohol Consumption". Quarterly Journal Studies on Alcohol, 35. 877-898.

OFFICE OF HEALTH ECONOMICS (1975). "Medicines which Affect the Mind". Office Health Economics, London.

RAE, J. and DREWERY, J. (1969). "Interpersonal Patterns in Alcoholic Marriages". Unpublished Study.

REGISTRAR GENERAL'S STATISTICAL REVIEW OF ENGLAND AND WALES (1974). H.M.S.O. London.

REGISTRAR GENERAL'S STATISTICAL REVIEW OF SCOTLAND (1974). H.M.S.O. Edinburgh.

SADOUN, R., LOLLI, G. and SILVERMAN, M. (1965). "Drinking in French Culture". Monograph 5. Rutgers Center of Alcohol Studies.

SCOTTISH ABSTRACT OF STATISTICS (1971). H.M.S.O. Edinburgh.

SCOTTISH DAILY RECORD AND SUNDAY MAIL LTD. (1975). "Survey of Individual and Household Income, Expenditure and Readership", Glasgow.

STEINGLASS, P., WEINER, S., MENDELSON, J.H. and CHASE, C. (1971). "A systems Approach to Alcoholism". Archives of General Psychiatry, 24. 401-408.

STEINGLASS, P. and WOLIN, S. (1974). "Explorations of a Systems Approach to Alcoholism". Archives of General Psychiatry, 31. 527-532.

TERRIS, M. (1967). "Epidemiology of Cirrhosis of the Liver, National Mortality Data". American Journal of Public Health, 57, 2076-2088.

WALLACE, J. (1970). "Who Drinks What in Britain". Third International Conference on Alcoholism and Addictions, Cardiff.

WHITTET, M.M. (1970). "Epidemiology of Alcoholism in the Highlands and Islands". Health Bulletin, 28.

SOCIAL CIRCUMSTANCES OF NON-CONVICTED vs CONVICTED DRUG USERS

CHARLES E. REEVES

SOCIAL SCIENCES RESEARCH COUNCIL

CALIFORNIA

SOCIAL CIRCUMSTANCES OF SOME DRUG USERS AT THEIR FIRST CONVICTION

Introduction

This paper deals with the criminal behaviour of a group of young drug users in the geographical area of Southern Hampshire, England. It is based upon data obtained in intensive personal interviews of a non-representative sample of one hundred and twenty-two drug users. It thus differs from most previous studies which have dealt mainly with the institutionalized drug users. Although studies on institutionalized users reveal a great deal about patterns of criminal behaviour associated with drug use, they reveal very little about what happens before and after the individual's institutionalization, and almost nothing at all about the individual's use of drugs and non-reported criminal behaviour, and the events that led to the criminal behaviour being brought to the attention of the various statutory bodies. Equally important, research conducted in institutions contains little or no information about many of the attitudes, values and other characteristics of a drug taking group with which this study is concerned.

It can be said that no one piece of research contains data about the whole spectrum of the attitudes, values and characteristics of individuals in a drug taking group. Only by talking directly to the 'users' and collecting data in an unstructured manner on a longitudinal basis using the techniques of participant observation is it possible to obtain realistic data on criminal and non-criminal behaviour associated with drug use and to try and discover how these patterns are related to individual behaviour. As far as is known, this is the first effort of this type to examine in detail over a period of time the patterns of behaviour leading up to criminal behaviour associated with drug use in the United Kingdom.

Impetus for this research project stems from a previous piece of work by the author: "Drug Use and its Management in South Hampshire". In that work, it was found that there was a large group of individuals involved in behaviour associated with drug use that can best be termed as criminal, the bulk of which was unreported. It was thought that it would be interesting to investigate a group of drug users and attempt to see why certain users were brought to the attention of the authorities and why other users appeared to escape attention. Various social variables were investigated and some rather interesting patterns emerged.

METHODOLOGY

When using the techniques for data collection mentioned above it is important at the outset to discuss the interactional elements present in this type of sociological interview; and, in particular, what has been termed the 'rapport' which exists between interviewer and respondent. An interview typography is frequently provided in terms of a continuum ranging from the scheduled, standardized interview for which the presentation of the interview is, in theory, similar for every respondent, to the non-scheduled non-standardized interview in which the interview is essentially a focused conversation.

The standardized form is generally preferred for quantitative 'hard' data; the unstandardized to get at hidden meanings. What is important in the present context is that methods of handling interactions vary considerably along this continuum. In this study spontaneous interaction between interviewer and respondent was encouraged - achieving good rapport was a major target of the interviewer. The success can be shown by the type of data that was collected.

Detailed information about drug and criminal activities was collected during the interview. At the completion of the field work period the investigator was given, by special arrangements, access to criminal records and these data were collated where possible with the data collected from the respondents. A few minor errors were discovered in the official records but they were obviously the result of what Hindess (1973) referred to as 'modifications and transformations' necessary in order to produce a specific final record or result. It was not known to the respondents that access to these records was available. It was found that in 17 cases (14%) there was evidence of systematic distortion. This appears to be of the same order of magnitude as the level of reliability reported by Ball (1967) in his work on narcotic users.

The respondents in this study fell into two distinct male groups*, those that: (1) had not been convicted for any offence (77 respondents, 63%), THE CONTROL GROUP; and (2) had been convicted for a criminal offence (45 respondents, 37%), THE EXPERIMENTAL GROUP.

It was thought that it might be possible to determine why one group of respondents were caught in their drug activities and convicted and another group did not come to the attention of law enforcement or social agencies, and, in fact, appeared to be almost immune from attention.

The following variables were applied to both groups to see if any significant patterns emerged :

(a) gender
(b) social class
(c) natal family intactness
(d) siblings' drug taking
(e) size of family
(f) age left home the first time
(g) age left school
(h) age of first sexual intercourse
(i) age of first drug experience
(j) injecting versus other methods of drug taking

There were other areas suggested by West (1969) as precursors of early delinquency that might be applied to a drug using group, such as :

(a) poor housing
(b) poorly kept accommodation
(c) inadequate income
(d) physical neglect
(e) conviction record of parent

It was found that these last five categories would yield either subjective data, or the numbers were too small to yield any significant data, so they were not used.

DESCRIPTION OF FIELD WORK

This paper is based upon a piece of empirical sociological research conducted in Southern Hampshire, England, during the period 1969 - 1974. The main feature of the research design was to

---

* Convicted females represented such a small proportion of the sample (four cases, 3.7%) it was thought practical not to include them.

interview a population of drug users and follow them up with re-interviews after periods of 18 months to see what changes, if any, developed over this period.

The study was primarily a 'participation observation study ' participant observation operationally defined as "Participant-as-observer" (Gold 1958).

The sample consisted of a non-randomly selected population of 122 drug users. The technique for building the sample was 'snowballing' - a technique described by Polsky (1971).

The interview and re-interview schedules were loosely structured and open-ended, but there was a check list of the main points of interest to ensure that they were systematically covered to provide data that could be coded.

One hundred and twenty-two initial interviews were conducted, one of which could not be completed because of the psychological condition of the interviewee. In addition, there were three outright refusals. After a period of 18 months it proved possible to achieve re-interviews with no less than 106 (86.9% of those originally interviewed). After a further period of 18 months, it was possible to re-interview 93 (76.2% of those originally interviewed).

It was from this population that evidence was gathered for this paper.

OPERATIONAL DEFINITIONS

It is essential at the outset to define what is meant by 'drug' and 'drug user' or 'drug taker'. For the purpose of this paper, a 'drug' is defined as any chemical substance or agent that affects physiological and/or social functions and is taken for comfort, stimulation or pleasure. A 'drug user'or 'drug taker' is any individual who takes these substances for the above purposes, or, in some cases, in order to re-live or create a 'mystical experience'. In this study all those interviewed had used a drug on at least three separate occasions. This criterion eliminated those individuals with only a passing interest in drugs.

In an earlier paper (Reeves 1972), the drug histories of the respondents were reported in detail. Briefly, the drug taking characteristics of the members of the sample reported in this paper are :

(a) Fifty-seven (47%) respondents who used the injection method of drug use on three or more occasions, and

(b) Only nine cases (7.4%) who were mono-drug users, the balance, 113 cases (92.6%) were poly-drug users. The mean number of different drugs used by the interviewed population was five.

CHARACTERISTICS OF THE RESPONDENTS

Social Class and Sex of Respondents

Different theories have been advanced regarding social class and drug taking of a quasi-casual kind; these suggest that the drug using population differs from the general population and from the criminal population. In social class terms there may be not one but two or more virtually separate drug populations with different aetiologies and, if this is so, different problems of control are posed.

Because of the youthfulness of the drug users in this population it would be unrealistic to assess social class on the basis of their own occupations. The classification is, therefore, of the fathers' occupations, based upon the Hall-Jones (1950) Scale of Occupational Prestige for Males.

The use of the fathers' occupation has the additional advantage of making it possible to compare the social class of female and male users. This is important in this study where 42% of the respondents are females.

Table I shows the distribution of the respondents by social class and sex. Because of the small size of the cells, the occupat occupations have been collapsed into two broad categories, manual and non-manual.

TABLE I : SOCIAL CLASS OF RESPONDENTS BY SEX

Social Class (based upon Fathers' Occupation)	Male	Female	Total
Non-Manual Class	39 (31.9%)	32 (26.2%)	71 (58.2%)
Manual Class	28 (22.9%)	15 (12.3%)	43 (35.2%)
Other (e.g. no father, does not know fathers' occupation etc.)	4 (3.3%)	4 (3.3%)	8 (6.6%)

AGE AND TYPE OF FIRST CONVICTION

The respondents were a relatively young group. Fifty-seven (46.7%) of them were 20 years of age or less at the time of the initial interview. The modal age of initial drug taking was 16 years. Fifty per-cent of the respondents had had their first drug taking experience by that time.

The age and type of first conviction of the experimental group is shown in Table II.

TABLE II : AGE AND TYPE OF FIRST CONVICTION

Age	Type of First Conviction	
	Drug	Non-Drug
10 years	0	3
11 years	0	1
12 years	0	0
13 years	0	0
14 years	0	2
15 years	0	2
16 years	3	3
17 years	2	4
18 years	1	6
19 years	5	2
20 years	1	1
21 years	3	2
22 years	1	2
23 years	1	0
Total (n = 45)	17	28

$X^2 = 2.86$, p 0.05, df = 2 (not significant)

As can be seen from the above table, in the experimental group there is no significant age difference between those who had had a first conviction for a drugs offence and those who had been convicted for an offence other than a drugs offence.

Various general variables were investigated in both groups (see Table II) and it was found that there were no significant differences between the groups. However, when more specific items were investigated, several revealing factors became apparent as can be seen in Table IV.

TABLE III

Variable	'Experimental Group'	'Control Group'	X^2
Gender (male)	30 (46.7%)	41 (53.4%)	2. 09, p .05, df 1*
Social class (manual) **	15 (33.3%)	28 (36.4%)	.119, p .05, df 1*
Natal family (intactness) ***	35 (79.5%)	56 (72.7%)	. 37, p .05, df 1*
Sibling drug taking	19 (51.4%)	27 (38.6%)	1.12, p .05, df 1*
Size of family:			
2 or less	15 (34.1%)	18 (23.7%)	
3 or 4	16 (36.4%)	38 (50.0%)	2.19, p .05, df 2*
5 or more	13 (29.5%)	20 (26.3%)	

The first four variables are Yates corrected

* not significant
** based upon parental occupation
*** both parents living at natal home

Table IV was developed in its particular format in order to demonstrate what was felt to be an extremely crucial period in the careers of the experimental group.

Referring to Table IV the variable 'age left home' clearly shows that the experimental group had left the natal home at a much earlier age than the control group. It is also apparent from the Table that the experimental group left school at an earlier age, and that this early school leaving was closely coupled with early home leaving. A further two variables closely linked by age to school leaving and home leaving were 'age of sexual intercourse'

TABLE IV

Variable	'Experimental Group'	'Control Group'	X^2
Age left home the first time (still at home: Ex: 11 Control: 24)			
Up to 16 years	18 (56.3%)	10 (18.9%)	13.31, p .05, df 2*
17 years	5 (15.6%)	10 (18.9%)	
18 years or more	9 (28.1%)	33 (62.6%)	
Age left school (up to and including 16 years)	35 (79.5%)	32 (43.2%)	13.38** p .05, df 1*
Age of first sexual intercourse (up to and including 16 years) ***	33 (82.5%)	31 (47.7%)	12.03** p .05, df 1*
Age of first drug experience:			
Up to 16 years	30 (68.2%)	31 (40.8%)	
17 years	9 (20.5%)	13 (17.1%)	9.75, p. .05,df 2*
18 years or more	5 (11.3%)	42 (42.1%)	
Injecting vs. other methods of drug use (injecting and other methods)	31 (70.5%)	24 (31.6%)	21.30** p .05 df 1*

* significant

** Yates corrected

*** Five respondents had not had sexual intercourse by the time of the interview; eight respondents either do not remember when or were evasive.

and 'age of first drug experience'. It would appear that the period up to sixteen years of age is of paramount importance. Although there were a few cases of respondents who had left home, taken a drug for a first time, etc. at a much earlier time in their careers, the majority of cases had experienced the variables shown in Table IV* in their fifteenth year. What can be inferred from these data is that upon leaving school at an early age (up to sixteenth birthday), the individual, in trying to establish his independence, seeks employment and accommodation away from the natal home, thus weakening parental and/or school control or regulation. With this independence he finds himself with more freedom to experiment with drugs and sexual activities. Due to the obvious lack of parental/school influence, the individual is able to spend more time in various public places that cater for young people. e.g. cafes, etc. These places are usually known to the law enforcement agencies and are not infrequently visited by them, the result being that any newcomer on the 'scene' would become immediately noticed as 'not having been around'. This in turn would increase the profile or visibility of the newcomer.

Several of the younger respondents, particularly those who had their own room or shared accommodation, frequently reported police harrassment, which usually took the following form: when a particular cafe closed, usually around 10 p.m., the young people leaving either to return home or to visit a friend, were frequently stopped, searched, asked their destination and had their names taken. When the police were asked the reasons for the search, the usual response was either "it is none of your business", or, more commonly, "just routine".**

A possible inference would be that with this sort of vigilance, young people (up to 16 years of age) would certainly be more at risk than their older associates. From Table II it can be seen that of the forty-five respondents with convictions, 23 (49 per cent) had received their first conviction between 14 and 18 years of age. A further factor that would place the experimental group at risk is that they have a higher incidence of injecting drugs. This form of

---

* Earliest first drug experience, nine years; earliest home leaving, 13 years.

** It should be noted that in the Chief Constable's Reports (1972, 1973) in the survey area it was reported that "street checks are made only when there are very real reasons to suspect the individual checked and that the individual can give a lead to association which merit even more serious investigation!" During 1973 in one of the cities in the study area, there were 57 street searches for drugs on individuals 18 years or younger. Only 20 of these searches were positive, that is, yielded drugs.

drug taking is much more difficult to conceal in that the injector will have visible marks about his body from injecting, paraphernalia associated with the preparation of drugs for injection, and be known to associate with those whose drug career is predominantly associated with injecting and who are usually known to the police or other social agencies. Table IV shows that almost three-quarters of the experimental group had injected drugs, whereas only about one-third of the control group had used this method.

In summing up the data collected from the two groups (experimental and control), it can be said of the experimental group that there was a tendency to leave school at a reasonably early age, seek accommodation and participate in behaviour that might not be possible whilst living at their natal home. In the new and less restrictive environment there would be a reasonably good chance that their behaviour would be brought to the attention of social control agencies; and most certainly because of its nature - that is, procuring supplies, methods of use, the environment in which it is used and the effects of its use, drug users, and in particular the injectors of drugs, would be more at risk and thereby increase the possibility of being caught and subsequently convicted.

COMMENTS TO THE METHODOLOGY

1. It has been reported that drug taking is "the problem that died"* Using participant observation and keeping a close view of drug taking in a community over a period of years, it was found that this was not the case. In fact, quite the reverse has happened. With increased police activities and increased penalties, drug users have become more covert in their activities and thereby there were fewer convictions in proportion to the increase in drug use and drug availability. Dr Robert L. DuPont, director of the National Institute on Drug Abuse recently reported (February 1976) that in the United States drug taking was again on the increase. This topic of increase in drug use is the subject of a paper now in progress.

2. Only that rapport which specifically relates to the roles of interviewer and respondent has been found to be of use in producing a valid response. In other words, only so far as the respondent understands what the interviewer is expecting of him, and only in so far as the interviewer is able accurately to convey this understanding and to interpret the respondent's replies correctly, will the response produced by the interview be valid. In terms of Kuhn's (1962) definition, such understanding is born of shared frames of reference - the sharing of language, values and general perspective. The author believes that this conclusion is

---

* Paul Harrison, in "New Society", August 15th 1975, reports that "the era of the drug problem is more or less dead".

fundamentally in accordance with the findings of Dohrenwend et al (1968) on the effect of social distance between interviewer and respondent on response errors. Rapport may be seen as a function of social distance. If social distance is great, there will be no common language and hence no rapport. If social distance is minimal, over-rapport, i.e. rapport in roles not related to the interview function, will result.

Finally it is necessary to emphasize that the author does not believe that a high degree of rapport is essential or even desirable for all types of interview survey, although rapport behaviour is inevitable. However, in this study it was undeniably a major factor in successfully obtaining this information.

3. If the results in this study are somewhat different from the conventional picture it may well be due to the fact that the data were derived from a close study of the drug scene in Southern Hampshire for five years. It suggests the importance of looking at the drug user in his natural habitat rather than through the focus of an institution.

REFERENCES

BALL, J.C. (1967). "Reliability and Validity of Data Obtained from 59 Narcotic Drug Addicts". American Journal of Sociology, 72, 650-654.

CHIEF CONSTABLE'S REPORT (1972, 1973). Hampshire, England.

DOHRENWEND, B.S. et al. (1968). "Social Distance and Interviewer Effects". Public Opinion Quarterly, 32, 410-422.

GOLD, R. (1958). "Roles in Sociological Field Observations", Social Forces, 36, 217-223.

HINDESS, B. (1973). The Use of Social Statistics in Sociology, The MacMillen Co. U.S.A.

KUHN, M.H. (1962). "The Interviewer and the Professional Relationship". In Rose, A.M. (ed.) Human Behaviour and Social Process, Houghton Mifflin, Boston.

POLSKY, N. (1971). Hustlers, Beats and Others, London, Penguin Press.

REEVES, C.E. (1972). "Motivation for Changing Methods of Drug Taking". Proceedings I.A.A. Conference, Amsterdam.

WEST, D. (1969). Present Conduct and Future Delinquency, International Universe Press, London.

THE PATHOLOGICAL AND THE SUBCULTURAL MODEL OF DRUG USE - A TEST OF TWO CONTRASTING EXPLANATIONS

KARL-HEINZ REUBAND

Zentralarchiv für empirische Sozialforschung

Universität zu Köln

1. Introduction

Until recently the explanation of drug use has usually been the domain of psychiatrists and other medically trained people. It was among them that the aetiology was mainly discussed and it was they who set the tone in other disciplines as well.* The dominant image of the drug user which they developed was that of an individual suffering from pathology. According to that view the motivation for drug use usually evolves from an attempt to escape from problems which the individual has to face. Drug use is seen as a flight from reality and the drug user is seen as a person deserving psychotherapeutic treatment.**

* The validity of the psychiatric model has usually been taken for granted in other disciplines as well and has been made a starting point for an explanation of its own. Thus in sociology for instance attempts were made to view drug use as a response to general conditions within society. According to that view drug use represents a structurally induced kind of retreatism (e.g. Merton 1957; Cloward and Ohlin 1960). The psychiatric model continues to be effective under this disguise since the drug use is still seen as a flight from unpleasant realities.

** Sometimes sociopathic, antisocial personalities are said to be prone to drug use as well. Within the context of the pathological model however this explanation seems to be of minor and moreover losing relevance (for some statements of the pathological model see e.g. the excerpts in Goode (1969), Schenk (1974a) and the discussion in Redlich and Freedman (1966), Kielholz and Ladewig (1973).

The changing social basis and the changing conditions of drug use which took place in the sixties have shifted this traditional conception somewhat away from its individualistic emphasis towards a more social orientation. Thus the existence of a pro drug ideology and the existence of friendship networks among drug users became recognized. However, generally seen, this recognition did not cause a major reorientation of the dominant pathological explanation, it only produced minor modifications of the theory. As a consequence it is still not uncommon today in the psychiatric and psychological literature to read that individual pathology and individual problems are the main reason for drug use. Social factors are said to be either relatively unimportant or mere surfaces phenomena which do not deserve a treatment of their own (Redhardt 1971, Scheidt 1976). There is no sign of a change towards a more social psychological or sociological explanation within the recent psychiatric and psychological literature in West Germany (and probably in other countries as well).

On the contrary: individualistic approaches seem to gain more prominence than ever (Haas 1974, Scheidt 1976), the social factors in explaining drug use are even said to have been over-estimated during the past (Täschner 1975). It seems due time therefore to address the research question whether individual pathology is the dominant reason for drug use and whether social factors are in fact of secondary importance. We need an empirical assessment of both approaches taken together since the relative value of an approach can only be found by strict methodological comparison. In the following we attempt such a comparison. We proceed from the traditional pathology model of drug use on the one hand and then contrast it with the subcultural model which takes social factors as a starting point for its explanation. Before entering into an empirical test of both approaches a brief discussion of both models is presented.

2. The pathological and the subcultural model of drug use

2.1 The pathological model of drug use

The most widespread explanatory model of drug use (especially in the medical field) has been that drug users have problems from which they are trying to escape. Drug use has accordingly been conceived as a symptom of individual pathology. A differentiation according to the type of drug user is seldom performed. Some authors already conceive those drug users as pathological who have tried drugs once or twice. Others - more or less implicitly - restrict this view to those who have gone beyond the phase of tasting the drug a few times (Kielholz and Ladewig 1973). Whatever the interpretation is, both sides agree on the fact that progression in drug use is linked to individual pathology, with unpleasant and frustrating experiences.

Despite its prominence the empirical proof of this hypothesis is rather dim due to the methodological and analytical deficiencies of past research. Namely:

(a) Type of date used: The most common approach of data collection has been to take drug users in clinics and counselling centres as the basis for analysis. In a strategy like this the danger of methodological artefacts cannot be denied, since drug users in both institutions seem to be a non-random selection of drug users within each category of drug use: drug users with personal problems are probably heavily over represented since the function of these institutions is not restricted to professional help on drug related problems alone. Counselling in other types of problems is rather widespread and is one major reason why people turn to such kinds of institutions (Berger and Zeitel 1976). A somewhat similar bias seems to operate even if the data collection takes place via snowball sampling: in this case those people seem to be especially interested in co-operation who hope to get help from the researcher in order to cope with their personal problems (Herha 1973, Kühne 1974). Random samples of users and non-users alike are one way to eliminate the bias of selection. Such kind of samples have gained prominence within the last years in drug research among medically trained researchers as well. A proper analysis, however, has hardly been completed: it has been customary simply to compare undifferentiated groups of users and non-users or at least dubious typologies of users (Sadava 1975, Schenk 1974b). An adequate understanding of the underlying dynamics of drug use after its beginning has therefore not yet been developed.

(b) Existence of frustrating experiences: Disturbed relationships - such as with parents and school - have usually been taken as an indicator for individual suffering (Schwarz 1972). A direct test of the hypothesis has rarely been carried out. Such strategy does not seem justified because of the possiblility (i) that disturbed personal relationships can have an effect on drug use by other mechanisms, such as by normative estrangement, for instance (Reuband 1976), and (ii) that the level of satisfaction is determined by a balance of weighted positive and negative experiences (Scheuch 1971, Reuband 1971), shifting and reweighting a dimension of satisfaction is henceforth always possible. A disturbed relationship for that reason does not necessarily imply an impact on the general level of life satisfaction. It needs to be measured as a separate variable.

(c) The determinators of causality: Even if we take the hypothesis for granted, that drug users have problems, the proof must still be made that the frustrating experiences are the cause and not the effects of drug use. Since the operation of the latter mechanism has been empirically documented (Wanke 1971) it has to be taken into consideration when interpreting a relationship. Past interpretations have too often been one-sided by neglecting either the causal or the effect dimension of drug use (Redlich and Freedman 1966).

(d) Magnitude of the observed relationship: Even if the problem of causality has been settled the problem of theoretical significance still exists: there is the danger of taking trivial relationships as a fundamental proof of one's hypothesis without realizing the weak basis of one's argumentation. The application of significance tests does not help further, since they are no measure of the strength of relationships. * Correlation coefficients must be computed in order to allow inferences about the theoretical significance of the relationship. Such a strategy is especially important when comparisons with other relationships are done. Past research on drug use has rarely taken this kind of strategy. It has been usual simply to assess differences and to make them the basis of theoretical conclusions.

Since the pathological model of drug use has not been tested adequately it is far from being proven as a valid and useful explanation of drug use. Nevertheless the possibility of adequacy does in fact exist. An empirical test has therefore still to be done. In the following we prefer to do so in comparison with the subcultural model which takes the social variables in the etiology of drug use into account.

2.2 The subcultural model of drug use

The subcultural approach - in contrast to the pathological one - takes social factors as explanatory factors in their own right. It starts from the basic fact that man is living in a society of people sharing a common culture. His behaviour is accordingly determined by societal expectations and internalized beliefs, values and norms. Since the process of socialization is not restricted to early or late childhood, but remains effective later on as well, (Brim and Wheeler 1966), man's behaviour can hardly be explained on the basis of purely psychodynamic factors alone. Societal influence has to be taken into consideration as well.

* It should be noted, moreover, that significance tests can only be applied under certain conditions, a fact, which has often been overlooked. For a discussion of significance tests see Morrison and Henkel (1970).

Societal influence - via expectations and internalized culture - does not derive from society as a global phenomenon itself, it is situated in the social groupings which make up the fabric of society. As long as these groupings share a more or less common culture essentially the same behaviour patterns are considered normal or abnormal within society. However, if existing or evolving groups develop a different perspective with regard to beliefs, values and norms it might happen that certain behaviour patterns are considered normal by one group and abnormal or deviant by other groups. Under this condition a subculture, i.e. a culture within the general culture, is said to exist.* If an individual comes under the influence of such a subculture his evolving behaviour might be a natural result of his subcultural participation, no psychopathological explanation is then needed in order to account for it. The behaviour is as normal as other kinds of prescribed behaviour patterns of the group and his learning of the behavioural norms is as normal as other kinds of learning processes.

Subcultures like cultures in general are man made but are not made by men acting individually. They evolve in a collective process which is determined by interaction and communication. Stability of beliefs, moreover is only possible, if some kind of social validation exists in day to day interaction.** Subcultures therefore usually have a social, interactive basis. This fact deserves especial mentioning, since it means that contact with a subculture can be established by interacting with its members without necessarily incorporating the extensive beliefs, values and norms of that subculture.*** In such a case the subculture's conduct norm **** is communicated by not necessarily the underlying beliefs, values and norms which give some kind of legitimization to it.

* For a discussion of subcultural theory see especially Arnold (1970) and Wolfgang and Ferracuti (1967).

** See Berger and Luckmann (1967) for a theoretical elaboration. For research supporting these conclusions see the group dynamics literature (e.g. Cartwright and Zander 1968).

*** For a discussion of this analytical separation see the remarks by Wolfgang and Ferracuti (1967: 102) and Johnson (1973: 10f.).

**** The term "conduct norm" has been developed by T.Sellin and later used by B.Johnson (1973: 9) in order to refer to behaviour that is expected within a group. The conduct norm in drug using groups would be to use drugs.

The question arising now is whether drug use today can be seen as linked up with a subculture of drug use. This question must be answered in the affirmative: from the very beginning of the recent drug wave - the middle sixties - drug use was backed up by a more or less elaborate ideology of drug use (see for instance Timothy Leary or the hippies ideology of drug use) and held together by intense interaction between drug users. This interaction was partially due to the ideology of drug use itself and partially due to other factors as well, such as the illegitimacy of drug use. Any explanation of drug use today therefore has to realize that drug use might be an outflow of sub-cultural participation and not an outflow from psychodynamic factors.

Due to the relevance of interpersonal expectations and the relevance of culturally mediated beliefs, values and norms on the one hand and due to the social and cultural basis of subcultures on the other hand, participation in the drug subculture can take place either interactively or symbolically. Interactively the individual is influenced by interpersonal expectations and the cultural elements which are usually transmitted in the interaction process as well. This kind of socialization is probably the most effective one since it involves a "double" process of socialization. Participation in the drug subculture might be purely symbolic on the other hand, if the participation is restricted to the beliefs, values and norms and does not include interaction with drug users. Such a kind of participation might be the result of various factors: it might be caused by reference group identification, influence by the underground mass media or even by traditional sources of information which have partially adopted elements of the subcultural beliefs of drug use.

The proposed hypothesis of subcultural influences on progressive drug use can thus be stated as follows: continuous drug use takes place with increasing participation in the drug subculture as measured by interaction with drug users and personal internalized beliefs, values and norms about drug use. In the following we want to restrict ourselves to the influences of these variables on the motivational genesis of progressive drug use. We do so for reasons of empirical testing: since we have no panel data exact behavioural consequences cannot be causally discerned by our kind of analysis. The influence of subcultural participation with regard to the access to drugs (Becker 1963) is therefore omitted from the discussion despite its evident relevance for actual drug use. Doing so we remain on essentially the same level as the pathological model which restricts itself to the motivation of drug use as well.

3. Methodology

A stratified cluster sample of the school population of Hamburg (West Germany) in 1975 constitutes the data basis of our research. The data were collected in the classroom setting by means of an anonymous questionnaire. Using an anonymous questionnaire usually has the effect that tabooed and sanctioned attitudes and behaviour patterns are more likely to be admitted (Hyman et al. 1954). Teachers were not allowed to be present in order to increase this feeling of anonymity. * Since the characteristics of an interviewer can have an influence on the responses (Hyman et al. 1954) an attempt was made to decrease the existing distance between students and interviewer by selecting relatively young interviewers (not older than 30 years of age) who were dressed informally during the interview. An intensive search of the questionnaires for any signs of non-co-operation and deception led to an elimination of 2% of the sample, leaving N = 5.426 for analysis.

The basis of the pathological model assumes that progression in drug use is determined by individual frustration. For that reason the optimal test does not compare users and non-users but users with different amounts of drug use. Since causality can only be determined if the observed variables have an appropriate temporal relationship to each other we restrict our sample of users to current users, including those who have taken drugs within the last six months. The latter strategy seems useful when recognizing the rate of change of relevant variables, such as life satisfaction for instance (Robinson 1969). The usual approach of simply correlating dependent and independent variables is ruled out or illegitimate, since the measured amount of drug use refers to the past for most of the drug users whereas the measured independent variables usually refer to the present (e.g. family relationships, life satisfaction). Even so, the recognition of the temporal dimension alone is not sufficient for a causal analysis. We need a more direct assessment of causality by ruling out recursive causal relationships. Thus merely correlating the amount of drug use with certain variables per se does not often help much in tracing causality since these variables could be either a cause or a consequence of drug use (e.g. family relationships).

* Teachers' presence seems to have an influence on answers - at least among young students - even if the questionnaire is anonymous (see Devereux 1970). This effect is probably not due to a fear of sanctions but due to an inner activation of cognitive dissonance.

Since we restrict our analysis to the motivational antecedents of continuous drug use the additional recognition of the willingness to continue drug use can be helpful in resolving the problem. Consequently we equally consider both aspects in order to discern the relevant determining factors in drug progression.

Progression in drug use is measured by summing up the average frequency of each of the mentioned drugs. The subsequent scale is partitioned into five classes according to the amount of drug use (N = 416 users). The drug use typology is presented in Table 1. (See next page). It shows that the variety of drugs used goes up with increasing drug use. Furthermore it shows that the committment to drugs increases with its usage: whereas the tasters hardly spend any money for drugs, most of the frequent drug users do. As a consequence the dependency on situational contingencies for drug use diminishes, the frequent user is more able to use drugs whenever he wishes than the less frequent user. This ability does not seem to remain on the level of possibility alone, it is in fact made use of : drug use on an individual basis increases with greater frequency of drug use. Since this trend is usually combined with social drug use as well it does not necessarily signalize a desocialization from other drug users as has been proposed by a number of writers. The most adequate interpretation would be that of an increasing diversification of drug use (Fisher 1974). Lone drug users are rarely represented even among those students who have used drugs more than a 100 times. Seen in the whole it can be stated that the chance of immediate motivational gratifications is greatest among our more frequent drug users. If there is a motivational link to drug use it should turn out in our data in any case.

The willingness to continue drug use is measured by asking current users for their differential willingness to stop drug use. The categories are then regrouped so that an ordinal scale evolves (N = 336 users).

4. A test of the pathological model

In the following we want to test empirically the pathological model of drug use. This is done in a number of steps. First we want to see to what extent disturbed relationships to subjective relevant persons and institutions do in fact correlate with drug use as it has been widely claimed. In a second step we want to see whether frustrations are decisive for the progression in drug use.

4.1 Disturbed relationships

Parents and school are considered the most important spheres of a young person. Both institutions absorb most of his time, both are relevant determinators of gratifications and sanctions as well.

TABLE I

DRUG RELATED CHARACTERISTICS OF THE SAMPLE (IN %)

	Amount of drug use				
	1	2-5	6-20	21-99	100+
Type of drug ever used					
Cannabis	83	94	93	94	100
Halluzinogens	-	5	9	35	60
Amphetamines *	8	13	33	50	60
Sedatives	-	-	1	2	-
Opiates	-	1	4	9	20
Inhalants	-	1	-	2	2
Not specified	13	5	3	2	3
Used more than two types of drugs **	-	1	6	21	49
No money spent for drug use	76	75	57	29	27
Social context of drug use ***					
Usually alone			2	3	-
About half alone, half in company of others			5	11	25
At first in company of others, later alone			4	6	8
(N=)	54	85	101	94	65

Annotation:

* Includes legally and illegally available amphetamines and stimulants.

** Type of drug as defined above.

*** Not ascertained for people who have used drugs less than five times.

We turn to the family first. A number of writers have argued that living in an incomplete family is one of the most important factors for drug use (Kielholz and Ladewig 1973). No substantial proof, however, is found for this hypothesis: 79% of the drug users have a family where neither divorce nor death has separated their parents. Although the proportion of complete families is somewhat lower among those who have used drugs more than a 100 times when compared to those who have used drugs only once (73% vs. 85%), a clear linear relationship between drug use and completeness of the family does not exist. The same holds true for the willingness to consume drugs. The relationship between drug use and completeness of the family is generally negligent in magnitude. Apparently the relevance of the broken home factor has been exaggerated, perhaps due to the sampling strategy used (clinics, counselling centres). If we turn to the relationship with the parents a majority among the drug users convey a rather positive impression of that relationship, an impression which is confirmed when other indicators of family life are considered (60% for instance state a positive relation to their mother, 46% do so with regard to the father. If the moderate relationships are included we get 93% respectively 77%), A majority of the drug users only stand in opposition to their parents when the identification with their parents' life style and beliefs is tapped. However, a majority among the non-users oppose the parental life style as well though to a lesser extent. This fact can therefore hardly be taken as a proof for the pathology hypothesis.

If we correlate drug use with indicators of disturbed relationships to parents and home life the correlations emerge as rather low and not each worth mentioning in most of the cases (Table II) (See next page). Where somewhat higher coefficients are achieved the question of causality is set into doubt such as in the case of the wish to reduce the contact to parents: if we take the willingness to continue drug use as an indicator, the correlation nearly vanishes. The only variable which has a noteworthy albeit small relationship according to both criteria is a variable which deals with the belief system and the life style of the parents ("I would like to be like my parents later on"). Henceforth it seems as if the emotional dimensions are of little value in explaining drug progression, only the normative identification seems to have some influence.

Similar low correlations are found when the relationships with the school are considered. A correlation greater than $\gamma = .10$ is only given according to both criteria with regard to school satisfaction in general, the relationship to teachers attitudes and one's class mates. There is not a single correlation greater than $\gamma = .20$, however, which satisfies both criteria. A moderate or even strong correlation cannot be found.

TABLE II

INDICATORS OF SOCIAL DISTURBANCE AND UNHAPPINESS AS RELATED TO DRUG USE (GAMMA CORRELATION COEFFICIENT)

	Actual amount of drug use	Willingness to continue drug use
Relationship to parents		
Mother	-04	-01
Father	-07	-20
Feels well at home	-06	-07
Wish for less contact with parents	22	02
Wants to be like parents later in life	-20	-18
Has different opinions than parents	10	06
Happy childhood	-08	-01
Relationship to school		
School (working place as to apprentices)	-11	-21
Teachers	01	-10
Has different opinions than teachers	11	10
Class mates	-14	-15
School achievement	02	-08
Repeater	23	07
Happiness		
Satisfaction with oneself and one's life	-06	-03
Feels often unhappy and sad	03	03

Annotation: The short description of the variables are phrased in the same evaluational direction as in the questionnaire. Negatively phrased items are therefore negatively phrased here as well. Where no evaluational direction is mentioned (e.g. "school") a negative correlation indicates that drug use increases when the relationship to the person or institution deteriorates.

4.2 Level of frustration and drug use

The core statement of the pathological theory of drug use refers to frustration as the decisive variable in explaining continuous drug use. Instead of following the common practice of solely using possible indicators of frustration (such as family relationships e.g.)we now turn to a more direct measurement of general frustration. If we do so, we can see that satisfaction with life is rather high among our drug users.* More important than that, however, is the fact that there is hardly any correlation between the amount of drug use on the one hand and the indicators of life satisfaction on the other. A similar finding has been shown in somewhat related studies as well (Schenk 1974a, Kandel et al, 1974). Since the correlation between drug use and life satisfaction is negligible in nature it seems rather unlikely that drug users are mainly using their drugs in a situation of depressive mood. The final and decisive test of the hypothesis therefore directly deals with the situation of drug use. People were asked for the situational mood in which drug use commonly takes place. According to this question a majority of drug users (58%) take drugs when no definite mood prevails. In 10% of the cases drugs are usually taken in a situation of bad mood and in 30% of the cases in a situation of good mood. If we break down the table according to the amount of drug use we can discern a trend towards increasing independence from definite moods. (Table III). (See next page). This trend seems to signalize increasing habitualization of drug use. As a consequence drug use in purely a good mood is decreasing with greater frequency of drug use, the same tendency - albeit a little less pronounced - can also be found with regard to drug use in bad mood. It might be concluded from these findings that there is little support for the traditional theory of drug progression. The correlations are rather low, if not negligible. Henceforth flight from reality cannot be the dominant motive of drug progression. The validity of this thesis must be limited to small minorities.

In view of these findings a major reorientation about drug effects seems warranted: instead of viewing the drug experience as a mere coping mechanism and the drug effect as satisfying only when a depressive mood prevails, it seems more reasonable to view the experience as pleasant in its own right: No psychopathological

* 50% of the drug users consider themselves as being strongly or moderately satisfied, only 19% admit being dissatisfied. A similar result can be obtained with regard to a statement which taps frequent feelings of unhappiness and sadness: 68% reject the statement as being valid for themselves.

TABLE III

SITUATIONAL MOOD AND USE OF DRUGS (IN %)

Mood	Amount of drug use				
	1	2-5	6-20	21-99	100+
If person is in a bad mood and wants to come into a better mood	11	16	11	3	10
If there is boredom	2	4	2	2	3
If person is in a good mood	39	31	32	28	21
There are no distinct situations, sometimes in bad and sometimes in good mood	48	49	55	67	67
(N=)	46	83	99	94	63

motivation is needed to explain why people experience it as pleasurable. Moreover the relevance of this effect for the continuation of drugs use has been documented (Becker 1963, Peterson and Wetz 1975, Zimmermann 1976). However, as Howard Becker has pointed out, learning to enjoy the drug experience is a necessary but not sufficient condition to develop a stable pattern of drug use. The user has "still to contend with the powerful forces of social control that make the act seem inexpedient, immoral or both" (Becker 1963). It is the question to which we want to turn to now.

5. A test of the subcultural model

According to the subcultural model drug use is not to be seen as pathological behaviour. It is seen as normal as far as it is due to the influence of a subculture which is centred around the activity of drug use. Drug use is accordingly seen as the product of an estrangement from traditional norms and their carriers and as a product of subcultural influence. The motivation for drug use is basically located in the cultural notions of drug use (and not in hidden motives of the personality) on the one hand and in a kind of conformity to relevant other persons on the other hand. We turn to the cultural components first and then to the interaction influences which partially encompass the cultural components and partially add additional weight to them.

5.1 Cultural orientation

As can be seen in Table IV (see next page) frequent drug users usually have less trust in the traditional media of information reporting on drug use than the seldom users. They have less negative views of the harmful effects of hashish use and perceive hashish as less dangerous than alcohol. They have apparently freed themselves from the conventional notions of drug use in which the negative effects are emphasized and the positive effects played down (Gaedt et al. 1976). With regard to the positive effects of hashish one can discern a trend among the frequent users to endorse the positive ascriptions more often. With regard to sociability and conflict reduction no such trend can be found: The belief in the sociability enhancing effects of hashish does not have an effect on the continuation of drug use and the belief in the conflict reducing effects is even contrary to that. This lends further credence to our notion that problems can hardly be the main reason for drug involvement. In the case of harder drugs than hashish a differentiation seems warranted. According to our data frequent users see less dangers in LSD than occasional users. No such relationship exists with regard to heroin. Perhaps we cannot find a similar relationship as in the case of LSD because of the users' differentiation between their own drug use and the use of heroin: heroin is probably seen as qualitively differing so that its rejection does not imply a rejection of one's own drug use. If we turn away from the definition of drugs to a definition of drug users and related aspects we find that progression in drug use is linked to other drug related attitudes as well. Frequent users for instance more often than occasional users plea for a liberation of the drug laws and for lesser sanctions applied to hashish dealers. Summarizing our results it can be concluded that progression in drug use generally depends on the extent to which the deviant perspective has been internalized. The correlations observed are usually stronger than those which have been observed with regard to disturbed relationships and life satisfaction.

5.2 Interaction partners

If we turn to the interaction partners as one of the most important agencies of stabilizing beliefs, values and norms on the one hand and of inducing conformity by interpersonal expectations on the other, we find that at least 92% of the drug users have current drug users in their friendship and acquaintanceship network. This proportion goes up to 100% with increasing frequency of drug use. At the same time as drug use increases the

TABLE IV

SUBCULTURAL INFLUENCES ON DRUG USE (GAMMA CORRELATION COEFFICIENT)

	Actual amount of drug use	Willingness to continue drug use
Negative effects of hashish		
Less dangerous than reported in the newspapers	31	43
Loss of self control	-25	-28
Addiction	-33	-32
Criminality proneness	-24	-35
Health hazards	-10	-27
Positive effects of hashish		
Satisfaction	17	11
Change of consciousness	17	19
Alleviation of sociability and contacts	-	03
Conflict reduction	-04	-15
Hashish vs. alcohol		
Less dangerous than alcohol	31	38
Less dangerous than alcohol when driving a car	14	19
Hard drugs		
LSD not dangerous under certain conditions	36	30
Heroin not dangerous under certain conditions	06	05
Drug related values and norms		
Drug user is a coward since he flees from reality	-22	-31
Right to determine one's fate and use drugs	20	13
Liberation of hashish laws	36	40
Stronger sanctions to hashish dealers	-36	-42
Interaction partners		
Proportion of drug users in the friendship and acquaintanceship network	50	42
Drug use among best friends	35	32

proportion of current drug users in one's environment increases as well. Thus we can find for instance that among the users who have taken drugs more than a 100 times 59% of them have a majority of drug users in their friendship and acquaintanceship network. A similar trend emerges if we do not consider the proportion of drug users but inquire whether the best friend or the best friends are using drugs. In this case we find essentially the same if smaller a trend. Thus 46% of the one time users and 77% of those in our most frequent group have good friends using drugs. There can be no doubt that the observed relationship between the frequency of drug use and having drug using friends is a reciprocal one: the drug using friends are a cause and a consequence of drug use (Wanke 1971, Johnson 1973). If we compare the correlations with regard to the actual amount of drug use and the willingness to continue drug use, however, it becomes evident the correlation with the willingness variable is only a little bit smaller than the correlation with actual drug use. Henceforth it can be concluded that the observed relationship is mainly due to the friends being the cause of continuous drug use and not vice versa. The correlations are rather strong, especially if compared with those which have been observed in our tests before. They are $\gamma = .50$ in the case of actual drug use and $\gamma = .42$ in the case of willingness to continue drug use.

6. Conclusions

The pathological model of drug use has found little confirmation in our data. Although the same trends were sometimes observed with regard to certain relationships as in past literature, a computation of correlation coefficients revealed that the observed relationships were usually low. In most of the cases they even were to be treated as negligble. We have to conclude, therefore, that past research has tended to overestimate the relevance of individual problems for drug usage, probably due to a number of methodological deficiencies in design and analysis. As a consequence the relevance of social factors in drug use has been underestimated.

According to our data it seems quite true to view drug use as an outflow of participation in the drug subculture. Drug use seems to be a rather "normal" kind of behaviour which usually does not need a psychodynamic explanation to account for it. The time is due for research to pay more attention to the role of subcultural influences on drug use, especially with regard to interaction patterns. *

* For some hopeful approaches into this direction see especially Johnson (1973), Kandel (1974), Plant (1975), Tec (1972), Goode (1970).

REFERENCES

ARNOLD, D.O., ed. (1970). The Sociology of Subcultures. Berkeley.

BECKER, H.S. (1963). "Outsiders" Studies in the Sociology of Deviance. New York.

BERGER, H. and ZEITEL, F. (1976). "Drogenberatungsstellen als therapeutische Instanzen. Zur Tätigkeit der Frankfurter Beratungsstelle "drop-in". K.H. Reuhand, ed. Rauschmittelkonsum. Soziale Abweichung und institutionelle Reaktion. Wiesbaden. 155-189.

BERGER, P.L. and LUCKMANN, T.(1967)."The Social Construction of Reality. A Treatise in the Sociology of Knowledge." Garden City, New York.

BRIM, O.G. and WHEELER, S. (1966). "Socialization after Childhood." New York.

CARTWRIGHT, D. and ZANDER, A. ed. (1968). "Group Dynamics." Research and Theory. 3rd Edition. London.

CLOWARD, R.A. and OHLIN, L.E. (1960). "Delinquency and Opportunity". New York.

DEVEREUX, E.C. (1970). "Autorität und moralische Entwicklung bei deutschen und amerikanischen Kindern." G. Lüschen and E. Lupri, eds. Soziologie der Familie. Sonderheft 14 der Kölner Zeitschrift fur Soziologie und Sozialpsychologie. Opladen. 353-379.

FISCHER, G. (1974). "Milieu of Marihuana Use." International Journal of Social Psychiatry, 20. 45-55

GAEDT, F., GAEDT, C. and REUBAND, K.H. (1976). Zur Rauschmittelberichterstattung der Tageszeittung der Tageszeitungen in der Bundesrepublik und West-Berlin. Ergebnisse einer Inhaltsanalyse. K.H. Reuband, ed. Rauschmittelkonsum. Soziale Abweichung und institutionelle Reaktion. Wiesbaden. 77-107.

GOODE, E. ed. (1969). "Marijuana" Chicago-New York.

GOODE, E. (1970). "The Marijuana Smokers". New York.

HAAS, E. (1974). "Selbstheilung durch Drogen." Frankfurt.

HERHA, J. (1973). "Erfahrungen mit Haschisch." Ergebnisse einer Befragung von 234 Konsumenten von Cannabis und anderen Drogen in Berlin (West). Medical Dissertation 1969/70, Berlin.

HYMAN, H., COBB, W., FELDMAN, J.H., HART, C.W., and STEMBER, C.H. (1954). Interviewing in Social Research. Chicago.

JOHNSON, B. (1973). Marihuana Users and Drug Subcultures. New York.

KANDEL, D. (1974). "Interpersonal Influence on Adolescent Illegal Drug Use." E. Josephson and E.E. Carroll, eds. Drug Use - Epidemiological and Sociological Approaches, New York. 207-240.

KANDEL, D., SINGLE, E., TREIMAN, D. and FAUST, R. (1974). "Adolescent Involvement in Illicit Drug Use : A Multiple Classification Analysis." Annual Meeting of the American Sociological Association, Montreal.

KIELHOLZ,P. and LADEWIG, D. (1973). Die Abhängigkeit von Drogen. München.

KÜHNE, H.H. (1974). "Motivationsläufe bei Rauschmittelgeschädigten." H. Müller-Dietz, ed. Kriminaltherapie heute. Berlin. 50-113.

MERTON, R.K. (1957). Social Theory and Social Structure. 2nd edition. Glencoe, Illinois.

MORRISON, D.E. and HENKEL, R.E. eds. (1970). The Significance Test Controversy. Chicago.

PETERSON, B. and WETZ, R. (1975). Drogenerfahrung von Schülern. Stuttgart.

PLANT, M.A. (1975). Drugtakers in an English Town. London.

REDHARDT, R. (1971). "Zur Psychopathologie der ideologischen und soziokulturellen Motivationszusammenhänge des Haschischmissbrauches". Zeitschrift fur Rechtsmedizin. 68, 57-72

REDLICH, F.C. and FREEDMAN, D.X. (1966). The Theory and Practice of Psychiatry. New York.

REUBAND, K.H. (1971). "Ein Modell zur Erklärung von Arbeitszufrie-denheit." Soziologenkorrespondenz. 2, 73-81.

REUBAND, K.H. (1976). "Normative Entfremdung als Devianzpotential. Über die Beziehung zwischen wahrgenommenem Lehrerverhalten und Bereitschaft zum Rauschmittelkonsum." Rauschmittelkonsum. Soziale Abweichung und institutionelle Reaktion. 17-40. K.H. Reuband, ed. Wiesbaden.

ROBINSON, J.P. (1969). "Life satisfaction and happiness." Measures of Social Psychological Attitudes. 11-43. Ann Arbor.

SCHEIDT, J.V. (1976). "Der falsche Weg zum Selbst." Studien zur Drogenkarriere. Munchen.

SCHENK, J. (1974a). "Zur Personlichkeitsstruktur des Haschisch-konsumenten." Wehrpyschologische Untersuchungen. Heft 1.

SCHENK, J. (1974b). "Ausmuss und Struktur des Drogenkonsums Jugendlicher in Deutschland." Wehrpsychologische Untersuchungen, Heft 3/4.

SCHEUCH, E.K. (1971). "Vorstellungen von Glück in unterschiedlichen Sozialschichten." H. Kundler, ed. Anatomie des Glücks, Köln. 71-85.

SCHWARZ, J. (1972). "Feststellungen zum Rauschmittelmissbrauch Jugendlicher in Schleswig Holstein." Grundlagen der Kriminalistik, 9. 196-208. Rauschgiftmissbrauch, Rauschgiftkriminalität. Hamburg.

SADAVA, S.W. (1975). "Research Approaches in Illicit Drug Uses: A Critical Review." Genetic Psychology Monographs. 91, 3-59.

TÄSCHNER, K.L. (1975). "Zur Frage gesellschaftlicher Ursachen des Drogenkonsums." Zeitschrift fur Sozialpsychologie, 6. 76-79.

TEC, N. (1972). "The Peer Group and Marijuana Use." Crime and Delinquency. 298-309.

WOLFGANG, M.E. and FERRACUTI, F. (1967). The Subculture of Violence Towards an Integrated Theory in Criminology. London.

ZIMMERMANN, R. (1976). "Zur Situation des ersten Rauschmittelkonsums. Über den Stellenwert situationsspezifischer Faktoren fur die Aufnahme und Fortsetzung des Konsums." Rauschmittelkonsum. Soziale Abweichung und institutionelle Reaktion. Wiesbaden. 63-75.

UNEMPLOYMENT AND SICKNESS ABSENTEEISM IN ALCOHOLICS

E. S. M. SAAD

DEPARTMENT OF PSYCHIATRY

BRIDGEWATER HOSPITAL, MANCHESTER

Industrial absenteeism due to sickness has reached alarming proportions and the Managers in industry are well aware of the fact that some staff members are more likely to have spells of sickness absence than others. No accurate estimates of absenteeism related to alcoholism are available for the United Kingdom and, therefore, the estimates of the cost of alcoholism to industry in Britain are guesses that point to the need for precise information.

Holtman estimated financially six forms of loss attributable to alcoholism; decreased life expectancy, unemployment, absenteeism, imprisonment, hospitalisation and car accidents. The list could be extended to include state and personal insurance benefits to alcoholics and dependents, treatment at home or at hospital outpatient departments, care from Social Services and voluntary agencies, police and other legal expenses including Court actions for separation or divorce.

This study examined the medically certificated incapacity and unemployment of a group of male alcoholics. Through the co-operation of the Department of Health and Social Security (D.H.S.S.) data were obtained of their state benefits over a 12 month period, the diagnoses on their medical certificates, and the weekly contributions to the State insurance scheme paid or credited over a five year period. The weekly contributions form a measure of regularity of employment. Each patient gave signed permission for D.H.S.S. to provide the data.

The subjects were 73 male alcoholics who were admitted to the Regional Addiction Unit in Chester or seen in its outpatient clinics in Chester and Liverpool between May and November 1973. All had developed physical dependence on alcohol, so that they were drinking to relieve withdrawal symptoms such as tremors, sweating, and anxiety.

The majority had suffered from a drinking problem for over ten years. Fifty two or 71.2% were inpatients and 21 or 28.8% were outpatients. There was no significant difference between the two sub-groups, perhaps because of the nature of the Regional Unit and its outpatient clinics. Forty-seven or 64.4% were in employment at the time of the interview and 26 or 35.6% were unemployed. There were significant differences between the two sub-groups.

The demographic data are summarised in the following tables:-

TABLE I : AGE DISTRIBUTIONS - MEAN 43.1 S.D. 9.5

Age Group	N	%
20 - 29	6	8
30 - 39	17	23
40 - 49	33	45
50 - 59	14	20
60 - 69	3	4
TOTAL	73	100

Age distributions, the mean is 43.1 and S.D. 9.5, 50 or 68.5% were in the most productive period of their lives (30 - 50).

TABLE II : MARITAL STATUS

	N	%
Married	42	57.5
Divorced	14	19.2
Single	9	12.3
Separated	7	9.6
Widowed	1	1.4
TOTAL	73	100

TABLE III : SOCIO-ECONOMIC CLASS ACCORDING TO THE REGISTRAR GENERAL'S CLASSIFICATION

Class	N	%
I	2	2.7
II	4	5.5
III	34	46.6
IV	21	28.8
V	12	16.4
TOTAL	73	100

Socio-economic class - 40 or 54.9% were skilled workers, there was a tendency to downward drift in social class. Five of the subjects who were in Class IV when interviewed, and two of those in Class V had belonged to higher strata.

TABLE IV : SOCIAL STABILITY SCALE 0 - 4 (STRAUS AND BACON 1951)

S.S.S.	N	Working	Out of Work
4	28	22	6
3	20	15	5
2	19	8	11
1	6	2	4
0	0	0	0
TOTAL	73	47	26

Social stability scale of 5 points, 0 - 4 according to Straus and Bacon. Forty-eight or 65.8% displayed fairly good social adjustment. Thirty-seven or 78.7% of those who were in employment were high in the scale while the pattern was reversed for the unemployed - only 11 or 42.3% high in the scale.

TABLE V : RECORDED PERIODS OF INTERRUPTION OF EMPLOYMENT (P.I.E.) IN THE YEAR BEFORE INTERVIEW (FIGURES IN BRACKETS FOR TEN ALCOHOLICS ALMOST CONTINUOUSLY AWAY FROM WORK FOR THE YEAR)

Work Status	N	P.I.E.	% of total loss	Average loss per person
Employed	47	3466	39.0	73.7
Unemployed	26	5416	61.0	208.3
	(10)	(3007)	(33.9)	(300.7)
TOTAL	73	8882	100	121.7

Recorded periods of interruption of employment in the year before the interview. P.I.E. are not working days lost for two reasons. Firstly, they are calculated on the basis of a six day week and a 52 week a year, regardless of holidays, giving 312 working days to the year even though most people are expected to work fewer days than this, perhaps 240 days each year. Secondly, the figures were inflated by 10 alcoholics who were in almost continuous absence from work and some of whom were unemployable. So 10 or 13.5% were responsible for 33.9% of the total. This is in accordance with the general finding that almost one third of all days of incapacity in D.H.S.S. records are incurred by a minority who are unemployable. On the other hand, spells less than four days are not counted in D.H.S.S. records which also do not include certified absences of members of the non-industrial civil service, the post office, or the armed forces, or absences which do not qualify for sickness or unemployment benefit because of inadequate contributions. Those who were in employment had an average loss of 73.7 working days in a year, while the loss for unemployed was 208.3 working days.

TABLE VI : CONTRIBUTIONS IN THE LAST FIVE YEARS OF THE TEN ALCOHOLICS IN ALMOST CONTINUOUS UNEMPLOYMENT

N	Paid	Credited
1	109	73
2	107	61
3	156	99
4	214	44
5	149	79
6	70	118
7	-	261
8	42	214
9	34	212
10	44	156

This shows the contributions of the ten who were in almost continuous unemployment. Five of them had been able to pay more than 100 weekly contributions throughout the previous five years rather than have their contributions credited for them by the State. The other five could be considered as unemployed.

TABLE VII : DISTRIBUTION OF RECORDED PERIODS OF INTERRUPTION OF EMPLOYMENT (P.I.E.) BETWEEN CERTIFICATED INCAPACITY AND UNEMPLOYMENT

Work Status	N	P.I.E.	Certificated Incapacity	Average loss per person	Unemploy-ment	Average loss per person
Employed	47	3466	2956	62.9	510	10.8
Unemployed	26	5416	3329	128.0	2087	80.3

Among the employed, the average loss due to certificated incapacity was 62.9 working days and only 10.8 working days were lost through unemployment. Those who were unemployed lost on average 128.0 days due to unemployment. One explanation could be the reluctance of the alcoholic to sign as unemployed.

TABLE VIII : DISTRIBUTION OF SPELLS OF SICKNESS ABSENCE

No. of spells	No. of Patients	Total No. of spells
1	21	21
2	15	30
3	11	33
4	4	16
5	3	15
6	4	24
8	1	8
Continuous	2	2
TOTAL	61	149

Distributions of spells of sickness absence. The majority have one to three spells.

TABLE IX : DISTRIBUTION OF SPELLS OF UNEMPLOYMENT

No. of spells	No. of Patients	Total No. of spells
1	11	11
2	7	14
3	4	12
4	1	4
Continuous	1	1
TOTAL	24	42

Distribution of spells of unemployment; again the majority here have one to three spells.

TABLE X : RELATIVE EFFECTS OF SICKNESS AND UNEMPLOYMENT IN THE YEAR BEFORE INTERVIEW

	Absence spells	Average spells per person	Recorded loss (days)	Average loss per person	Average loss per spell	% of total loss
Sickness	149	2.0	6285	86.1	42.2	70.8
Unemployed	42	0.6	2597	35.6	61.8	29.2
TOTAL	191	2.6	8882	121.7	46.5	100

This shows the relative effects of sickness and unemployment in the year before the interview. The national rate of sickness spells for males during the twelve months terminated in June 1971, was less than 0.5 per person. The average here is two spells per person. So the increased sickness incapacity of alcoholics is the result of more spells of sickness rather than of markedly longer spells. The average yearly loss through certificated incapacity of 86.1 days is considerably more than among the general population whose loss was calculated in the North West for the year ending June 1972 as 19.8 days. The average length of sickness absence spell was 42.2 working days, only slightly larger than the national average of 36.5 for the year ending June 1971 - (Office of Health Economics 1973).

TABLE XI : STATE BENEFITS PAID IN THE YEAR BEFORE THE INTERVIEW

Work Status	N	State Benefits	Average per Person
Employed	47	£ 8,074.30	£ 171.79
Unemployed	26	£10,360.50	£ 398.48
TOTAL	73	£18,434.80	£ 252.52

The total of sickness, unemployment, and supplementary benefits paid by the state to the group in the year before the interview amounted to £18,434.80 with an average of £252.52 per alcoholic. The average for the unemployed was £398.48, over twice the average of £171.79 per person in the year paid to those in employment.

TABLE XII : DIAGNOSES ON MEDICAL CERTIFICATES

	No.	%
1. Psychiatric		
Nervous Debility	22	
Neurasthenia	6	
Anxiety State	5	
Functional Nervous Disorders	8	36.2
Depression & affective Disorders	4	
Psychiatric illness and Psychoneurosis	4	
Alcoholism	4	
D.T.	1	
2. Hospitalisation	24	16.1
3. Respiratory Infections		
Bronchitis	13	
Asthma	2	
Sinusitis, Sore Throat & 'Flu	4	14.1
Coryza and Pleurisy	2	
4. Digestive Disorders		
Enteritis	6	
Gastritis	6	
Dyspepsia and Diarrhoea	6	
Colic and Vomiting	3	
5. Accidents	15	10.1
Bruises, lacerations, head injury, back injury, broken ribs and recent fracture		
6. Musculoskeletal Disorders	7	4.7
7. Circulatory		
Myocardial ischaemia	1	0.7
8. Miscellaneous Causes		
Virus infection	3	
Medical Investigations	2	4.0
Defective Vision	1	
TOTAL:	149	100.0

The diagnoses supplied on medical certificates to D.H.S.S. minimised the absences that were in reality due to alcoholism. Though it was not possible retrospectively to determine the precise proportion of sickness absences that should have been attributed to alcoholism, it was clear from the patients' descriptions that the great part of their incapacity was actually due to their drinking.

The findings confirm the considerable extent of sickness absenteeism and unemployment among alcoholics. Employers and Trade Unions therefore, should press for industrial alcoholism programmes aimed at early identification and treatment of alcoholic personnel.

ALCOHOL PROBLEMS IN WOMEN

A.BALFOUR SCLARE

DEPARTMENT OF PSYCHIATRY

DUKE STREET HOSPITAL, GLASGOW

Women are "news" these days in a political and sociological sense. We have recently completed International Women's Year and on 29th December 1975 the Sex Discrimination Act and Equal Pay Act became law in this country. Half of all our married women in Britain are now employed whole-time or part-time. At the moment of birth females have a longer mean expectancy of life than males - 72 years compared with 68 years.

It is against this background of social change in the role of women that we must view any changes in their attitudes towards alcohol. Women are emerging from their socio-political purdah. It is mistaken, however, to assume that this emergence is complete. "Double Standards" continue to flourish conventionally regarding behavioural norms for men and women.

Certainly in Scotland imbibing alcohol and drunkenness have become traditionally associated with toughness and sociability among men, as indicated in the University of Strathclyde study by Davies and Stacey (1972). A recovered alcoholic lady in Glasgow recently remarked, "the Calvinistic attitudes which still exist in Scotland are largely to blame. A drunk man is reasonably acceptable, but a drunk woman is the end of the line and a drunk mother is completely beyond the pale so far as the public is concerned".

During the 1960's the much neglected woman alcoholic became an object of some interest in the world of psychiatry. At that stage it was the medical interest rather than the phenomenon itself which was new. Much informal evidence is available to suggest that gin drinking and sherry swilling among members of the fair sex was not unknown during the economic depression of the 1930's - and before.

In Victorian times, as Charles Dickens has portrayed in his midwives, women in the lower socioeconomic groups drank along with their menfolk.

Pemberton (1967), working at Dumfries in Scotland, began to indicate some specific features in the alcoholic female, e.g. a more rapid course of the illness and a generally poorer outlook. Winokur and Clayton (1968) undertook a comparison of female and male alcoholics. An alcoholic type of drinking commenced at an average of 34 years in the females and at 26 years in the males. Psychiatric problems were commoner in the parents of the females than of the males.

Mayer and Green (1967) described their therapeutic efforts, employing group techniques, with alcoholic women ex-prisoners in Massachusetts: these patients were grossly disturbed and impulsive.

THE GLASGOW STUDY

This study compared the clinical attributes of 50 randomly selected female alcoholics with those of 50 male alcoholics (Sclare 1970). Among the females there was a significant peak in the 30-39 age group. The females generally commenced their drinking later than the males but often reached the stage of alcohol dependence within 5 years - compared with a generally slower build up in the male. Thus the woman alcoholic frequently seems, as Pemberton (1967) also suggested, to follow a more dramatic collision course.

The alcoholic patients in this investigation had all been hospital inpatients. Domestic stress was the most significant reason for admission to hospital in the case of the female alcoholics, while a critical situation at work was outstandingly common among the males. This difference probably relfected the life situation in which each sex is chiefly engaged.

The women alcoholics generally consumed only whisky or vodka at home, while their male counterparts usually drank spirits with beer in the public house. Quantity of alcohol intake proved difficult to evaluate precisely but there was an impression that the female drinkers tended on the whole to ingest rather less alcohol than the males. As time proceeded, both sexes tended to gravitate to wine when cash supplies ran short.

It was found surprisingly that delirium tremens (D.T.'s) occurred as often in the women alcoholics as in the men. Perhaps there should not have been surprise. There is no real reason why alcohol dependence and withdrawal phenomena should be physiologically different in the two sexes.

No substantial difference between the alcoholic women and men was found in regard to the amount of their offences against the law. However, there was a discrepancy between them in the nature of their offences. The women displayed a pattern of recklessness, e.g. 'drunk and incapable', while the men more often behaved aggressively while drunk, e.g. convictions for theft.

The family history of alcoholism likewise revealed no quantitative difference between the two groups. But significantly more of the alcoholic women had an alcoholic spouse, while more of the alcoholic men had an alcoholic father. The alcoholic ladies who had alcoholic husbands had a competitive, spiteful attitude.

The physical complications were considered. Iron deficiency anaemia was significantly commoner in the women alcoholics - doubtless a reflection of dietary insufficiency. On the other hand, liver damage and amnesic spells were much less frequent in the females, possibly owing to their lower total level of alcohol consumption.

The women patients more often reported sexual problems than did the males. The main difficulty comprised a lack of sexual interest and drive. Such an occurrence may stem from what one psychiatrist (Kinsey 1968) has described as a lack of preparation for adult roles in the female alcoholic.

Quantitatively there was no appreciable difference in the occurrence of psychiatric disorder between the two groups. The ladies, however, - within the field of psychiatry - displayed a marked excess of primary depressive illness antedating the onset of their alcoholism. They also, more often than the men, reported having had an unhappy childhood - suggesting a vulnerability to subsequent stresses.

THE PAST FIVE YEARS

What is new about alcoholic women during the past five years or so ? Information is becoming available that female problem drinkers are being more frequently referred or self-referred for help. In 1969 clients coming to the attention of the Glasgow Council on Alcoholism had a male/female ratio of 4.1/1 but by 1974 this ratio had become 2.9/1. Looking at the admissions of alcoholic patients (ICD 303) to Scottish psychiatric hospitals and units: in 1965 male alcoholics outnumbered females by 3.0/1 but by 1973 this figure had dropped to 2.7/1. Put differently, the figures show that male alcoholic admissions increased by 77% during the eight-year period, while females showed a 138% increment.

Evidence is also available from the Merseyside Council on Alcoholism of an increase in female referrals. In the working year

1973/74 the male/female ratio was just over three to one, but this dropped to just below three to one by the year 1974/75. At a certain seaside resort in the north of England, this ratio is now 2.5/1.

In short, although men with alcohol problems continue to outnumber the fair sex, the gap between them has recently been closing. The graph for both sexes continues to soar, but the lines are tending to converge. The question of interpretation of this fact arises. Are more lonely, shameridden women drinkers now coming to attention in the new climate of female emancipation ? Or is there a true increase in the incidence of alcoholism in women ?

Possibly both of these factors are operative, although it is difficult to be certain. Further research is required. In a population, such as ours, in which the annual alcohol intake per head is gradually increasing, it is worth bearing in mind de Lint's (1974) estimate that anyone who is drinking more than 15cl per day is likely to develop alcohol-related health damage.

What are the sources of alcohol for the lady drinker ? These are not substantially different from those of her male counterpart. Women are now more confidently consuming alcohol in public houses, lounges, restaurants and in the open air. To some extent, however, women with alcohol problems furtively add a quarter bottle of gin or whisky to their shopping basket in the supermarket or licensed counter now appearing in certain chain stores.

Schuckit (1972) has competently summarised the data from 28 studies of the feminine alcohol abuser. This pooled information shows clearly that alcoholic women present special features of clinical course and family history. These patients first abuse alcohol between 28 and 39. The stages of alcoholism tend to become telescoped: the alcoholic woman tends to become hospitalised by the time she is 40, i.e. about the same age as her male counterpart who began abusing alcohol much earlier.

Women alcoholics, according to the amalgamated data, report more depressive symptoms and have a higher incidence of suicide attempts than do males. Depression is also more common in the female relatives of women alcoholics. Feelings of sexual dissatisfaction are commonly reported. The divorce rate in two American series is 44% (Schuckit et al 1969; Winokur et al 1970).

PARTNERS OF WOMEN ALCOHOLICS

Fox (1968) has reported that the male partners of female alcoholics are less accepting of the problem than are the wives of alcoholic men. Accordingly they are more liable to reject their addicted spouses, offering little support and understanding.

THE INFANTS OF ALCOHOLIC WOMEN

As long ago as 1900, Sullivan observed a high still-birth rate and infant mortality rate in the offspring of alcoholic women examined in prison at Liverpool. In subsequent years paediatricians and social workers have recognised that the infants of alcoholic women are especially at risk in terms of neglect, malnutrition and battering. However, there has been comparatively little research into the quality of life experienced at a somewhat later stage of childhood by the offspring of alcoholic mothers.

Special attention has recently been drawn (Jones et al 1973; Jones and Smith 1973; Jones and Smith 1975) in Seattle to a group of 11 infants suffering from the "foetal alcohol syndrome". This condition is characterised by the following features :-

PERFORMANCE:	Prenatal growth deficiency Postnatal growth deficiency Mental retardation
CRANIOFACIES:	Microcephaly Short palpebral fissures Epicanthal folds Maxillary hypoplasia Cleft palate Micrograthia
LIMBS:	Joint anomalies Altered palmar crease pattern
OTHER:	Cardiac anomalies Anomalous external genitalia Capillary haemangiomata Fine-motor dysfunction

The alcoholic mothers were drinking during the time they were pregnant with these infants. The American workers suggest that, owing to the magnitude of the risk, consideration should be given to termination of pregnancy in chronically alcoholic women.

IN CONCLUSION

The author's current thinking is that three groups of alcohol abuse in women can be discerned :

(1) Girls in their teens or early twenties who drink heavily and often become intoxicated in the company of their drinking boyfriends at weekends. Alternatively they empty the gin or vodka bottle and may replace it with water, while babysitting.

(2) Women in early middle life whose children are growing up and who spend their new-found earnings in a jolly extraverted manner which is a facsimile of the traditional male drinking style; and

(3) Women in early middle life who have a basic depressive condition arising from the loneliness and embitterment of divorce, separation, widowhood or neglect by a husband who is more devoted to his profession or another woman.

There may be some who consider that the highlighting of femine alcoholism constitutes in itself an example of male chauvinism. The author does not share such a view. There would appear to be specific features associated with women alcoholics and it is scientifically appropriate to take account of these matters.

REFERENCES

DAVIES, J.B. and STACEY, B. (1972). Teenagers and Alcohol, H.M.S.O. London.

FOX, R. (1968). "Treating the Alcoholic's Family". Alcoholism, R.J.Catanzaro (ed.) Thomas, Springfield, Illinois, U.S.A.

JONES, K.L. and SMITH, D.W. (1973). "Recognition of the Foetal Alcohol Syndrome in Early Infancy". Lancet, 2, 999-1001.

JONES, K.L. and SMITH, D.W. (1975). "The Foetal Alcohol Syndrome". Teratology, 12, 1-10.

JONES, K.L., SMITH, D.W., ULLELAND, C.W. and STREISSGUTH, A.P. (1973). "Pattern of Malformation in Offspring of Chronic Alcoholic Women". Lancet, 1. 1267-1271.

KINSEY, B.A. (1968). "Psychological Factors in Alcoholic Women from a State Hospital Sample". American Journal of Psychiatry, 124, 1463.

de LINT, J. (1974). "The Epidemiology of Alcoholism". First International Medical Conference on Alcoholism. N.Kessel, A.Hawker and H.Chalke (eds.) Edsall, London.

MAYER, J. and GREEN, M. (1967). "Group Therapy of Alcoholic Ex-Women Prisoners". Quarterly Journal of Studies on Alcohol, 28, 493.

PEMBERTON, D.A. (1967). "A Comparison of the Outcome of Treatment in Female and Male Alcoholics". _British Journal of Psychiatry_, 113, 367-373.

SCHUCKIT, M. (1972). "The Alcoholic Woman: A Literature Review". _Psychiatric Medicine_, 3, 37-43.

SCHUCKIT, M., PITTS, F.M., REICH, T., KING, L.J. and WINOKUR, G. (1969). "Alcoholism: I. Two types of alcoholism in women". _Archives of General Psychiatry_, 20, 301-306.

SCLARE, A.B. (1970). "The Female Alcoholic". _British Journal of Addiction_, 65, 99-107.

WINOKUR, G. and CLAYTON, P.J. (1968). "Family Studies IV: Comparison of Male and Female Alcoholics." _Quartery Journal of Studies on Alcohol_, 29, 885-891.

WINOKUR, G., REICH, T. and RIMMER, J. (1970). "Alcoholism: Diagnosis and familial psychiatric illness in 259 alcoholic probands". _Archives of General Psychiatry_, 23, 104-111.

THE AETIOLOGY OF DEPENDENCY

FRANK A. SEIXAS

NATIONAL COUNCIL ON ALCOHOLISM INC.

NEW YORK

First of all it must be recognized that it is not idle to say that human beings are part of one another. Interdependence is so great that almost every action - large or small - affects fellow beings in this world (human and animal) in a variety of ways, and the ripples of each action extend depending on position and others who take this same action - perhaps to infinity. So that dependency can be said to be part of the human state - the complete helplessness of the infant graduating to the seemingly total independence of self-motivated adults. Therefore, when looking for aberrancy, in this system, therefore one must look for aberrancy not in dependence, but in independency - what the Greeks would have called hubris - in this case the kind of self-swelling that denies interdependence. Two psychological theories of the genesis of alcoholism reflect this, the one of Howard Blane (1968) which focuses on the failure of our social scheme to provide outlets for dependency needs in males, and that of McClelland et al. (1972) which tells us of the reverse - the unfulfillable drive to personal power somehow short-circuited by alcohol.

Alcoholics Anonymous instinctively (or perhaps not so instinctively) recognizes this as it says, "We grew to recognize that we were powerless over alcohol". Surrender of the hubris is the first step towards recovery.

But interestingly enough Alcoholics Anonymous bring in a specific substance - the substance that is the trouble for most people over most of the world, and it is necessary for us to consider the qualities of this substance in relationship to the type of dependence (if you like this term) that we are specially considering.

At the outset before one can overuse this substance, it must be available- and such availability has been unsuccessfully restricted by law in the U.S.A. for two decades - and in other countries for variable periods, only to have these laws repealed.

Another way to restrict availability, although not to eliminate it, is to have a restrictive tax policy on alcoholic beverages. Despite the impressive statistics in favour of this, the author's clinical experience shows that it is so much of a game to avoid taxes and so easy to ferment anything and make alcohol beverages, that such laws would not succeed in efficiently curbing alcoholism. There is another and rather pernicious school of thought which says that society prescribes the 'role' of the alcoholic to people and they try to learn that role. As McAndrew and Edgerton (1969) say, "Within limits we get the drinking comportment that we deserve". One of the reasons that this is pernicious is that it tells us "unlearn what we have learned to treat alcoholism". It is not a sickness or disease but defective education! And its aficionados are so intense that Eugene Leblanc (1976) recently reminded us in an elegent paper on the Biochemistry of Tolerance, that when the biochemistry of memory is better understood, we will have more insights into the biochemistry of tolerance and vice versa.

McAndrew & Edgerton (1969) had said, though within limits, we govern our behaviour with drunkenness; and likewise LeBlanc (1976) said, that "within limits our learning and memory processes may enter into similar transactions to that of tolerance". So finally, all who deal with alcoholism must come to grips with the fact that the course of progression of this disease has a relationship to the facts contained in the interaction of a specific drug with the mind and body of man. It is only since this chemistry has been looked at carefully that we have begun to learn the parameters of this interaction.

A social system in which a person has the freedom of choice to take or not take alcohol, is being looked at. In most of the United States, and probably England and elsewhere, most people are exposed to the opportunity to drink. Many, (even those with difficult emotional problems, not drinking heavily) begin to drink excessively and it can be said that what governs that choice has to be determined - is there something in the body that either gives or withholds permission for that?

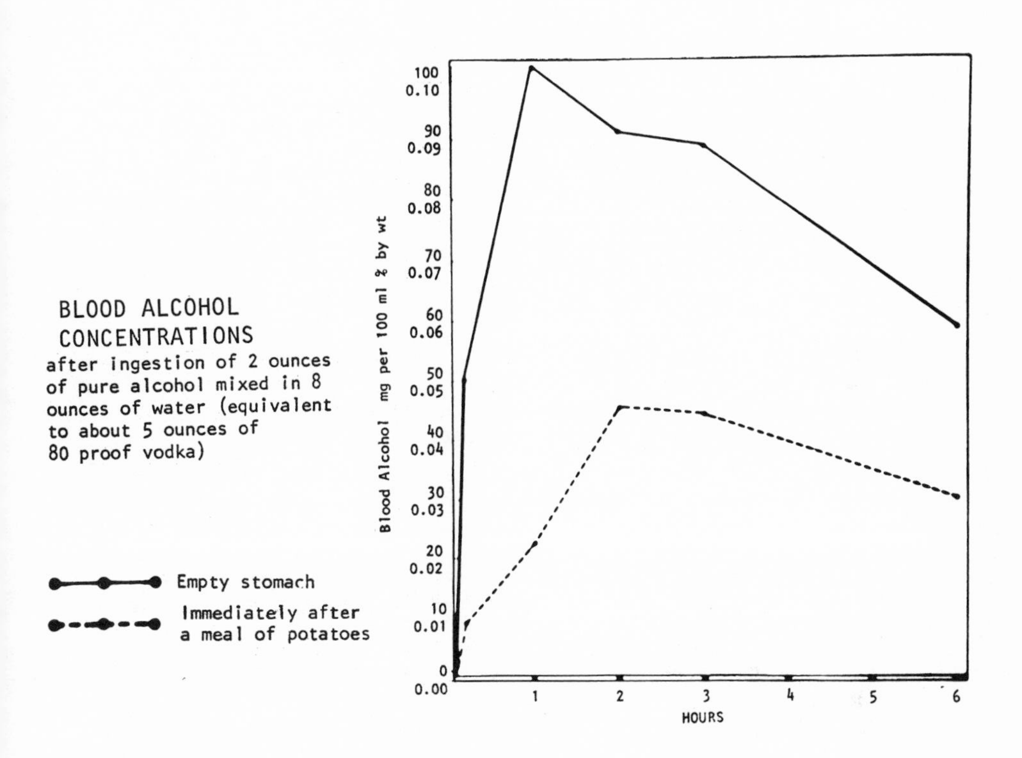

FIGURE 1

Figure 1. How does the body handle alcohol. When taken in, it is absorbed in the stomach and intestines. Food delays the passage through the stomach. After absorption it is oxidized in the liver by alcohol dehydrogenase.

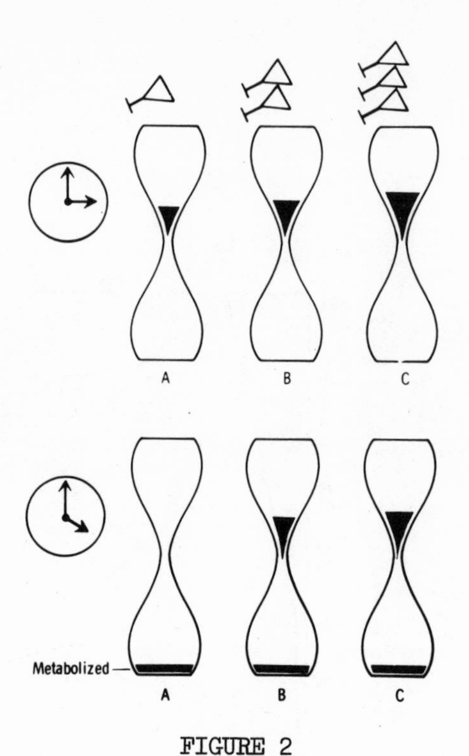

FIGURE 2

Figure 2. This reaction proceeds at a constant rate and this rate is not increased by adding more alcohol at the same time. Thus blood alcohol levels may build up.

CORRELATION OF BLOOD ALCOHOL LEVELS

Blood Alcohol Levels Mgm%	Effect
50	Loss of Inhibitions
100	Ataxia Dysarthia
150	Difficulty in Standing
200	Periods of Sleep
300	Stupor
400	Coma
500	Death in Respiratory Paralysis.

FIGURE 3

Figure 3. shows the actions of alcohol in the brain in the naive drinker.

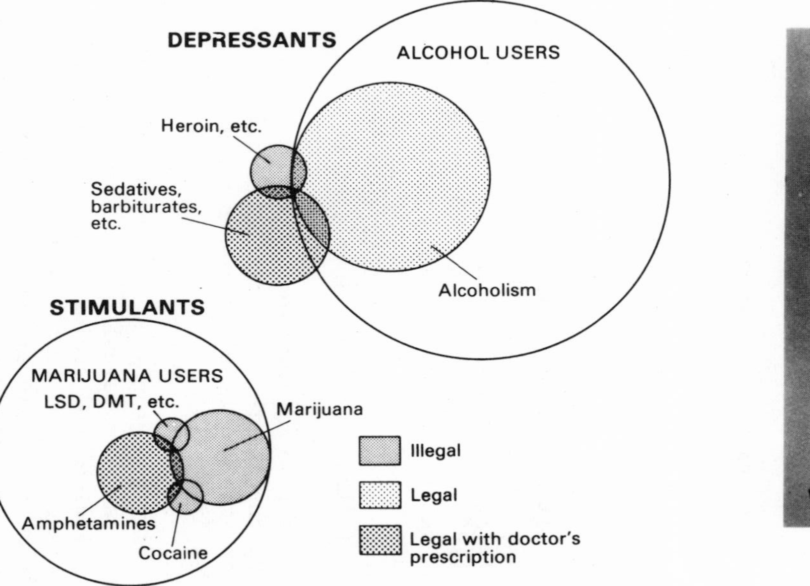

FIGURE 4. Alchol is a depressant drug like opiates and the other sedatives. It still holds that whereas there may be psychological habituation with the "uppers only" the "downers" produce physiological habituation, which requires the development of tolerance and the withdrawal syndrome. Tolerance can be understood when one recognises that thousands of people are arrested every year for driving under the influence of liquor.

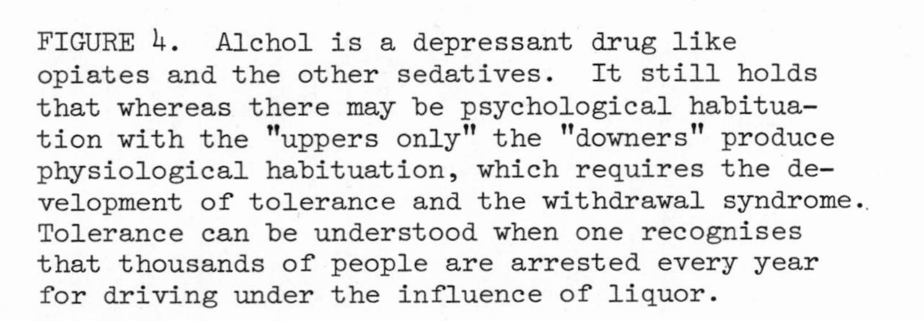

FIGURE 5. This requires for an 180 lb. man 11 ounces of 90 proof whiskey in two hours after a meal. Most people would be unable to open the door of a car after taking this amount.

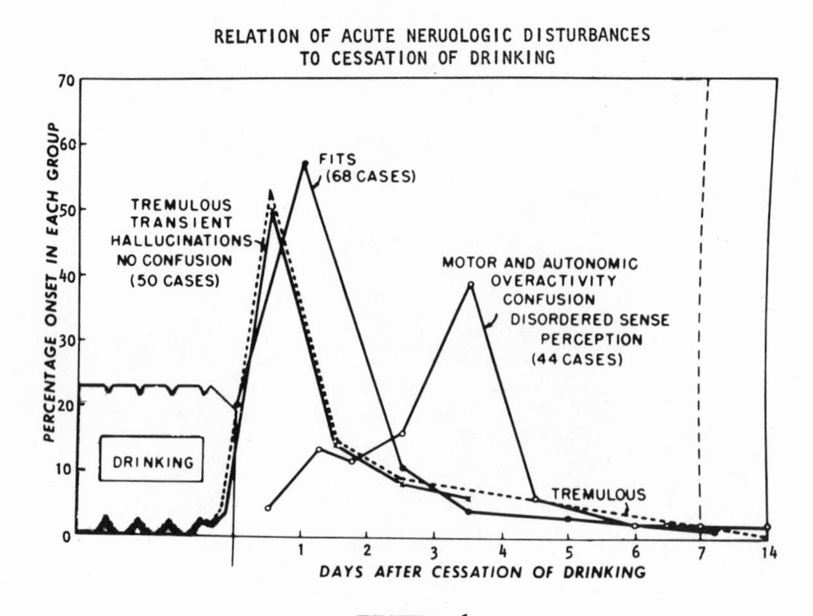

FIGURE 6

Figure 6. The withdrawal syndrome is familiar.

It is this habituation, this change in the reaction of the brain to alcohol, which is referred to when talking of alcoholism, and it is found that the characteristic blackout periods are explained by this and the pattern of continued drinking to ward off the withdrawal syndrome is also explained. Why one should drink again under these circumstances after having stopped is not explicable in this rubric; this is the so-called craving. Mello et al. (1968) have respectively shown an "alcohol addiction memory" exists which is characterised by the more rapid development of withdrawal in alcoholics who resume drinking after having stopped, and Gitlow (1976) has shown that tolerance is re-acquired more readily in those who have developed it before.

Ludwig (1974) has presented an ingenious explanation of the resumption of drinking by showing that the alcoholic may be responding to internal or external cues, misinterpreted as similar to withdrawal features.

Only the tolerant individual can find it possible to drink enough to develop the late stage physical consequences of alcoholism and its psychological and social disequilibriums. The mental and social disturbances arise from the hubris of the single-minded determination to obtain alcohol, and from alcohol's action on the brain, with alternating somnolence and irritability.

How is it then that some become alcoholic and some do not? The major work of Dr. Donald Goodwin et al. (1974) about the alcoholism inherited by people adopted at birth, whose natural parents were alcoholic is not the only clue. There is substantial other work showing that the ability to develop alcoholism has an hereditary component - and it seems to me this component relates to the ease of developing tolerance. The unconfirmed work on alcohol intolerance, involving a marked flushing response to alcohol ingestion, may relate to this concept.

Dr. Lieber's (Hasamura et al. 1975) recent study showing that acetaldehyde levels in dry alcoholics subjected to a single dose of alcohol are twice that of controls, probably shows an effect not a measure of the cause of alcoholism - but it needs further study.

Thus, in the aetiology of alcoholism, we must first have a person in whom it is biologically possible. This possibility when screened through anthropological, sociological and emotional influences can result in the full development of the disease. For opiates and other sedatives we know that similar conditions of tolerance and dependence exist. Buf if we look at the victims, they do not have to pejoratively be called dependent, in any way - more than is implied by the concept that they require this medication. They turn out to be real human beings in whom we can find a worthy aim to help them get away from alcohol, and other drugs they may be taking, and return to a world in which honest interdependence can be healthy!

REFERENCES

BLANE, H.T. (1968). The Personality of the Alcoholic: Guises of Dependency, Harper & Row.

GITLOW, S. (1976). Unpublished conference paper.

GOODWIN, D.W., SCHULSINGER, F., HERMANSEN, L., GUZE, S.B. and WINOKUR, G. (1973). "Alcohol Problems in Adoptees Raised Apart from Alcoholic Biological Parents". Archives of General Psychiatry.

HASUMURA, U., TESCHKE, R., and LEIBER, C.S. (1975). "Acetaldehyde Oxidation by Hepatic Mitochondria: Its Decrease after Chronic Ethanol Consumption". Science, 189, 727-729.

LeBLANC, E. (1976). National Drug Abuse Conference, New York City.

de LINT, J. (1974). "The Prevention of Alcoholism". Preventive Medicine, 3, 24-35.

LUDWIG, A.M. (1974). "The Irresistible Urge and the Unquenchable Thirst for Alcohol". Fourth Annual Alcoholism Conference of the National Institute on Alcohol Abuse and Alcoholism, 3-22.

MACANDREW, C. and EDGERTON, R.B. (1969). Drunken Comportment, A Social Explanation, Aldine Publishing Co., Chicago.

McCLELLAND, D.C., DAVIS, W.N., KALIN, R. and WANNER, E. (1972). The Drinking Man, The Free Press, New York.

MELLO, N.K., MENDELSON, J., and SOLOMON, P. (1968). "Small Group Drinking Behaviour, An Experimental Study of Chronic Alcoholics". The Addictive States, Association for Research in Nervous and Mental Disease, XLVI, 399-430, Williams and Wilkins Co., Baltimore.

Section III

Treatment

A PILOT CONTROLLED DRINKING OUT-PATIENT GROUP

DOUGLAS CAMERON

THE ALCOHOL RESEARCH AND TREATMENT GROUP

CRICHTON ROYAL, DUMFRIES

This paper describes a Pilot Controlled Drinking Group which was set up using six male alcoholic out-patients with the explicit goal of reshaping their drinking behaviour towards moderate intake.

METHODS

1. Group Therapy

Because of the author's experience with group therapy it was decided that this would be the major treatment modality. It was anticipated that there would be two themes - firstly drinking behaviours, the problems encountered and controlling stimuli; secondly life problems, those problems which are often dealt with by drinking but which could be coped with in other ways.

2. Didactic Instruction on Drinking Behaviour

Sobell et al (1972) have demonstrated differences in baseline drinking behaviour between alcoholics and normal drinkers, albeit in a mock-up hospital bar. They have shown that alcoholics tend to drink a greater quantity, of more concentrated beverages with gulps rather than sips, and larger time-gaps between gulps than is the case with non-alcoholic controls. It may well be that, as the subjects knew they were under observation, these are differences in people's stereotypes of drinking styles rather than actual differences; but nonetheless, shaping towards non-alcoholic drinking stereotypes is probably desirable. It was decided to instruct and suggest to the clients that they try to emulate normal drinking patterns. They were encouraged to forsake neat whisky and to take beer and other less potent beverages instead.

3. Personal Intake Goal Setting

No judgements were made of the "how much is bad for you" type; and the subjects were allowed to decide how much, and how often they wished to drink per week. This goal was scrutinised in the group, and the subject's spouse asked to be a party to the final decision. In this way, goals would be set for session intake and weekly intake.

4. Facilitating Social Dysjunction

Bacon (1971) has hypothesised that problem drinkers find a peer group whose drinking behaviours resemble and support their own. It was hoped to increase insight into the role of the subject's peers and to facilitate social dysjunction where appropriate. This might mean simply suggesting a change of pub, or of drinking time or of the location within a pub.

5. Social Reinforcement

Hunt and Azrin (1973) have shown that with a vigorous system of social reinforcement, it is possible to modify alcoholics' behaviour towards drinking less, working more, spending more time with their families and out of institutions. It was decided to offer support and practical help contingent upon the maintenance of appropriate drinking behaviour. Furthermore, such behaviour would be reinforced by approval from three agents - the group, the spouse and a drinking companion. The drinking companion would be a member of the group, who would serve as an observer and a reinforcer of appropriate drinking in the subject's preferred drinking environment.

6. Drinking Practice

This would happen, when deemed suitable, as a group activity and as a paired pursuit, involving the client with his participant observer companion. Information on the success or otherwise of drinking experiences would be reported back to the group.

PROCEDURE

It was decided to operate a closed group of finite length (three months of twice weekly meetings after an initial week of daily meetings) of around six subjects per group. At first, subjects would be visited at home and a stylised history would be taken, including a detailed account of drinking habits since starting to drink.

At the first meeting, a goal of total abstinence for the first week would be set and, thereafter, group discussion along the lines indicated would take place. Spouses would attend a "couples meeting" once during the first week, and afterwards, meetings of the subjects alone, or accompanied by their wives, would be held alternately.

During the first week subjects would define their drinking goals, which would be agreed by the spouses, after which the subjects would be free to drink, along with a group member (including the therapists) to their declared session and weekly intakes, and such drinking would be called "legal". These social constraints would remain until the "legal" drinking pattern was quite stable. After that, a subject would be able to drink in the company of his own choice but would continue to keep in regular contact with the group until its closure.

Following the three-month programme, a regular follow-up of the subjects would be undertaken.

SUBJECTS

Six male out-patients participated in the project, all of whom would qualify for the diagnosis of "alcoholism" if one accepts the World Health Organisation definition.

Subject 1 was a 39 year old married insurance supervisor with two children, who thought that his drinking had been a problem for three years and who had a history of alcoholic amnesias, hangovers and minor hallucinosis.

Subject 2 was a 30 year old electrician, divorced, remarried, whose drinking had been a problem for under a year, who had a history of hangovers, alcoholic amnesias and passing-out while drinking.

Subject 3 was a 49 year old, married, railway worker with one son, whose drinking had been a problem for 20 years and who had a history of hangovers, alcoholic amnesias, early morning drinking and delirium tremens.

Subject 4 was a 45 year old man, married with three children, whose drinking had been a problem for between one and two years and who had a history of hangovers, alcoholic amnesias, passing-out while drinking and early morning drinking.

Subject 5 was a 22 year old estranged storeman, cohabiting, with two children, unemployed, who reported that his drinking had been a problem for 15 months. He had a history of hangovers, alcoholic amnesias and the shakes.

Subject 6 was a 41 year old roofing contractor, divorced, remarried with one child. He said his drinking had been a problem for between one and two years and had a history of alcoholic amnesias and the shakes.

All these men were currently living with some other person, normally the wife and the co-operation of both the man and his spouse was requested before inclusion in the group was possible.

Detailed information on the changes in drinking patterns were elicited from subjects. Some of the clients were well aware of the many factors precipitating changes in the pattern, e.g. departure of wife, change of job, leaving the forces, and so on. Others had great difficulty in commenting about their drinking in terms other than "It's just got gradually worse over the past five years"; "oh yes, I did drink more when I drove the brewery lorry!".

CHANGES IN PROCEDURE

Alcoholics present erratically to the out-patient services, predominantly at times of crisis. They tend to require help rapidly and thus suitable clients for the controlled drinking programme emerged sporadically. The structure of the group changed to accommodate them by becoming an open-closed group of indefinite length. This has made it more difficult to obtain "tidy" data; but this is essential if the group is to maintain an adequate number of clients and provide a valued service. Clients introduced after the group had started were seen three times during the first week, once at home for the original interview, and twice in the group, once along with their spouses. This less regular contact may in part explain why they ("S.4" and "S.6") had more difficulty in achieving a week's abstinence.

Clients are not ready for discharge from the group after three months. This may be because the group only meets twice weekly or because the clients are very disturbed people or because we are poor psychotherapists. But it is thought likely that a group of this kind would function best as a continuous self-help support system - a non abstinent "Alcoholics Anonymous".

The original procedure did not allow for "wives only" groups. This made the wives feel a little jealous, deprived, or left behind in the group proceedings. Accordingly instituted irregular "wives only" groups have been instituted. One function of these in the early days of the group was to allow the wives to speak out about the past misdemeanours of their husbands - a need which subsides with time.

The discussion of drinking behaviour and suggested changes has produced few problems. The clients tended,(with approval) to revert from spirits and/or beer to beer only. This personal-intake-goal-setting presented little difficulty, most clients opting for around seven pints of beer per week, with a session intake of two to three pints. It is not known why they opted for this quantity; perhaps it is easier during a drinking session to stop at three pints rather than stop at seven !

It may be relevant to note that half of the clients drank regularly in the same public house. They came to the group independently - not by personal recommendation. This poses the question, whether that particular establishment is a morbid drinker's pub. ("S.3" still drinks there, "S.4" who has just come back to us still drinks there, but "S.2" has opted to drink elsewhere).

The amount of general social support requested has been slight; but one of the therapists is telephoned regularly (once or twice a week) by various clients or spouses, for practical help or for help with inter-personal difficulties. This is provided, contingent upon, previous group-approved behaviour, althoug because we do not wish clients to "drop out", there is never total rejection. This means that it has not been possible to provide a rigid contingency management regime, even though that is what the clients are told is the aim.

It has not been possible to restrict drinking to "accompanied by a group member only", in part because clients come from up to 20 miles away, but also because numbers are small. Thus, "legal drinking" has been redefined as drinking of the declared beverage, at or below declared session intake and frequency, in the company of a group member, spouse or someone else in the family aware of the member's involvement in the group and personal intake goal.

This modification is believed to be unsatisfactory, but inevitable in a scattered rural area such as this. Obviously, all drinking is illegal during the 'dry' week. A week of abstention remains the expectation until it has been met. It may be of some interest that the subjects who have done best are those who achieved the week's sobriety at the first attempt, although they also had the help of meetings every evening ("S.1" and S.2").

Attendance at Meetings. This was an area in which difficulties were anticipated. As the therapists were dependent on group pressure and personal motivation for attendance, it was thought likely that it would be erratic. If people failed to turn up and had not told the group that this was going to happen, up to two attempts were made to re-establish the contact. The attendance at the couples group has been more consistent than that at the all-male group; but attendance rates have been very satisfactory.

It is a group rule that all turn up to meetings sober, although very occasionally clients have arrived intoxicated. This is usually obvious and the group readily confront the offending person. Nobody has been so drunk as to be group-disruptive; and nobody has been asked to leave because of his inebriated state. One subject ("S.6") consistently denied being intoxicated when his behaviour made his condition obvious to others. This was more of a nuisance - a red herring - than it was destructive.

DRINKING BY ALCOHOLIC CLIENTS

This, of course, was the area of our greatest anxiety. Like most alcohologists, we were brought up on the belief that alcoholics could not consistently control their drinking and were anxious lest resumption by our clients of their drinking careers should prove disastrous. As can be seen from the figure, this does not seem to have been the case.

The data on weekly intake presented here is self-report retrospective; collected as a group exercise. It is, therefore of dubious validity depending, as it does, entirely upon trust.

This puts it in the same calibre, and subject to the same bias and error as reported intake prior to the start of the programme (see the obliquely crossed columns on the left of the histograms). Some of this self-report data was simply not believed, notably that of "S.6". Apart from him, wives have reported general agreement with the amounts cited. It is refreshing to find how open clients are about drinking misdemeanours when they know that the group's (and therapists') ethic is not one of total sobriety.

The majority of "illegal" drinking shown on these histograms is so called because it has been undertaken outwith the company of family, group member, or therapists. Nonetheless, wives report that "he wasn't drunk like he used to be when he came back from the pub". However, all that can be claimed with confidence is that the reported alcohol intake of four of the clients has fallen dramatically since first contact with the group.

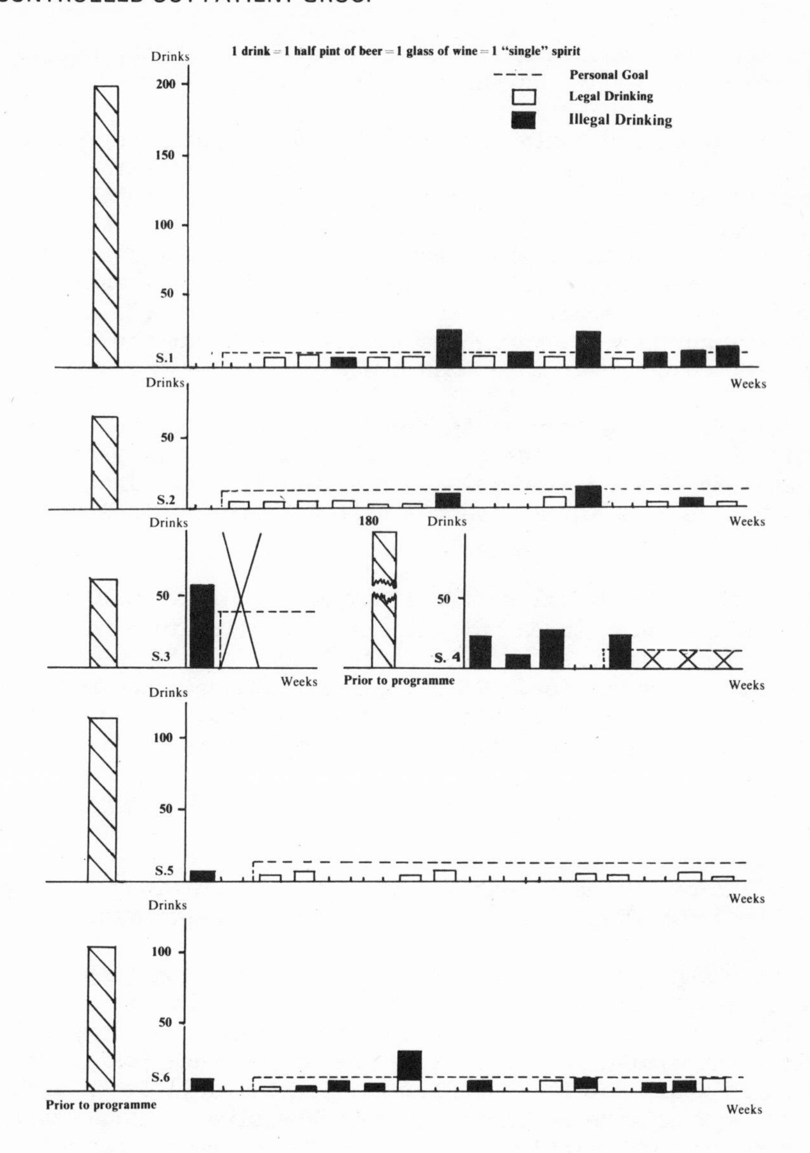

FIGURE 1. Weekly Alcohol Intake of Controlled Drinkers

The two group-drinking sessions that have taken place presented no problems, but instituting the companion participant/observer system has not been consistently successful. When two members of the group have gone drinking together, neither has reported a failure to maintain intake below declared session intake. When a therapist has accompanied a client, session intake has been maintained. It is of note that when clients have been drinking in pairs, they have chosen not to go to their old haunts, but have patronised some anonymous "neutral" establishment.

Some clients clearly differentiate two types of drinking; controlled drinking, within intake and socially constrained limits, and uncontrolled drinking. Both types are present in the repertoires of our more successful clients. They are quite capable of going out with their wives and having two pints of beer, and no more. They are also capable of getting drunk. Presumably, that makes them no different from "normal social drinkers", who are all at risk of getting drunk now and again. No client has had a drinking session extending over more than one day.

Only "S.2" has been given licence by the group to extend his drinking circle beyond the group and the family. This has caused more anxiety for his wife than we had anticipated. Thus far, he claims his session intake has not exceeded his original declared intention.

The therapists feel sufficiently confident at this stage to state that allowing these out-patient alcoholics to continue drinking within group determined social constraints has not been a dangerous or disastrous enterprise. Of course, clients have not maintained their drinking totally within prescribed limits, but most of these people's drinking has usually been moderate and socially appropriate.

DISCUSSION

No claims are made that the information contained in this paper is of greater validity than that of clincial observation and anecdote.

The author is well aware that self-report data (albeit corroborated by spouses) is of highly dubious validity. But if one is dealing with people in the real-life situation as opposed to people incarcerated in hospital or elsewhere, much must be taken on trust. It would not be possible, nor indeed would it be wished, to set up a spy network sufficiently comprehensive to obtain more objective data on drinking behaviour. But there are more fundamental problems in the study. If contact with the group has coincided with with a great reduction in alcohol intake by a number of subjects to levels well within the range of what one may call "moderate drinking", then can one claim to have been successful ? Within this highly limited focus, the answer is probably 'yes'. Alcoholics Anonymous claims success if continuing weekly contact with the organisation helps to maintain sobriety.

But the psychotherapeutic efforts of the group have been towards opening up the possibility of other methods of problem solving. There is no data upon how successful the enterprise has been. If the focus of treatment is taken away from alcohol consumption then measures of such variables as trust, ability to cope with new situations, openness within the marriage, quality of life and relationships need to be incorporated into the assessment of efficacy. In other words an assessment of psychotherapeutic progress independent of drinking behaviour must be made, and a meaningful assessment of efficacy of psychotherapy is a difficult job.

There remains the problem of a lack of control groups; so the study is limited to comparing the outcome of this study with either a group which is not strictly comparable or with currently held views of what happens if alcoholics resume drinking.

A further obvious conclusion is that this kind of programme is not suitable for all clients. It would have been valuable to have been able to predict who would benefit and who would not. Thus far, there are two clinical pointers. Firstly, drinking history. If a client could perceive his drinking as having been environmentally responsive, i.e. to have changed contingent upon change in social and emotional circumstances, then we suspect that he is likely to respond to contact with the group. If a client cannot see his drinking in these terms it is suspected that he will do badly; the reason being that the clinical rating of personal progress of clients in contact with the group correlates with the number of changes in drinking pattern reported at the initial interview. Secondly, if he were able to maintain a week's abstinence in his home environment at the beginning of the programme, it is suspected that the client is amenable to this kind of group. Clients who are construed as doing well, managed a week's abstinence at the first attempt ("S.1" and "S.2"). The one client who dropped out at an early stage ("S.3") failed to achieve this goal; and thus, the week's abstinence may act as a useful selection mechanism.

REFERENCES

BACON, S.D. (1973). "The Process of Addiction to Alcohol". Quarterly Journal Studies on Alcohol, 34. 1-27

HUNT, G.M. and AZRIN, N.H. (1973). "A Community-Reinforcement Approach to Alcoholism". Behaviour Research and Therapy, 11, 91-104.

SOBELL, M.B., SCHAEFER, H.H. and MILLS, K.C. (1972). "Differences in Baseline Drinking Behaviour between Alcoholics and Normal Drinkers". Behaviour Research and Therapy, 10, 257-267.

PROGRAMMING ALCOHOLISM TREATMENT: HISTORICAL TRENDS

RAYMOND M. COSTELLO

ALCOHOLISM TREATMENT UNIT

SAN ANTONIO, TEXAS

The presentation is prefaced by the comment that the work being looked at is not so much the author's but observation of work done by others. It is compiled of 20+ years work looking for elusive pieces to an intriguing and complex puzzle. With the exception of one document, the author's own, the articles for discussion belong to others. Fifty-eight manuscripts dated from 1951 to 1973 and representing the work of investigators in seven countries, have been collected. Thirty-four from the United States, 12 from England, six from Canada, three from Scotland and one each from Ireland, Netherlands and South Africa. The articles all represent descriptions and simple outcome evaluations of alcoholism treatment projects. Some of the projects are grand attempts at comprehensive treatment programming involving large numbers of clients, much money, intense staff effort and multiple therapeutic components. Others are projects of lesser scale involving few clients, limited staff, and a particular therapeutic strategy of special interest to the investigators. The empirical methodologies ranged from widely detailed overworked statisitcal ventures to cursory frequency tabulations.

No attempt has been made to write yet another position paper spelling out the deficiencies of past work and issuing standards of required competence for future efforts. Others have done this work previously, and sufficiently adequately that these works have become historical landmarks which must be studied by any serious student the area, (Hill and Blane 1967, Pattison 1966).

Instead conclusions have been drawn from a reasonable, although not rigorous methodology. The work was begun with the hope that certain robust findings would become obvious despite the various methodologies previously employed by individual investigators. Many of the findings that were revealed seem very reasonable but some of them are likely to be considered still equivocal.

The author's interest in alcoholism treatment programming and evaluation began in 1971, when a psychiatrist and himself assumed the responsibility of implementing a comprehensive, community based alcohol treatment programme, one of the first funded by National Institute on Alcohol Abuse and Alcoholism and based on the comprehensive community mental health centre model. Although they both had had experience in comprehensive community mental health centres, they had no experience in alcoholism treatment and had no guidelines other than those provided by the required services outline of NIAAA. They began collecting and reading descriptive and evaluative articles as they were implementing the programme. While doing so it became obvious to them that there was no adequate collation of the tremendous output in this area. They had collected more than 200 references, none of which provided them with a crystallized pattern.

Focus was placed on the 58 articles which had been summarized in a previous paper, (Costello 1975). Graphic displays of empirical observations with regard to the procedures employed by investigators as relating to subject selection or screening and to choice of therapeutic components will be studied. The aim will be to arrive at a common conceptualization of management strategies, differentiating programmes of various outcome classes and to itemize salient observations as they were reported originally.

Five facts should be noted before proceeding: (1) only studies reporting outcome at one year will be reported. (2) Outcome parameters as reported originally were accepted whenever possible. If, however, the investigator exploited a "floating n", outcome was recalculated. A "floating n" results from using a denominator figure to calculate success rates that is smaller than the original number of subjects and results in an inflated success estimate. If an investigator sampled his original pool with adequate technique, no exploitation was indicated. If subject attrition or "experimental mortality" provided the sampling technique, results were recalculated. (3) Each study was described by a pattern of four scores (percent of cohort: A. dead, B. problem drinking, C. sober or no problem drinking, and D. lost). Sobriety, extended over the full follow-up year, was not required for classification in the success category, although this was usually found to be the case. (4) A point-prevelance estimation method, (MacMahan and Pugh 1970), was applied wherever possible. This method examines outcome at a single point in time, for example, on the exact one-year anniversary date following formal termination of treatment. As alcoholism is a condition of high morbidity, this epidemiological tactic is sufficient for homogeneous clusters on the basis of outcome score patterns by a statistical technique of heirarchial grouping, (Ward, 1962).

RESULTS

Table I displays the group percentage means for each outcome index. Note that a five group clustering solution was accepted and the groups were ordered from Best, to Good, to Intermediate, to Poor, to Poorest, on the basis of success rates. The number of studies in each cluster is relfected in the Table, and shows that each group contains a sufficient number such that generalizations should be reliable. The percentage of patients reported as deceased is also reflected in the Table. Although the better groups report a higher death rate the explanation is that these studies tended to be more comprehensive in their follow up and actually looked for patients in Vital Statistics. The poorer programmes tended to ignore this follow up resource.

TABLE I : GROUP PERCENTAGE MEANS FOR EACH OUTCOME INDEX: ONE YEAR FOLLOW-UP STUDIES

		Outcome Index			
Group	N	Dead	Problem Drinking	Success	Lost
Best	15	1	44	45	10
Good	11	1	27	35	37
Intermediate	12	2	47	30	21
Poor	13	0	79	18	3
Poorest	7	0	60	12	28

Table II displays the twelve patient exclusionary criteria most often reported. These are common criteria but they are used unsystematically, rendering selection bias as an obvious confounding variable and suggesting alternate competing explanations to treatment effects when studies are compared. (See next page).

Table III displays the percentage of studies in each group which reported use of at least one exclusionary criterion. Note a perfect linear function from Best to Poorest such that a perfect rank order correlation with the success rates displayed in Table I would be found if calculated. Obviously, the studies reporting higher success rates are more likely to utilize a management strategy of poor-prognostic subject exclusion. Although no Table depicts this additional phenomenon, the more successful studies also reported a higher average number of exclusionary criteria than did the poorer outcome studies. Only one study in the Poorest group reported use of an exclusionary criterion, and their study used only one such criterion.

TABLE II : EXCLUSIONARY CRITERIA

1. psychosis
2. organic brain syndrome
3. inability to pay fees
4. physical disabilities
5. sociopathy
6. contraindications for or unwillingness to take antabuse
7. vagrants
8. unmarried
9. no contact with relative in past year
10. geographical (state or county) residence
11. other "obvious" psychiatric problems
12. outstanding police problems

TABLE III : PERCENT OF CASES IN EACH GROUP REPORTING USE OF EXCLUSIONARY CRITERIA

Group	Percent
Best	80%
Good	45%
Intermediate	25%
Poor	17%
Poorest	14%

Table IV displays an overview of the treatment components reported as available to the patients in the various groups. Note that an inpatient component was reported as available by most of the studies. This is only to say that one-year evaluative efforts are most likely to be generated by inpatient programmes than by other types of programmes. However, the therapeutic milieu or therapeutic community concept was not normally reported. Twenty-seven percent of the Best studies, 27% of the Good studies and 17% of the Intermediates reported such a component. None of the Poor and Poorest studies reported such a component. The percentages relfected in the Table were calculated by using as a denominator figure the number of studies in each group reporting an inpatient component, rather than the total number of studies in each group.

The rate of reporting of involvement of collaterals such as

the spouse, other family members, friends or employers seemed to follow a straight linear function from Best to Poorest. Likewise so did the rate of reporting of use of aggressive follow-through or after-care programmes. The rate of reporting of use of Antabuse also followed this linear trend, save for one reversal in the better programmes. No linear trend was detected for rate of reporting with regard to use of group therapy, behaviour modification techniques, nor Alcoholics Anonymous.

TABLE IV : OVERVIEW OF TREATMENT COMPONENTS REPORTED BY STUDIES IN EACH OUTCOME GROUP-IN PERCENTS

Component	Best	Good	Intermediate	Poor	Poorest
In-patient	80	55	100	83	67
Milieu	33	50	17	0	0
Group Therapy	73	73	91	46	57
Collaterals	60	45	17	8	0
Antabuse	40	27	33	8	0
Behaviour Modification	27	18	25	0	14
Follow-through	67	64	50	0	0
Alcoholics Anonymous	26	64	58	8	29

Table V displays a 2X2 factorial summary of expected success rates based on the historical review. Programmes with multiple treatment components available and a patient class of good prognosis might be expected to achieve a 50% success rate at one-year follow-up. Conversely, programmes with limited treatment resources and a patient class of poor prognosis might expect to achieve a 15% success rate at one-year follow-up.

TABLE V : 2X2 FACTORIAL SUMMARY: EXPECTED SUCCESS RATES

		Client Prognosis	
		GOOD	POOR
Treatment Scope	BROAD	50%	30%
	LIMITED	30%	15%

In order to become better acquainted with history and not be doomed to repeat it, the following comments are offered. These comments are not always exact quotations but the original source is offered for those who might be interested. Each of the comments has a quality of timelessness such that if they were written tomorrow they would have just as contemporary a ring as they must have had when originally published. The comments have been arranged in chronological order, by decades, with reference to year of publication.

1950's

The following profile emerged of the patient with favourable outcome: intermittent drinker of good previous personality, with satisfactory work record, with close personal ties to at least some one person, married, seeking help for the first time, rating one or more for social stability (Straus and Bacon Scale), who continues to accept the aid afforded by disulfiram and Alcoholics Anonymous. Characteristics which had no prognostic significance were those of sex, age, source of reference, length of drinking, intellectual level, type of work, social class and nationality. (Davies, Shepherd and Myers 1956).

Group meetings in which the therapist is not actively "therapeutic" have no therapeutic value for most of the participants and may result in harmful changes. (Ends and Page 1957).

With continued visits patients develop a good relationship with the therapist, and should they slip, they may consequently feel guilty. Their attendance at the clinic often is stopped when feelings of guilt cause them to fear rejection if they return to the therapist. Here it becomes necessary for the therapist to adopt aggressive efforts to obtain the patient's return. (Epstein and Guild 1951).

Ideally, psychotherapy should supplement any physical treatment. It is difficult to effect this on an outpatient basis, however, for the following reasons: (a) Most patients refuse psychotherapy and the suggestion of the psychic origin of their maladjustments. (b) Psychotherapists are not available to all alcoholics who present themselves for treatment. (c) Patients tend to put their entire faith in the disulfiram tablets and express confidence that once sober they can carry on independently. (d) Not all alcoholics require formal psychotherapy to make an acceptable adjustment. (e) Most patients become threatened by deep psychotherapy and flee unless they are in a closed hospital. (Epstein and Guild 1951).

Many members of the local Alcoholics Anonymous group who are successful in remaining "dry" with the A.A. programme alone have become hostile to those taking disulfiram - a fact we deplore. These may conceive of disulfiram as a threat to themselves and the group. Consequently, it is stressed to new patients that we feel that AA is important in their continued successful readjustment and that it is in no way minimized by the disulfiram therapy; our aim is essentially that of AA - to aid alcoholics to re-establish themselves in society - and any measures which might seem helpful are welcomed. (Epstein and Guild 1951).

Disulfiram is an effective weapon in the medical armamentarium presently available to combat alcoholism. A total approach to the problem is necessary, in which disulfiram should be considered as merely one factor. (Epstein and Guild 1951).

Faced with the anxiety of increasing inadequacy of adjustment, most alcoholics revert to drinking before breakdown and psychosis appear. (Epstein and Guild 1951).

A larger percentage of women than of the men, and of the spree drinkers than of the plateau drinkers responded favourably. The younger the patient, the shorter the duration of alcoholism, and the happier the childhood history, the more earnest and co-operative was the attitude during treatment. Inpatient treatment followed by continued therapy in the outpatient clinic yielded the best results. Faithfulness in attendance at group meetings, especially in company with the spouse or another family member, was a prominent feature of the group whose outcome was classified as good. The combined chemopsychotherapeutic programme appears to offer more advantages than any other method so far advocated for rehabilitation of the alcoholic, provided the patient can be motivated to follow through with a reasonable degree of regularity. Mandatory participation in such a programme under legal supervision may offer a solution for at least a part of the problem. (Fox and Smith 1959).

Community living with other alcoholics rather than the official group-psychotherapy sessions seems to be responsible for the change of heart and attitude which relatives of patients notice after only a few weeks. This change of heart seems to begin more rapidly in men than in women, perhaps partly because the women are fewer and lack the opportunity of living in a community of alcoholics, though they attend the group meetings, which are "mixed". (Glatt 1955).

Disulfiram is a valuable aid in the management of alcoholic patients, however, it is not a substitute for sound psychiatry. It provides the patient with a simple objective act to reinforce his decision and to protect him against weakening in the next twenty-four hours. A resolution which can be given immediate objective expression is easier to keep than one which must remain a subjective hope. Disulfiram shortens the period of hospitalisation and allows greater freedom while in residence. It improves the relationship between patient and doctor. Disfulfiram is a valuable adjuvant to a total psychiatric programme of diagnosis and rehabilitation. Only in such a context can its real value be appreciated. (Moore and Drury 1951).

This inpatient facility is serving several important functions in the overall treatment programme. It is providing services to patients whose immediate needs cannot be met on an outpatient basis. It provides opportunities for the application of adjunctive therapies for which hospital supervision is indicated. It is also serving continuing clinic patients for whom crises may arise and is thereby helping to sustain sobriety and the treatment process when both are threatened. Still another group of patients, who do not recognise the real nature of their condition and enter the hospital seeking only help in terminating a drinking episode, are exposed to different attitudes and to an approach which frequently has led to the initiation of sustained therapeutic contacts resulting in ultimate improvement. (Walcott and Straus, 1951).

These facts speak clearly in favour of disulfiram treatment whenever it is medically permissable and desired by the patient. Obviously nothing can be accomplished, even with the use of disulfiram, without the patient's co-operation. He can (and many do) stop taking the drug and return to drinking a few days later. It appears, then that the patient's willingness to take disulfiram indicates, in many cases, a strong motivation to conquer his alcoholism. It is no wonder, then that disulfiram-treated patients are frequently the most successful ones; they were better prospects for any treatment in the first place. (Wexberg 1953).

Real help for alcoholics of all types would require a boarding home, a hospital ward, an institution for alcoholics requiring internment, in addition to outpatient clinics. (Wexberg 1953).

1960's

Apart from improvements in the therapeutic techniques themselves, the patient's degree of involvement in his treatment will influence the effects of therapy in preventing relapse. (Blake 1965).

Our data would seem to support the thesis that drinking and activities attendant to it become a way of life for the alcoholic. As a corollary, if changes occur in the drinking pattern, then changes must be expected (and were observed here) throughout the individual's characteristic life style. The incidence of symptom replacement in abstinent alcoholics was very low. (Bowen and Androes 1968).

Even when abstinence cannot be maintained, the treatment

programme can have beneficial spin offs in many other areas of social functioning. (Bowen and Androes 1968).

The act of leaving treatment prematurely, whatever the reason offered, seems to be followed by an inevitable return to drinking. It appears that the earlier the departure, the more severe the drinking problem. It is difficult not to conclude that the dropouts are those who have already decided to continue to drink. They are unable to concieve of a world without alcohol, without fear and trepidation, and soon return to a life centred about alcohol. (Bowen and Androes 1968).

Remaining sober seems correlated to the extent that an individual can relate to others. (Bowen and Androes 1968).

The mercurial character of many alcoholics, the waning of initial enthusiasm for treatment, and mounting social pressures combine to produce a progressive attrition in the numbers of successfully treated patients. What, at first blush appears to be a successful therapeutic methodology, often results in being a commonplace, mediocre therapy when measured by the dual criteria of double-blind observations and long-term follow-up. Shorter periods of observation are less meaningful as many alcoholics have demonstrated their ability spontaneously to maintain abstinence for periods of three months or less without benefit of any treatment, only to relapse shortly thereafter. (Charnoff et al 1963).

The indiscriminate use of psychopharmacologic agents by themselves does not constitute sufficient nor definitive therapy. Neither tranquilliser nor antidepressant medication has been shown to be superior to placebo when given to an unselected group of alcoholics. (Charnoff et al 1963).

Most patients do not utilize outpatient aftercare. Most patients that relapse do not utilize aftercare programmes. Many successful patients do not utilize aftercare programmes following discharge from an inpatient setting. Further research is necessary to determine how aftercare may be more efficiently used. (Dubourg 1969).

The most striking finding in the present study is the degree to which initial social stability was related to final outcome. It can be said with a fair degree of certainty that a patient with high social stability will do well, and a patient with low social stability will do badly, and the accuracy of prognosis is little impaired by ignoring personality factors. That good results can be achieved with socially stable patients is gratifying, but that bad results are obtained with the socially unstable, presents a perplexing problem. (Edwards 1966).

The similarity of inpatient and outpatient results, in this study, could mask particular subgroups of patients for whom one or other treatment is more suitable, and there is a need for clinical trials which aim at investigating the specific indications for one or other type of treatment. There are, in all probability, patients who will benefit from two to three months respite in hospital with forced removal from drinking and who would not recover if left in the disorganised social setting with which they can no longer cope; and in future analyses the effects of hospital admission per se and of psychotherapy should be separately assessed. We do not believe that our findings undermine the position of the specialist inpatient units, but rather suggest that they point to the need for the alcoholism treatment systems to be expanded so that these units form part of a properly comprehensive service. (Edwards and Guthrie 1967).

While it is widely agreed that off-hour emergency calls are not therapeutically justified for psychoneurotics, in the case of alcoholics a nightly phone call can often make the difference between intoxication and sobriety. (Jensen 1962).

Ethnic background per se does not appear to be a prognostic factor except in so far as it relates to social stability or instability. (Kissin et al 1968).

The time is well past when alcoholics should be indiscriminately assigned to a treatment programme without consideration of the needs and potential of the patient. (Kissin, Rosenblatt, and Machover 1968).

The treatment of alcoholics with the focus on underlying psychological problems may be less effective than simply providing the patient with a period of rest and diversion. In the highly structured total push programme, the patient may readily perceive treatment as something being done to him and his seeming participation in therapy may simply be a cover for inactivity. (Levison and Sereny 1969).

With the opportunity for medical care, adequate nutrition, group discussion of their problems, exposure to an anti-drinking but supportive environment, and, most important, the sudden cessation of drinking and the perpetration of this sobriety for at least one month, the vicious circle of chronic alcoholism tends to be disrupted, enabling the patient to regain some self confidence and gain some perspective on the personal and social consequences of his drinking behaviour. Return to drink does not imply sustained, continued drinking behaviour thereafter or even a serious disruption of gains made in other areas of performance. Most patients are able to carry on most of their other social tasks at least at a higher level than that noted on hospital admission while engaging in periodic drinking. (Ludwig et al 1969).

Assignment of patients to take disulfiram on an outpatient basis following hospital discharge does not seem to make any difference in terms of treatment outcome. (Ludwig et al 1969).

The overwhelming, consistent, empirical findings force us to conclude that the various LSD procedures, as used in this study, do not offer any more for the treatment of alcoholism than an intensive milieu therapy programme. (Ludwig et al 1969).

The conclusions are that the outcome in alcoholism depends very little upon the treatment given but largely upon individual factors relating to each patient and upon the natural history of the condition. The cost of establishing and running the type of special unit which caters mainly for the treatment of alcoholics with social and behavioural characteristics associated with good prognosis may not be justified. More attention should be given to the provision of a range of facilities to suit the management of alcoholics with less favourable attributes who are likely to continue drinking in spite of psychiatric treatment. (McCance and McCance 1969).

In both sexes alcoholism frequently followed a marital disruption (death of spouse, separation or divorce). The difference was that the males in this situation appeared to have sufficient interests and outlets of their own to make rehabilitation relatively easy. On the other hand the females had been dependent upon their husbands for interests and contacts outside the home, and so when they were on their own they were lost and lonely. (Pemberton 1967).

The impulsive young alcoholic with a severe personality disorder will demand more detailed study and the development of different treatment techniques. (Ritson 1969).

A patient who is reluctant to discuss his alcoholism with friends and relatives and who refused to disclose the diagnosis to his employer, often has a poor prognosis. It may be that his desire to conceal his addiction from others represents an extension of his own denial of the illness. It was the clinical impression that involvement of others often marked a turning point in the patient's progress. (Ritson 1969).

During the period of treatment by psychotherapy, both individual and group, difficulty of handling patients full of tension was experienced and it was soon realised that a person in a state of "tension" does not think clearly, appears incapable of making rational decisions, and so explored the possibility of hypnosis to relieve this tension. This process of relaxation is very simple to develop, and greatly increases the value of psychotherapy, especially if the session starts with the relaxation. It was quickly realised that relaxation was not, in itself, adequate treatment, and is now looked upon as a useful adjunct to other therapy. (Smith-Moorhouse 1969).

Patient control of aspects of the hospital organisation is designed to lead the patient to perceive that he is capable of responsible behaviour which is in fact expected of him. Doors are not locked, there is no bedtime hour, a newly admitted patient is shown the hospital by a fellow-patient, patients elected by their fellows run the hospital shop, etc. Rules are kept to a minimum, so that patient autonomy in the hospital's social climate can be fostered (with technical participation of the staff to modify the social interactions into a treatment process). The staff orientation and co-ordination, together with patient autonomy, are remarked on by patients in terms conveying security. At the same time angry reproach is repeatedly expressed that not enough is being done, i.e., no active treatments applied of which patients can be passive recipients and the ineffectiveness of which, presumably, patients can demonstrate as soon as they leave the hospital. (Walton 1961).

1970's

The most important finding is that total abstinence was associated with the patient's mother being deceased. For those with mothers living, relatively infrequent contact was associated with abstinence. (Bateman and Petusen 1971).

The addition of LSD to the therapeutic regime pays off in terms of greater short-term (i.e. three months) gain. Patients seem to become more self-accepting, to show greater openness and accessibility, and to adopt a more positive, optimistic view of their capabilities to face future problems. This gain washes out, however, over the long-term (i.e. one year) as patients report a waning of the initial inspiration, euphoria, and good intentions with the former stresses and difficulties in their lives. (Bowen at al 1970).

Although some alcoholic patients respond favourably to traditional group therapy based on conventional psychodynamic principles or on behavioural concepts, the therapy cannot pierce the armour of many others. Many alcoholic patients are socially remote and inaccessible men, who seem to be encapsulated within defences of denial and rationalisation. One way to break through hardened mechanisms of self-deception and denial is by aggressive and direct confrontation of psychological defences in encounter groups - the Daytop Village concept. Frontal assault is not the only, and perhaps not the most powerful, method of penetrating defences. Marathon group therapy provides opportunity for gentler and more subtle means of "breaking through". (Dichter et al 1971).

The use of hallucinogenic drugs has not been of striking

benefit in the treatment of chronic alcoholism. Unless a new treatment philosophy, different from the psychologic concept is shown to be efficacious to sustain the initial improvement, hallucinogenic drugs, including LSD, DET and DPT, appear to have limited therapeutic value in the treatment of chronic alcoholism. (Faillace et al 1970).

Whether or not a patient completes the treatment programme provides some indication of his subsequent adjustment; if he completes it, he is considerably more likely to attain a satisfactory community adjustment. Programme completers are somewhat more likely to become sensitive to the re-emergence of drinking problems at an earlier stage than are non-completers. Despite completers' generally better community adjustment, their re-admission rate for alcoholism equals that of the non-completers, suggesting a greater tendency to make use of a previously helpful resource at an earlier time in a disruptive drinking phase. A higher dropout rate among women can be anticipated during their first treatment attempt. This drop-out rate does not appear to have the same significance as in men. While in men, non-completion on first admission is a rough predictor of non-completion in a second treatment attempt, for women this predictor cannot be used as effectively. Many of the women who are good prospects for later treatment efforts drop out of the programme during their first try. The reasons for these sex differences certainly appear worthy of further investigation. (Fitzgerald et al 1971).

We conclude that marital couples group therapy is the treatment of choice at this time for married alcoholic patients. The denial and projection mechanisms, exaggerated in the alcohol-marital problem, are more easily approached and treated in a group. The spouse of the patient helps to keep "pulling" the patient back to treatment, and drop-out rate is lower as a result of the spouse's co-operation. The therapist obtains a more realistic view of the home life of the patient by seeing some of the marital interactions in the group, forcing the patient to use less denial and distortion. (Gallant et al 1970).

With similar programmes, differing markedly in amount of group therapy but producing essentially the same results, one is forced to question seriously the value of group psychotherapy in these programmes, especially when a direct comparison within a single programme (as that at Ford Meade) finds no demonstrable effect. These results do not, of course, provide a definitive indictment of the efficacy of group therapy but do suggest the need for serious controlled study of this therapeutic modality. (Kish and Hermann 1971).

It is clear that social and even more so, psychological factors play an important role in patient acceptance of two of these treatment modalities. Socially and psychologically intact alcoholics accept psychotherapy and reject inpatient rehabilitation, while the opposite appears true for socially and psychologically unstable alcoholics. (Kissin et al 1970).

By designating individuals as controls we may have created a feeling of rejection and may actually have lowered the natural recovery rate in this group. (Kissin et al 1970).

These data indicate that different types of alcoholics do better under different programmes, and that the overall success rate in the treatment of alcoholics could be improved by offering a variety of treatment programmes and allowing the patient some choice in the treatment he receives. (Kissin et al 1970).

Although it is reasonable to emphasize the desirability of sobriety, thinking should not be rigidly forced into two such categories as "abstinence equals adjustment" and "return to alcohol equals maladjustment". (Rohan 1970).

Harmful drinking is better viewed as a type of activity rather than as a sickness or disease. Emphasis is on overt activity. In this way harmful drinking itself is seen as a special or defective way of acting and its role in the functioning of the individual is assessed. (Rohan 1970).

In comparing our results to other studies many obstacles are evident. Various programmes select different types of subjects in terms of the severity of the alcohol problem and the extent of social and behavioural deficits. Treatment differs in kind and duration. Also very important was an exclusion factor in reporting success or failure rates. Exclusion of certain subjects from final statistics increases sobriety percentages but the reasons for exclusion across studies were inconsistent. Often excluded were those who didn't complete treatment, could not be found at follow up, or were transferred to another ward because of disruptive behaviour. This problem in comparing similar studies is avoided by using the total admission sample as a reference point. Sobriety percentages then reflect that portion of the admission sample reporting complete sobriety within a specified period of time after discharge. Percentages, inflated by exclusion, seem to minimize certain problems such as failure to stay in treatment, which could be interpreted as a measure of a programme's efficacy rather than a subject characteristic. (Rohan 1972).

In current thinking, the individuals whose drinking results in extreme physical and social damage are labelled "alcoholic" and placed in a qualitatively separate category distinct from "social drinkers". This traditional dichotomy suggests a drinker is either an "alcoholic" or he is not, and it is assumed that if an "alcoholic" drinks at all, he drinks at his maximum limit. This simplistic formulation reflects a conceptual model which accounts for damaging drinking within a "disease" frame of reference. It is contradicted by the diversified drinking patterns found among "alcoholics". In fact, return to drink is highly variable on the dimension of time spent drinking and amount of alcohol ingested. It would be more useful to conceptualise alcohol ingestion within a continuum of amount over time. Within the drinking population the range extends from non-damaging low rate to severly damaging high input. (Rohan 1972).

Disregarding, for the moment, the possibility of physiological factors, it makes good sense to assume that social drinking is a skill which derives its popularity from the rewarding consequences it has. Those who do not have this skill cannot experience these rewarding consequences. Like all skills, drinking in a socially acceptable manner should be trainable as long as relevant behavioural parameters for such behaviour are known. (Schaefer 1972).

REFERENCES

BATEMAN, N.I., and PETERSON, D.M. (1971). "Variables related to outcome of treatment for hospitalised alcoholics". International Journal of the Addictions, 6(2), 215-224.

BLAKE, B.G. (1965). "The application of behaviour therapy to the treatment of alcoholism". Behaviour Research and Therapy, 3, 75-85.

BOWEN, W.T. and ANDROES, L. (1968). "A follow-up study of 79 alcoholic patients: 1963-1965". Bulletin of the Menninger Clinic, 32, 26-34

BOWEN, W.T., SOSKIN, R.A. and CHOTLOS, J.W. (1970). "Lysergic acid diethylamide as a variable in the hospital treatment of alcoholism: A follow-up study". Journal of Nervous and Mental Disease, 150(2), 111-118.

CHARNOFF, S.M. KISSIN, B., and REED, J.I. (1963). "An evaluation of various psychotherapeutic agents in the long term treatment of chronic alcoholism; results of a double blind study". American Journal of Medical Science, 246, 172-179(1-2).

COSTELLO, R.M. (1975). "Alcoholism treatment and evaluation: in search of methods". International Journal of the Addictions, 10(2), 251-275.

DAVIES, D.L., SHEPHERD, M. and MYERS, E. (1956). "The two-years' prognosis of 50 alcohol addicts after treatment in hospital". Quarterly Journal Studies on Alcohol, 17, 485-502.

DICHTER, M., DRISCOLL, G.Z., OTTENBERG, D.J. and ROSEN, A. (1971), "Marathon therapy with alcoholics". Quarterly Journal Studies on Alcohol, 32, 66-77.

DUBOURG, G.O. (1969). "After-care for alcoholics - a follow-up study". British Journal of Addiction, 64, 155-163.

EDWARDS, G. (1966). "Hypnosis in treatment of alcohol addiction controlled trial, with analysis of factors affecting outcome". Quarterly Journal Studies on Alcohol, 27, 221-241.

EDWARDS, G. and GUTHRIE, S. (1967). "A controlled trial of inpatient and outpatient treatment of alcohol dependency". Lancet, 1(1), 555-559.

ENDS, E.J. and PAGE, C.W. (1957). "A study of three types of group psychotherapy with hospitalized male inebriates". Quarterly Journal Studies on Alcohol, 18, 263-277.

EPSTEIN, N. and GUILD, J. (1951). "Further clinical evaluation of tetraethylthiuram disulfide in the treatment of alcoholism. Quarterly Journal Studies on Alcohol, 12, 366-380.

FAILLACE, L.A., VOURLEKIS, A., and SZARA, S. (1970). "Hallucinogenic drugs in the treatment of alcoholism: A two-year follow-up study". Comprehensive Psychiatry, 11(1), 51-56.

FITZGERALD, B.J., PASEWARK, R.A. and CLARK, R. (1971). "Four-year follow-up of alcoholics treated at a rural state hospital." Quarterly Journal Studies on Alcohol, 32(3), 636-642.

FOX, V. and SMITH, M.A. (1959). "Evaluation of a chemopsychotherapeutic programme for the rehabilitation of alcoholics: observations over a two year period". Quarterly Journal Studies on Alcohol. 20, 767-780.

GALLANT, D.M., RICH, A., BEY, E. and TERRANOVA, L. (1970). "Group psychotherapy with married couples: A successful technique in New Orleans alcoholism clinic patients". Journal of the Louisiana State Medical Society, 122(2) 41-44.

GLATT, M.M. (1955). "Treatment centre for alcoholics in a mental hospital." Lancet, 268, 1318-1320.

HILL, M.J. and BLANE, H.T. (1967). "Evaluation of psychotherapy with alcoholics; a critical review". Quarterly Journal Studies on Alcohol, 28, 76-104.

JENSEN, S.E. (1962). "A treatment programme for alcoholics in a mental hospital". Quarterly Journal Studies on Alcohol, 3(Suppl. 1), 315-320.

KISH, G.B., and HERMANN, H.T. (1971). "The Fort Meade alcoholism treatment programme: a follow-up study". Quarterly Journal Studies on Alcohol, 32(3) 628-635.

KISSIN, PLATZ, & SU. (1970). "Social and psychological factors in the treatment of chronic alcoholism". Journal of Psychiatric Research, 8(1), 13-27.

KISSIN, B., ROSENBLATT, S. and MACHOVER, S. (1968). "Prognostic factors in alcoholism". Psychiatric Research Reports, 24, 22-43.

LEVINSON, T. and SERENY, G. (1969). "An experimental evaluation of 'insight therapy' for the chronic alcoholic". Canadian-Psychiatric Association Journal, 14(2) 143-146.

LUDWIG, A.M., LEVINE, J., STARK, L.H. and LAZAR, R.A. (1969). "A clinical study of LSD treatment in alcoholism". American Journal of Psychiatry, 126(1) 59-69.

MACMAHON, B. and PUGH, T.F. (1970). "Epidemiology: Principles and Methods" Boston: Little, Brown and Co.

McCANCE, C. and McCANCE, P.F. (1969). "Alcoholism in Northeast Scotland: its treatment and outcome". British Journal of Psychiatry, 115, 189-198.

MOORE, J.N.P. and DRURY, M.O'C. (1951). "Antabuse in management of chronic alcoholism". Lancet, 261, 1059-1061.

PATTINSON, E.M. (1966). "A critique of alcoholism treatment concepts with special reference to abstinence". Quarterly Journal Studies on Alcohol, 27, 49-71.

PEMBERTON, D.A. (1967). "A comparison of the outcome of treatment in female and male alcoholics". British Journal of Psychiatry, 113, 367-373.

RITSON, BRUCE. (1968). "The prognosis of alcohol addicts treated by a specialized unit". British Journal of Psychiatry, 114, 1019-1029.

RITSON, BRUCE. (1969). "Involvement in treatment and its relation to outcome amongst alcoholics". British Journal of Addiction, 64, 23-29.

ROHAN, W.P. (1970). "A follow-up study of hospitalized problem drinkers". Diseases of the Nervous System, 31, 259-265.

ROHAN, W.P. (1972). "Follow-up study of problem drinkers". Diseases of the Nervous System, 33, 196-199.

SCHAEFER, H.H. (1972). "Twelve-month follow-up of behaviourally trained ex-alcoholic social drinkers". Behaviour Therapy, 3, 286-289.

SMITH-MOORHOUSE, P.M. (1969). "Hypnosis in the treatment of alcoholism". British Journal of Addiction, 64, 47-55.

WALCOTT, E.P. and STRAUS, R. (1952). "Use of a hospital facility in conjunction with outpatient clinics in the treatment of alcoholics". Quarterly Journal Studies on Alcohol, 13, 60-77.

WALTON, H. (1961). "Group methods in hospital organization and patient treatment as applied in the psychiatric treatment of alcoholism". American Journal of Psychiatry, 118, 410-418.

WARD, J.H. (1962). "Multiple linear regression models in H. Borko (Ed.)". Computer Applications in the Behavioural Sciences, Englewood Cliffs, New Jersey: Prentice-Hall.

WEXBERG, L.E. (1953). "Reports on government-sponsored programme". Quarterly Journal Studies on Alcohol, 14, 514-524.

A YOUNG PROBLEM DRINKERS PROGRAMME AS A MEANS OF ESTABLISHING AND MAINTAINING TREATMENT CONTACT

BRIAN COYLE and JAY FISCHER

CHARING CROSS CLINIC FOR DRUG AND ALCOHOL PROBLEMS

GLASGOW

Despite the increased frequency with which young problem drinkers are presenting themselves for treatment (Smart and Finley 1975), almost no work has been reported concerning specialised treatment services for persons age 30 or under who abuse alcohol. The usual practice in Britain has been for these patients to be included in treatment where the predominant age range of patients is 35 to 50. Glasgow is no exception; specialised treatment services for this clinical subgroup is non-existent within the national health service.

The practice of interspersing younger patients with a majority of older drinkers ignores the possibility that the young problem drinker might have different or more severe emotional problems (Rosenberg 1969) and that these could best be dealt with by treating young patients separately. Also, in the prevailing practice already noted, we are undermining a basic feature of specialised treatment, which is to foster group cohesiveness; this in turn might help to bring about beneficial therapeutic effects. Instead, the possibility is increased that the young problem drinker will become alienated from the remainder of the treatment group, due to the often considerable variance in patient age. In short, rather than fostering group cohesiveness, we may be inadvertently creating group isolates.

Yalom (1975) has noted that the group isolate, or patient markedly deviant from the remainder of the treatment group, tends to terminate treatment prematurely. Given the inability of many young problem drinkers to establish and maintain treatment contact and the tendency for these patients to become isolated from the mainstream of the treatment programme, the question then becomes one of how best to create group cohesiveness, while reducing feelings of deviancy. It was felt that in establishing a separate treatment programme for young problem drinkers, we might obviate the problem by reducing feelings of deviancy which would facilitate group cohesiveness and more frequent treatment contact.

The aims of this paper are:

1. To describe the establishment of the specialised treatment programme for young drinkers, hereafter called the young persons programme (Y.P.P.).

2. To discuss changes in the young persons programme that have been made since its inception, which relate to the issue of continued treatment contact.

3. To test the assumption that the Y.P.P. will be effective in developing group cohesiveness and helping patients to establish and maintain treatment contact.

ESTABLISHMENT AND STRUCTURE OF THE PROGRAMME

Selection of Patients

In the first instance, approximately forty general practitioners and six social work departments in the Glasgow area were informed by letter of our intent to establish a separate treatment service for young problem drinkers within the Charing Cross Clinic. In the letter, the nature of the programme as well as general information regarding the clinic was given and referrals were requested. Additionally, existing case notes were scrutinised for suitable patients.

A structured intake interview of thirty-five minutes duration was arranged for each patient, in which length and type of problem were examined, as well as suitability for treatment. Clinical information regarding the patient's interpersonal, work and psychological functioning was also elicited.

Patients were considered unsuitable for treatment if, during the course of the intake interview, any of the following were noted :

1. Psychotisism
2. Low level of motivation
3. Self-injecting drug abuse
4. Barbituate use

A low level of motivation was defined as unwillingness to accept the need for treatment and the need to attend two meetings per week.

Patients were accepted for treatment in whom alcohol abuse was secondary to other problems, including drug abuse, except in categories 3 and 4 listed above. It seemed reasonable to include patients who abused alcohol in conjunction with other drugs, as multiple substance abuse is common in patients of this age group (Wechsler and Thum 1973; Rosenberg 1969).

During the intake interview, patients were told about the nature and rationale of the programme. Potential group members were also informed about what they might expect from the treatment process and that group therapy did not offer an easy solution to their problems. In relation to this, patients were informed that, at times, they might feel worse rather than better. Explaining the nature of the treatment process, particularly in relation to patient expectations, follows from the work of Fischer (1976), who noted the importance of patients having a clear and undistorted idea about the treatment process.

PATIENT POPULATION

Prior to the first meeting of the Y.P.P., thirteen male patients were interviewed. One was excluded and four failed to attend the first meeting. During the first nine months of the treatment programme, an additional twelve males and three females were interviewed. Of these additional fifteen patients, two were excluded and four failed to attend. Four of these patients were interviewed while in-patients for detoxification.

GROUP STRUCTURE

A closed group was created with the first eight patients who accepted the offer of treatment. Meetings lasting seventy-five minutes were held on Tuesday evenings and Saturday mornings. It was thought particularly important to have the group meet during the weekend, as this is often the period of heaviest drinking in Glasgow.

Within the group, patients are encouraged to express and explore feelings about themselves and other group members. Considerable emphasis is given to exploring present problems, as opposed to past history. Such an approach offers maximum possibility for modifying present maladaptive behaviour (Mintz 1971). This is not to minimize the importance of past experience, but rather it is our clinical judgement that undue emphasis on the past is often a means of avoiding present interpersonal or intrapsychic difficulties.

In addition to exploring current problems, the group also attempts to give advice and support in the handling of drinking and drug-taking situations.

TREATMENT GOALS

Treatment goals for each patient were individualised at the time of the intake interview. We have found that two main types of patients present themselves for treatment. The first type of patient, labelled the early alcohol abuser, presents himself when his pattern of drinking changes from moderate to heavy alcohol consumption. Included in this group are the patients who have a period of heavy alcohol abuse within a year prior to the intake interview and who have since reduced alcohol consumption but have increased the use of other drugs. For the early alcohol abuser, the treatment goal, where possible, is a return to moderate drinking with a cessation of other drug abuse, if applicable.

The second type of patient, labelled the advanced alcohol abuser, is one who has a history of heavy drinking of several years duration. Often these patients exhibit "loss of control" drinking (Jellinek 1960). Because of their repeated inability to drink in moderation, the treatment goal for the advanced alcohol abuser is abstinence.

Treatment goals are tentatively formulated at the intake interview. Modification of these goals is possible and is done within the context of the group.

MODIFICATIONS OF THE Y.P.P.

Admissions of New Out-patient Referrals

Originally, it had been our intention to admit new patients to the programme following the intake interview, as space became available. This however proved unwise, since some patients failed to attend more than one session. These premature terminations were discouraging for other group members, who felt that group time was being spent unwisely on patients who were unwilling to make a commitment to regular group attendance. In the first instance, the existing group was often quite willing to devote a considerable amount of time to the immediate concerns of a new member and, in so doing, to delay the exploration of ongoing issues. Repeated incidences of new patients entering the group and terminating prematurely meant that the exploration of ongoing issues could be delayed for two to three sessions.

In response to problems posed by premature termination, the group suggested that following the intake interview, new patients should be required to attend a basic discussion group called the "survival group", which is normally attended by all incoming patients in the clinic. The survival group is an open discussion group which meets twice weekly. It stresses the cessation of alcohol consumption and how best to avoid drinking situations. Patients referred to the young persons programme are now required to attend the survival group for three weeks before being admitted to the Y.P.P. Since the survival group is not closed and is rather large in comparison to the young persons group (often above fifteen persons), the problems posed by premature termination of some members will have less effect on the group as a whole.

Admission of In-patient Referrals

Originally it had been thought possible to include patients in the Y.P.P. who were still in hospital, but who had been detoxified and who were judged by hospital staff as being able to benefit from a specialised programme. The need to include such patients is caused by the absence of specialised services within the hospital setting. The argument already discussed, as to why the young alcohol abuser would derive little benefit from non-specialised out-patient treatment, would appear to be applicable to in-patient settings as well. Many young persons who are hospitalised for alcohol abuse report difficulty in identifying with the concerns of older patients. Similarly, hospital staff indicate that these patients often do not make a satisfactory adjustment to the treatment regime in the hospital. Also, it was hoped that establishing treatment relationships with patients while in hospital would increase the likelihood of their subsequent clinic attendance (Demone 1963). None of the in-patients (four) attended more than three sessions.

In light of the poor response by in-patients, it was decided that they would be eligible for the Y.P.P. only after discharge from hospital. Furthermore, as in the case of other patients, they would still be required to attend the "survival group" for three weeks before being admitted to meetings of the Y.P.P.

The Intake Interview

In relation to the above modifications in the admissions policy, prospective patients were told at the intake interview that they would have to attend the "survival group" for three weeks before being admitted to the Y.P.P. The nature of the survival group was explained, as was the reasons for their attending it.

Follow-Up Procedure

The last point in this section concerns an addition to the programme as opposed to a modification. Originally, due to lack of staff, it was felt that it would not be possible to follow up patients once they had ceased to attend the group. However, in the course of the first ten meetings, it became readily apparent that patients were concerned about those who ceased to attend the treatment group. Patients recognised that a number of premature terminations would threaten the stability and cohesiveness of the group, particularly in its early stages of development (Yalom 1975). In response to this problem, patients suggested that they seek out members who did not attend. At first, this endeavour was treated with caution as it was felt that it was beyond the capability of patients just entering treatment.

In keeping with the policy of individualisation, each case of non-attendance was discussed in the group, with respect to the most appropriate method for encouraging the non-attending patient to return. In two instances, non-attenders were visited by members of the group; in two other cases, the group elected a member to draft a letter to the non-attending patients; and in the remaining instance, the non-attender was telephoned by one of the group members.

Results

One of the goals of the Y.P.P. was to establish group cohesiveness, here defined as identification with, or expressed concern about, the members of one's treatment group. The development of group cohesiveness, because of its ambiguity, is difficult to measure directly. However, we can infer that it has developed from the following patient comments and behaviour :

1. During the course of a meeting, members will often comment on differences between the Y.P.P. and other clinic groups.

2. Patients in the Y.P.P. comment on differences between themselves and other clinic patients, maintaining that it is difficult for older patients to understand them and vice versa.

3. Patients indicate that, if not for the Y.P.P., they would have ceased to attend the clinic.

4. Patients suggesting that they follow up non-attending group members.

5. Patients choosing to drink tea amongst themselves, before and after group meetings, rather than mixing with other clinic patients.

The following results, except where noted, are based on seventeen patients aged 30 and under (henceforth called "younger patients") who attended the clinic for at least one session following the Y.P.P. intake interview, during the period May 1975 through January 1976, and seventy-two patients over age 30 (henceforth called "older patients") who attended the clinic from May 1975 through November 1975. Chafetz et al. (1962) indicated that five or more sessions is the minimum time necessary for the alcoholic patient to establish effective treatment contact. Eleven younger patients attended the clinic for at least this number of sessions following the Y.P.P. intake interview. This contrasts with only two younger patients who attended the clinic for the requisite number of sessions in the nine months preceding the Y.P.P. and represents more than a five-fold increase.

The percentage of younger and older patients who attended the clinic for at least five, or fifteen sessions is presented in Table 1.

TABLE I : PERCENTAGE OF YOUNGER AND OLDER PATIENTS ATTENDING THE CLINIC FOR FIVE OR FIFTEEN SESSIONS

	Younger patients N = 17	Older patients N = 72
Five or more sessions attended	64.7	52.8
Fifteen or more sessions attended	37.5*	21.0

* N = 16

If five sessions are accepted as the minimum time necessary for patients to establish treatment contact and fifteen sessions as indicating the maintenance of the treatment relationship, then Table I shows that a higher proportion of younger patients, as compared to older patients, established and maintained treatment contact.

DISCUSSION

Before discussing the results, let us consider in more detail two aspects of the Y.P.P. One of the tenets underlying the young persons programme has been to individualise treatment goals. While this has always been a basic psychotherapy postulate, it has not, with any frequency, been adopted by alcoholism treatment programmes, who still adopt abstinence as the main treatment criterion. One reason for this is the fear that differential goals, particularly as they relate to drinking behaviour, might cause disruption within the treatment programme. It is often felt that the patient whose goal is abstinence might not be able to accept the fact that other patients have a goal of moderate drinking. The prevailing practice of adopting a unitary treatment goal, irrespective of patients' ability or need to meet it, increases the likelihood that patients will become frustrated with the treatment process and terminate treatment prematurely. To date, the acceptance of differential treatment goals by the patients has not posed a problem. However, patients are encouraged to explore feelings of resentment in the treatment group.

As mentioned previously, we were initially reluctant to allow patients to follow up those who ceased to attend. Experience to date shows that this innovation has been highly successful. Three of the five patients were encouraged to resume continued attendance. Also, it was found that patients following up other patients helps to increase group cohesiveness.

In terms of the present paper, development of group cohesiveness was seen as a goal of the Y.P.P. as were establishment and maintenance of treatment contact. This follows from the work of Yalom (1975) and Madden and Kenyon (1975), who noted that cohesiveness and continued treatment contact are related to successful treatment outcome. The authors are aware that it is possible for patients to maintain continued treatment contact without showing improvement in social functioning or cessation of alcohol consumption (Rosenberg and Amadeo 1974). Because of this, future papers will report on changes that patients show in these other areas.

The data presented concerning patients' comments and social interaction while attending the clinic, although tentative, do seem to indicate that group cohesiveness was established when young problem drinkers were treated separately from older patients. The authors are limited in the conclusions that can be drawn from this data, since it was not possible to observe a control group of young problem drinkers treated in conjunction with older patients.

One explanation that might account for the increase in young problem drinkers who attended five or more sessions since the inception of the Y.P.P. is that specialised young persons treatment reduces the likelihood that young problem drinkers will perceive themselves as group isolates, which in turn will be efficacious in encouraging treatment contact. The relationship between the extinction or diminishing of a patient perceiving himself as a group isolate or deviant and continued treatment contact is consistent with Rathod et al (1966), who note that the patient who is markedly different from the remainder of the treatment, e.g. the young problem drinker, is more likely to terminate treatment prematurely.

Another possibility is that Y.P.P. patients were encouraged to become involved in treatment, explore personal problems, express hostility and have a clear idea about the treatment process. These aspects of treatment milieu have been shown to be related to positive alcoholism treatment outcome (Fischer 1976).

The 550% increase in the number of younger patients attending the clinic since the inception of the Y.P.P., while encouraging, should be regarded with caution. It is possible that the emergence of a young persons programme may have resulted in a substantial increase in the number of young problem drinkers referred to the clinic. If this were the case, it is possible that the proportion of these patients attending the requisite number of sessions might have remained constant and any increase in numbers could therefore be attributed to increased referrals. Taking into account any attraction the Y.P.P. had as a referral source, it is unlikely that the rate of referrals increased sufficiently to account for the above findings.

The difference in attendance between younger and older patients, indicating that a higher proportion of younger patients established and maintained treatment contact, while not significant, is encouraging, given that younger patients are often thought to have a poorer prognosis with respect to treatment contact (Miller et al. 1970). Again, we must view the above results with caution. Since younger patients are given an intake interview in which suitability for treatment is assessed, it is possible that any differences between younger and older patients could be partially attributed to factors within the selection process, namely, patients most likely to remain in treatment could be those judged to be most suitable (Whyte 1975). The above criticisms notwithstanding, the evidence presented is encouraging enough to warrant continuation of specialised treatment for young problem drinkers in our clinic and it is suggested that the approach offered here may be found to be the treatment of choice in other alcoholism treatment programmes.

SUMMARY

The establishment of the first specialised alcoholism treatment programme for young people within the N.H.S. is described in detail. The results for the first nine months of the treatment programme indicate that, as compared with older patients (over age 30), a higher proportion of younger patients (age 30 or under) established and maintained treatment contact. This is contrary to the literature which suggests that younger patients have a poorer prognosis with respect to continued treatment contact. It is suggested that specialised treatment services for young problem drinkers deserves further exploration.

REFERENCES

CHAFETZ, M.E., BLANE, H.T., ABRAM, H.S., GOLNER, J., LACEY, E., McCOURT, W.F., CLARK, E. and MEYERS, W. (1962). "Establishing Treatment Relations with Alcoholics". Journal of Nervous & Mental Diseases, 134. 394-404.

DEMONE, H.W. (1963). "Experiments in Referral to Alcoholism Clinics". Quarterly Journal of Studies on Alcohol, 24. 495-502.

FISCHER, J. (1976). "An Evaluation of the Importance of Staff-Patient Relationships and Treatment Milieu as Process Variables in Alcoholism Treatment". Ph.D. Thesis. University of Edinburgh.

JELLINEK, E.M. (1960). The Disease Concept of Alcoholism. Hillhouse, New Haven.

MADDEN, J.S. and KENYON, W.H. (1975). "Group Counselling of Alcoholics by a Voluntary Agency". British Journal of Psychiatry, 126. 289-291.

MILLER, B.A., POKORNY, A.D., VALLES, J. and CLEVELAND, S.E. (1970). "Biased Sampling in Alcoholism Treatment Research". Quarterly Journal of Studies on Alcohol, 31. 97-107.

MINTZ, E. (1971). Marathon Groups: Reality and Symbol, Appleton, Centry-Crofts, New York.

RATHOD, N.H., GREGORY, E., BLOWS, D. and THOMAS, G.H. (1966). "A Two-Year Follow-Up Study of Alcoholic Patients". British Journal of Psychiatry, 112. 683-692.

ROSENBERG, C.M. (1969). "Young Alcoholics". British Journal of Psychiatry, 115. 181-188.

ROSENBERG, C.M. and AMADEO, M. (1974). "Long-Term Patients seen in an Alcoholism Treatment Clinic". Quarterly Journal of Studies on Alcohol, 35. 660-666.

SMART, R.G. and FINLEY, J. (1975). "Increases in Youthful Admissions to Alcoholism Treatment in Ontario". Drug and Alcohol Dependence, 1. 83-89.

WECHSLER, H. and THUM, D. (1973). "Teenage Drinking, Drug Use and Social Correlates". Quarterly Journal of Studies on Alcohol, 34. 1220-1227.

WHYTE, R. (1975). "Psychiatric New-Patient Clinic Non-Attenders". British Journal of Psychiatry, 127. 160-162.

YALOM, I.D. (1975). Theory and Practice of Group Psychotherapy. 2nd Edition. Basic Books, New York.

FAMILY FOCUSED TREATMENT AND MANAGEMENT: A MULTI-DISCIPLINE TRAINING APPROACH

JERRY P. FLANZER and GREGORY M. St. L. O'BRIEN

SCHOOL OF SOCIAL WELFARE

UNIVERSITY OF WISCONSIN, MILWAUKEE

FAMILY FOCUSED APPROACH FOR THE MANAGEMENT OF ALCOHOLISM PROBLEMS: THE MODEL AND IMPLICATIONS TO PROFESSIONAL EDUCATION

Training for professional/paraprofessionals who work in the field of alcoholism has been grossly inadequate. Those programmes that do exist tend to focus on training for working with the alcoholic individual, ignoring the context of his environment and particularly the family. The authors intend to present consideration of a multidisciplinary curriculum aimed at highlighting work with the alcoholic family using it as a means to working with the designated alcoholic.

The concept of alcoholism as a disease, rather than as a symptom of immorality, weakness, or self-indulgence has come of age in the American legal, medical, and social service system. Probably, no man is more responsible for this than E.Jellinek (1960). His efforts greatly influenced the 1956 American Medical Association's action to regard alcoholism as part of the preview of medical practice. His works, no doubt, influenced the two monumental 1966 legal system's decisions declaring alcoholism a medical problem - a disease.* (See next page.)

The political awareness of the electorate's increasing concern is evidenced by the rush of the federal and state legislatures and the appropriate departments of health, education and welfare to promote the creation of direct services, train personnel and extend research in the alcoholism arena. The re-ordering of the National Institute of Mental Health to the Alcohol, Drug Abuse and

Mental Health Administration clearly suggests this. Perrow's (1967) point of the mental health domination of treatment, education and health research to the exclusion of other less organized and endowed agencies is clearly understood by the financial push in the alcoholism direction. Most concretely, recent times have seen systematic changes in the laws affecting (1) workmen's compensation, (2) insurance benefits and (3) commitment referral procedures to medical facilities rather than to jails (Hirsch 1967).

SCOPE OF THE ALCOHOLISM PROBLEM - INDIVIDUAL ORIENTATION

It is difficult to compute the number of alcoholics in the United States. The National Centre for Prevention and Control of Alcoholism using a regional sample survey estimates a population of adult drinkers in the United States at 80 million, 5.6% or 4½ million are influenced with alcoholism (Alcohol and Alcoholism 1972). At one time Jellinek's formula using the total number of deaths from diagnosed cirrhosis of the liver was the standard being used to estimate the alcoholic population in a given area (Seeley 1960). Statistical evidence and expanding problem definition has forced abandonment of this standard. Thus, although the numbers clearly show a problem, comparable, valid, reliable and annual demographic data remain unavailable. This definitely has affected planning, evaluation and policy procedures.

There have appeared to have been more male alcoholics than women. Recent willingness of alcoholic women to seek treatment and the increasing awareness by agencies of their needs have improved the women/men ratio. One wonders if the previous sex ratio statistics were culturally biased or valid ? The alcoholic is a risk to himself and to others. Alcoholic cirrhosis and alcoholic psychoses account for .8% of all deaths (U.S. National Centre for Health Statistics 1966). The alcoholic dies earlier, due to a long list of associated illnesses.

In 1950 it was estimated that yearly losses due to alcoholism totalled $ 765 million in the United States with wage, crime and

---

* (See previous page)
In one, the U.S. Court of Appeals for the Fourth Circuit overturned the public drunkenness conviction of a North Carolina man on the grounds that it was unconstitutional to punish a person for acts he could not control. Similarly, the U.S. Court of Appeals for the District of Columbia ruled chronic alcoholism is not a crime. The District of Columbia ruling held that proof of chronic alcoholism is a defence against a drunkenness charge because the defendant "has lost the power of self-control in the use of intoxicating beverages".

accident loss composing the bulk of this figure (Thompson 1956). Since then, an increasing alcoholism rate coupled with a more acute awareness by industry, police and the public have served to revise estimates demonstrably upwards: specifically (1) the direct loss per year due to traffic accidents of over 50,000 lives, 2,000,000 disabling injuries and costs over $ 7 billion in property, wage losses, medical expenses and overhead costs of insurance (National Safety Council 1966); (2) the direct cost to American industry due to loss of manpower inefficiency, replacements, fringe benefits and invested training expenses of a minimum of 25% of each alcoholics salary - or naturally $ 2 billion/year; (3) the direct cost to the taxpayer in police and corrections service is not obtainable. The number of acts incurring physical and/or property damage caused by intoxication is astronomical. One of the major reasons, for the push for designating alcoholism as a disease and not as a crime, has been to reduce the costs of "handling" public intoxication, disorderly conduct and vagrancy. The cost of American taxpayers for the arrest, trial and maintenance in jails of these excessive drinkers could no longer be tolerated. One author study showed alcoholism to be a large factor in 43% of convicted felons in one prison setting (Guze 1972).

THE TYPES OF STAGES OF ALCOHOLISM

Most classifications of alcoholic types have failed to prove reliable. Over and over again, alcoholics have proven to be heterogeneous resisting typology according to an alcoholic personality or causation. Alcoholism has multiple causations effecting multiple personalities, suggesting multiple treatments. Phases in the development of the alcoholism condition, however can be documented ranging from the early alcoholic dependence to the emerging severe physical complications. As the alcoholic progresses through these stages his relationship to his family leads to crisis, more crisis and despair.

"MAKING THE LOOP" - THE AGENCY CRISIS

Golden opportunities for early detection are missed almost daily in hospitals, courtrooms, social agencies (Plaut 1967). Those who practice in the helping professions see alcoholism every day, but seldom do anything about it. Approximately 15% of the alcoholic population is receiving treatment for alcoholism. Another 25% are patients or clients somewhere being examined, probed and treated for all sorts of emotional and physical disorders - but not for alcoholism.

The physician, the social worker, the police officer, etc. can serve as a case finder of alcoholics; yet until physicians, or their attitudes and knowledge about alcoholism, undergoes change, many alcoholics and pre-alcoholics will not receive treatment until the disease is so far advanced that treatment becomes highly complicated, therapeutical goals limited and necessary resources extremely limited and expensive .... (Blane et al. 1963).

Corrigan's (1974) study reveals that present statistics of the number of alcoholic patients in treatment is misleading and exacerbated; for many detected alcoholic patients pass right through sources of care. They receive a preliminary diagnostic work-up only to be referred to another institution to start all over again. Wiseman's (1968) study of the skid row population again shows that the alcoholic agency statistics and treatment realities are ballooned as the alcoholic "makes the loop", the shunting from institution to institution - differentially treated by race and/or socioeconomic class.

The need for diagnosis, case finding awareness among the disciplines and the need for case management among collaborating agencies is a major motivation behind development of University of Wisconsin-Milwaukee's alcoholism curriculum.

THE PROBLEMS OF ALCOHOLISM FOR FAMILIES

Incalculable are the intangible effects felt by non-alcoholic family members; the broken homes, the physical and emotional abuse, the lost income, and self-respect. Alcoholism in the family poses a situation defined by the culture as shameful, yet the few prescriptions which are effective and permit direct action are in conflict with their cultural prescriptions (Jackson 1954). "Alcoholism is shameful and just shouldn't occur". Family members are influenced by the cultural definition of alcoholism as evidence of weakness, inadequacy or sinfulness, and by the cultural values of family solidarity, sanctity and self-sufficiency.

Alcoholism is one of the intrafamily stressor events that leads a family to become crisis-prone. Because an intrafamily event reflects on the family's internal adequacy it is more disorganising than would be an intrafamily stressor event (Hill 1965).

Because the family repeatedly fails in its efforts to cope with the problem of alcoholism and becomes progressively more disorganised, its vulnerability to crisis is increased. For many families being in a chronic state of crisis becomes a life style in itself. The crisis and the response to it are not maladaptive but, in fact, are an attempt at adaptation serving to ward off deep unconscious depression, anxiety, or even underlying psychotic processes (Rappaport 1970). These families are "stuck" in a downward spiral landing to disaster.

The family's attempt at coping are described by Jackson's (1958) seven stages :

1. Attempts to deny the problem (doesn't exist).

2. Attempts to eliminate the problem

3. Disorganisation

4. Attempts to reorganise in spite of the problem

5. Efforts to escape the problem

6. Reorganisation of part of the family

7. Recovery and reorganisation of whole family (separation, divorce or family intact).

Family efforts seldom get beyond the fifth stage, "efforts to escape the problem" without external intervention help and support. Often family breakups may lead to the creation of two families in distress in place of one. Professional intervention does decrease the chance of repetitive marital families (Meeks et al. 1970).

No-one denies that the spouse of the alcoholic is seriously disturbed by the time she reaches an identifiable source of help. However, there remains the basic question as to whether her disturbance antedates the partner's alcoholism or stems from it (Bailey 1961). Much clinical evidence points to the fact that in many instances the wife of an alcoholic unconsciously because of her own needs, seems to encourage her husband's alcoholism (Futterman 1953). When the alcoholic becomes sober, the spouse decompensates and begins to show symptoms of neurotic disturbance (Whalen 1953; Price 1945). Transactional-interaction evidence points to the fact that the behaviour of the spouse of an alcoholic is largely a function of the changing pattern of marital interaction (Lemert 1960). The alcoholic marriage then is a symptom pointing to either an expression of underlying psychopathology of both partners (Bailey 1963) or of faulty communication and interaction patterns. And indeed, Gorad's (1971) study shows the alcoholism partnership locked into a competitive "one-up-menship" competition over control and co-operation in intimacy avoidance. Steiner (1971) illustrates the family transactional pathology so well in Games Alcoholics Play.

As part of these transactions, the alcoholic wife finds herself in distinctly conflicting roles alternating between "rescuer", "persecutor" and "dummy" roles (Albertson and Vaglum 1973);

the bewildered spouse is caught in the crisis prone cycle. In addition, the spouse must face the "hard facts". She must take an increasing economic-control position to maintain child-rearing and family maintenance responsibilities. She must do this in a spiral of increasing costs due to family inefficiency and economic disparity towards women (Gage 1975). Simply stated the alcoholic family is poorer than its normal counterpart and a full 33% poorer (Chafetz 1971) and the spouse must manage.

IMPACTS ON THE CHILDREN

Clearly, children of alcoholic parents are subject to a high risk of developing alcoholism in their adult years. Attention in recent years has increasingly focused on the alarmingly high incidence of emotional and behavioural disorders among this group (Chafetz 1971; Fine et al. 1975). Life in the home of an alcoholic parent can be chaotic, confusing and unpredictable: it frequently involves parental neglect and even physical abuse of the children. Such children often have a poor self-concept; low frustration tolerance; poor school performance; and have numerous adjustment problems during the adolescence years. Twenty-five to fifty per cent of all alcoholics have had an alcoholic parent or close relative (Bosma 1975; Schuckitt 1973; Fox 1972). The 25 million children of alcoholics are twice as likely to become alcoholic as the children of their non-alcoholic counterparts (Globetti 1973; Booz et al 1974). Besides the increased likelihood of a birth defect due to pregnant alcoholic women, children of alcoholics are more likely to suffer from psychoses due to an increased likelihood of prolonged separation from parent(s) and inconsistent parental substitutes. As Hindman writes:

> "Contributing to the emotional problems of such a child (child or an alcoholic family) is the fact that the behaviour of the alcoholic parent, and often that of the non-alcoholic spouse as well, tends to be erratic and inconsistent. The focus of family life is on the alcoholism; children are often ignored or neglected, disciplined inconsistently, and given few concrete limits and guidelines for behaviour. In addition, the family is generally isolated from other members of the community. Because of the alcoholism problem there are few family outings or group activities at home. Friendships are often avoided by both the non-alcoholic spouse and the children because they are ashamed of the presence of alcoholism in their family (Hindman 1975).

FAMILY INVOLVEMENT IN MANAGING ALCOHOLISM PROBLEMS

Only two approaches are taken by professionals who are working with alcoholics and their families. One approach, noting that the alcoholics's family's behavioural interactions have positive and negative effects on the alcoholic and his drinking, advocates involving family members in the treatment process in order to help the alcoholic with his problem. The other approach, noting that the alcoholic alone has the power to resolve his problem, advocates treating the alcoholic independent of the family. This latter in effect relieves family members of the painful feelings of responsibility. At first glance, these two approaches appear to be in conflict; and it is no wonder, then, that professionals chose to remain with one or the other. Both points of view are valid, so that approaches need to be developed which incorporate both.

The family focused approach to managing alcoholism problems accepts both the family and the individual orientation. However, the initial and primary focus is placed on the family for four reasons: (1) the family's basic needs come before an individual member's basic needs (to be discussed further on); (2) family members must be able to establish their separate identities before true family reintegration may occur; (3) much less knowledge, research and training time has been spent on working with the alcoholic family, then the alcoholic himself; (4) an alcoholic's chances of success are greatly increased when consistent family pressure creates a demand for cessation of drinking.

The family will be able to grow and accept help when its basic needs show signs of being met; when the integrity of each individual family member is safeguarded; when relief from day-to-day crisis decision-making gives way to beginning realistic steps in planning for the family and family members; and when, the extra-family resources permit growth and integration into the community.

When this occurs, the alcoholic will be enabled to accept help; for he is released to take responsibility while his family may begin letting go emotionally. It means giving up attempts to control the drinking and providing protection from the consequences. This forces the alcoholic to decide if he wants to be part of the family and places the onus of responsibility where it belongs.

A FAMILY FOCUSED MANAGEMENT APPROACH

The family focused management approach strives to (1) provide the family with skills to cope with high stress situations; (2) provide professionals with the skills to create a range of carefully linked, knowledgeably administred community resources readily available to help the alcoholic families as well as the alcoholic; (3) provide the family with skills to realise and change their status as co-deviants (Wiseman 1975) in the alcoholism game;

(4) provide the family members with the opportunity to develop "parts of self" independent from the family unit; including the development of skills that guide them towards economic, vocational, educational, and social self-sufficiency; (5) provide simply and efficiently the information needed to accomplish one through four. Wiseman (1975) refers to a group of wives of alcoholics from Finland who create totally independent existences for themselves while remaining married. They increase work and hobby skills, make social plans apart from husband; and schedule their time to avoid contact with their alcoholic husbands completely. Many of these husbands responding to the new family rules find their attempts to sobriety complicated by a wife who is torn between marital duty and this separate individually successful life-style. The family focused management approach has been developed with the realisation that a major crisis point is the reintegration of the total family, and professional helping skills must be especially provided for this crucial stage.

FAMILY FOCUSED MANAGEMENT AND FAMILY THERAPY

Clearly, an emphasis on family-focused management, as that phrase is used here, is conceptually broader than the usual family therapy approach which views the whole family as a patient, to be treated in respect to its internal dynamics and the effect such dynamics have on the problem drinker or alcoholic. The usual thrust of such treatment is to help the family become more understanding and supportive of the drinker. Although such understanding and support has its potentially humane aspects, it can entrap the family in an alcoholism-dominated system, with subsequent serious adverse social, economic, physical and psychological effects for the spouse or parents, children or siblings, and members of the larger kinship network.

Such a system can induce a vicious downward spiral for all concerned, and its effects can be often felt in succeeding generations. Family-focused management that views the family not only as an internal system, but as a system interacting with larger external systems, can concentrate on assisting family members not to be trapped in the "alcohol system" by providing a range of carefully linked, knowledgeably administered community resources readily available to problem drinkers, alcoholics and their families. A leading question in the family focused management approach is the inter-relationship between the family and individual system.

THE FAMILY NEEDS AND INTERVENTIONS

Need Level

The underlying assumption to the family-focused approach to managing alcoholism is that families have a hierarchy of basic needs. Maslow's (1954) hierarchy of an individual's basic needs has received wide acceptance. Building upon Maslow's need hierarchy concept for the individual, we have postulated a hierarchy of basic needs for the family; one which lends itself readily to understanding the specific hierarchical needs of the family of the alcoholic. Just as in Maslow's individual needs concept, the family basic needs concept make the assumption that each basic need is built on the previous and supports the next. Principle 1 : Each basic need cannot be truly fulfilled until the need levels upon which it is based have already been met.

TABLE 1 : HIERARCHY OF BASIC NEEDS FOR THE FAMILY

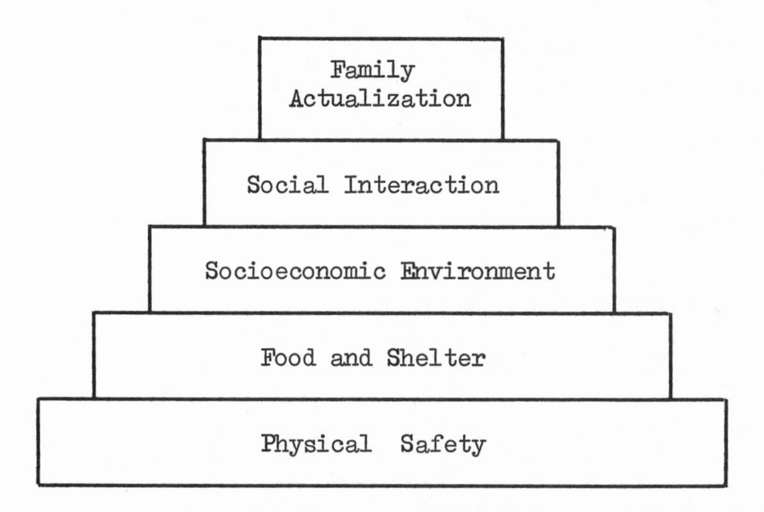

An important derivative of this principle is that family given level of need must be met before the individual alcoholic's need can be met. A family might be addressing needs of one level higher than those being addressed by the individual alcoholic(s). The reverse, (the alcoholic member addressing needs of a higher level than those of his family) cannot occur. Working with the alcoholic family to meet their need at a given level prior to addressing those same needs for the individual, allows for functional growth to all parts of the family system.

The alcoholic, as his condition worsens, tends to slide to more basic needs. As he does this, the family struggles to maintain their position. The family, however, cannot maintain itself more than one need level above the alcoholic and still maintain its family status. For example, while the alcoholic person begins to address the need of finding shelter for the night, his family might be faced with the needs of a surmounting monetary crisis. The family and its alcoholic members can be on the same level - often, with an "uneasy" equilibrium. If both the parties slide to the most basic need level of physical safety, they cannot survive as a family unit. This would further engage society in a myriad of problems derivative of isolation, separation of children, repetitive marital failures.

INTERVENTION STEPS

Primary intervention, then, must be aimed at working with the family to stablise - not to slide to more basic needs. This allows for a diagnosis of need-level for the family and the individual. The next interventive step is to help the family members reach and meet needs on the next higher level. This in effect, creates the freeing of the environment described earlier, allowing the alcoholic to take needed responsibility for self. In addition, it frees the individual family members to work at achieving the need-balance between the family group and the individual's own self-development.

Principle II : There must be professional specialization to need level.

The specific interventions, are of course dependent upon the diagnosis of present need level and the next higher need level to which the family could address itself. The skills needed for each level differ; our society has discovered that services and resources are most efficient when focused on one need level of a problem at a time. Organising services to meet multi-need levels for the entire problem has worked best when the professionals within them focused on one particular problem-need level. This has presented the necessity of a team approach. Principles three and four follow.

Principle III : Professional specialisation must have content specific to need level.

Principle IV : Continuum of care must be provided to help families meet their current and aspired needs.

As will be illustrated, in order to create a continuum of care, each professional specialisation must have direct knowledge of the content and issues of one need level above and one below their own particular need level of specialisation. A working inter-relationship with the professionals oriented to the need level one above and one below is essential for a maximal quality of care continuum.

The following is a description of the basic needs for families with suggested interventions. Selected professional specialisations at each need level will be presented: specifically, examples utilising criminal justice, home economics, family financial management and graduate social work will be highlighted as they fit the programme at our school. An example of their needed competencies and resultant corresponding curriculum will be given to illustrate the direct inter-relationship of the needs of families of alcoholics and the university curriculum.

No matter what need level orientation, all helping professionals must have in their repertoire a wealth of knowledge of human behaviour, skills in human interaction, and an awareness of "self". It is presumed that a specialisation in alcoholism, will be built on such a generic base, no matter which need is the target of intervention.

I. PHYSICAL SAFETY

The guarantee of the safety of each family member is primary. This includes protection from child abuse and neglect, wife battering and other physical assaults. The cardinal rule for the existence of any group is the guarantee of safety from physical harm. The concept of physical safety also includes the physiologically damaging effects the alcoholic incurs due to his/her problem.

Intervention at this need level includes provision of the myriad of police corrections and protective services oriented to restrict individuals from harming others and themselves. The gathering of evidence, knowledge of the law, the effect of polydrug (alcohol) abuse or behaviour and self-defence skills are among the specified knowledge needed by different professionals operating to fulfill needs at this level; similarly a conceptual base for the integration of these skills is required of the men and women who are working in or preparing for the criminal justice system in the areas such as prevention, enforcement, corrections and courts must have special knowledge and skills to deal with alcoholics and alcoholism. Such knowledge and skill base must include :

1. A knowledge base in social and behavioural sciences to better understand the relationship between alcoholism and criminal behaviour, its consequences for the victims of crime - including family members - and society's response to it.

2. A familiarity with the various mechanisms, approaches and programmes which combine to make the criminal justice system, including areas such as prevention, diversion, enforcement, adjudication, corrections, probation, parole, etc. responsive to the alcoholic problem.

3. Skills in the analysis of programmes and policies which guard and direct the operation of the criminal justice system and which might be focused upon the improvement of the system on behalf of alcoholics.

In order to achieve these competencies, the professional must have learned specific information which inter-relates the problems of families of alcoholics to the criminal justice system. Specifically :

1. The inter-relationship of Crime and Alcoholism: its control and social context; this includes the economic and social impact on families of alcoholics and associated criminal behaviour, societal reaction and cultural variations.

2. Criminal Justice Policy and Implementation: issues of law, regulation and social control in criminal justice as it relates to alcoholics and their families; this includes labelling and legal institutions response to the entire family system.

3. Alcoholism and the Criminal Justice Inter-System Dynamics: issues of programme design and development, programme implementation and co-ordination. This includes utilisation of planning, management and manpower development; enforcement and treatment techniques for the alcoholic and his family.

4. Methods of Inquiry into Criminal Justice Effect on the Alcoholic Family: this includes applied research methodology involved in programme development, implementation and assessment.

5. Field Experience: direct supervised work with families of alcoholics or alcoholic-related agencies to integrate and apply the above.

The person working in the criminal justice system is often in the position to gather evidence to make an assessment as to the potential of the family of the alcoholic to reach the next need level. His knowledge of the fields serving the need of food and shelter is essential to reorganising the potential and making possible the referral needed for an integrated continuum of care.

For example, the police officer must have first hand knowledge of the agencies and professions which provide emergency housing services in his community. Implementation of the law, in itself, often can exacerbate an alcoholic family's dilemma. The trained police officer making an assessment that temporary alternative housing might protect a family better than waiting for a technical breaking of the law in an alcoholic family dispute. He, then, might help the family reach these alternative services. This often takes the form of crisis intervention counselling. Thus, he has guaranteed a continuum of care aimed at preventing family disintegration/ disorganisation. In effect, he has added a social service component to serving the law. He has allowed for a continuum of care, because of his generic knowledge base and his knowledge of the need level for the family one above his own area of specialisation. (And we might add, he validates the need for advanced training and/or college education for police officers).

II. FOOD AND SHELTER

Alcoholic excess is liable to undermine health, for one reason, because the person who consumes much alcohol often does not eat properly and fails to get enough of the proteins, minerals and vitamins (particularly vitamin B) needed (Williams 1951). His family is also not likely to have the proper nutritional habits due to economic, stress related or informational reasons. Such families often do not have the necessary nutritional knowledge. In response to family stress, too often they lack the necessary concepts of time and structure to "sit down" to plan meals. Nutritional imbalance adds stress to the stressful emotional situations, depressing the physiological capacities of coping.

Nutritional problems are further complicated because alcoholic consumption not only gives rise to body deficiencies, but such deficiencies also give rise to alcoholic consumption. It is thought by some, that genothrophic diseases - a disease of nutritional deficiency resulting from inherited personalities in the chemistry of metabolism - might be responsible for as much as 25% of the known alcoholics (Woods 1955). Being inherited all family members may be vulnerable.

Shelter-housing looms as a basic problem need of many alcoholic families. Getting behind in the rent, loud disturbing quarrels, visibly deviant behaviour by the alcoholic members, rejection by one's neighbours etc., all contribute to a pattern of frequent eviction, and lowering housing standards. Quite clearly, intervention at this need level includes professions concerned with nutrition and emergency housing-shelter care needs.

The men and women who are working in the field of Home

Economics in areas such as nutritional education and foods, home management and hygiene also must have the knowledge and skills to deal with the alcoholics and his family's particular needs. Among the things a home economist must know in helping to meet these special needs are :

1. the variety of ways to meet a nutritional balance ;

2. the food combinations and quantities necessary for each family member (all who are at different stages of the life cycle, making sure that a proper distribution of food to family members is occurring);

3. the distribution of food to fit the family's pattern of expenditure (not forgetting helping the family reallocate money for basic foods that originally went for alcohol or for protecting the alcoholic member) ;

4. the pattern of physical availability of the food and the pattern of consumption (promoting these families to have a structured meal and time orientation of meeting home health tasks ;

5. the consumer aspects of buying nutritionally (correlating nutritional needs to money supply and best "nutritional buys" per unit cost).

These knowledge competencies translate into needed "back-up" of curriculum inputs - specifically such course areas as :

1. Nutrition: the physiological, biochemical properties of food; nutritional significance of food; the needs of the body machinery; methods of preparing foods guaranteeing minimal nutritional loss.

2. Food Supply: the consumer aspects of the food supply; specific nutritional needs correlated to best value for the money.

3. Home management: strengthening of structuring of home life to meet basic nutritional and home health needs.

The home economist while helping the family of an alcoholic must ascertain whether or not the basic safety needs have been met. At the same time, he is looking for opportunities to relate his clientele to the next level of need - the socio-economic environment. The home economist must know something about manpower, job training, and income maintenance possibilities to help the families make this link.

III. SOCIO-ECONOMIC ENVIRONMENT

Job instability and financial insecurity plague alcoholic families (Gage 1975). Needed financial resources for family members at key times in the families life cycle increase the crisis atmosphere, and undermine the meeting of basic needs. This makes relocation of the higher need levels improbable. The lack of job skills by most spouses limits opportunities for independent action (Alcohol and Alcoholism 1972).

Professionals specialising on this need level must deal with manpower training, job training, income maintenance and family financial management. The financial integrity of the family must be maintained and improved; so that the resources will be there as the need arises. The family financial manager often needs to re-educate the distressed families of alcoholics who are manifesting dysfunctional spending patterns. The family financial manager, then, must be able to :

1. evaluate relationship of financial priorities to spending pattern ;

2. demonstrate specific expenditure areas, explicate marketing techniques so as to increase rational consumer decisions :

3. articulate and explain process of credit granting, collection and the legal implications of becoming involved in the credit system ;

4. link clientele to community resources to assist financially distressed families ;

5. differentiate common consumer problems from unique alcoholic-related consumer problems, linking lack of assertiveness to poor money management.

Specific curriculum in the area of identifying a family's spending priorities, expenditures, the money-credit system and consumerism must be obtained. In addition, specific financial habits of alcoholics and their families must be learned.

Of course, family financial management is meaningless if the client does not have food to eat. Proper decision-making rests on "nutritional power". The family financial planner must be aware of the existence of a nutritional problem if he hopes to have the client think beyond gratification of immediate needs.

The interaction of a family discussing their spending habits creates a plethora of information useful for diagnosing specific problems in social interaction. Helping the family see how their inability to make appropriate financial plans lies partly in their ability to trust and relate co-operatively - if that be the case - helps to prepare the client family to begin to address the next higher need level. This is an important function for specialists focused on financial concerns. This function cannot be achieved without first understanding the content and the professional specialisations associated with the need level of social interaction.

IV. SOCIAL INTERACTION

Alcoholic families have difficulties in meeting four inter-related sub-needs of social interaction: (1) provision of a normal growth and development environment; (2) establishment communications and relationship models; (4) mastery on interpersonal relationships - particularly in the area of intimacy attainment (Steiner 1971; Futterman 1953; Rapapport 1970; Jellinek 1960).

The counselling and psychotherapeutic professions, the rehabilitation and family education professions; the group services and recreation professions, all can focus on the social interaction of the families of alcoholics. Social interaction oriented interventions vary from socialisation modelling techniques to the role analysis of game playing to "insight" learning of intrapsychic problems. No other profession other than social work is equipped to train students to concentrate in all of the above sub-needs, services or interventions.

Simply stated, the graduate social worker specialising in working with alcoholics and their families may develop skills on one of three levels: (a) skill in providing family oriented treatment and rehabilitation services to alcoholics and their families; (b) skill in managing and administering alcoholism prevention, treatment and rehabilitation programmes; (c) skill in planning programmes and organising community resources to most effectively create agency delivery systems concerned with enhancing social interaction needs.

Social workers providing direct family oriented treatment and rehabilitation building on generic social treatment skills, must know: (a) specific alcoholic family interactions, the alcoholic games and conflicting roles (Steiner 1971; Albertson and Voglum 1973); (b) techniques for unmasking symptomatic behaviour; and (c) the workings of Al-Anon, Alateen, Alcoholics Anonymous and other such support groups, as well as, the workings of antabuse programmes and therapeutic communities.

Thus, graduate social work curriculum specialising in meeting social interaction needs of the families of alcoholics must especially contain:

1. Knowledge of the various theories on causation of alcoholism ;

2. knowledge of the alternative treatment modalities;

3. knowledge application through direct supervised experience ;

4. knowledge of the special issues facing provisions of services to alcoholics and their families, such as, federal and local alcohol-specific legislation, insurance decisions and family isolation.

The MSW with a family focused management alcoholism speciality will be on the "look-out" to see that the family's socio-economic environment is stable, permitting the future orientation and released energy for the self-development needed. As the clientele progress in meeting the social-interaction needs, the social worker helps the family begin to actualise itself as a family, interfacing into society, learning to live in its value framework.

V. FAMILY SELF-ACTUALISATION

As the family of the alcoholic begins dropping its label and becomes the 'X' family, groping to grow and develop as an independent entity, making its own decisions and taking full responsibility for its actions, the family as a unit has begun to reach the self-actualisation level. The needs of this level are never fully achieved. As the family begins trying to meet this need, it mounts efforts to interface with society. It socialises as a family with other families; it creates a self-help inter-family system; it assimilates into the community's activities and neighbourhood; and, the family no longer needs professional intervention. The family has the skills to learn and grow on its own.

The graduate social work curriculum is designed as an inter-related matrix of social issues, areas and methods specialisations. With careful faculty and peer advising, the student is able to develop a course of study that builds on individual experience, interests and strengths. Beginning with a general core curriculum and moving to specialisations and integrative experiences through field work, projects and seminars, students develop an academic programme to achieve the integration of knowledge and skills required for professional competence.

The competencies and specialisations of advanced social work practitioners are becoming increasingly complex and demanding. They require a sound basis in knowledge, skills, theories and research. The knowledge and skill base, drawn from an ideological commitment to human rights, provides students with an understanding of the varied service fields and methodological approaches of the social work profession. Behavioural theory and research courses give students an empirical foundation for the analysis of individual and social problems and the study of intervention strategies. This core serves as the starting point for specialisation within the matrix curriculum.

Currently there are four social issues areas defined within the curriculum, and three methods specialisations.

The social issues areas are :

Mental and Physical Health
Family Development and Family Substitutes
Economic and Distributive Systems
Crime and Delinquency

The methods specialisations are :

Social Treatment
Community Organisation and Planning
Human Services Administration

MSW TRAINING WITH ALCOHOLISM SPECIALTY

Consistent with the family-focused philosophical orientation discussed above, students obtaining a Master in Social Work degree elect a concentration in the Mental and Physical Health issues area - with a specialisation in alcoholism - and take a number of electives from the Family Development and Family Substitutes issues areas. Specific alcoholism related courses are: Adaptation in Alcoholism; Social Treatment for Alcoholics; Social Issues of Alcoholism; Special Issues Affecting Alcoholic Families; Family Development over the Life Span - with special emphasis on alcoholic families. In addition, each student through his thesis seminar project in alcohol will engage in a programmatic effort related to alcoholism.

UNIVERSITY OF WISCONSIN-MILWAUKEE

The four professions mentioned were chosen because they are offered within the School of Social Welfare at the University of Wisconsin-Milwaukee, Milwaukee, Wisconsin, U.S.A. The School of

Social Welfare is the first mutli-disciplinary school in the United States and it provides a continuum of education in the fields of social work, criminal justice and home economics. In addition, it offers a family financial management specialisation as part of its social work undergraduate programme. As part of a major urban university, the School has access to the resources of a number of other school/departments, including psychology, sociology, nursing, urban planning, and education. It has a working relationship with local schools of medicine and law and it has close ties with a substantial number of community agencies engaged in planning and providing services to alcoholics, problem drinkers, and their families.

The basic core of the alcoholism specialty lies within the graduate social work curriculum. The criminal justice and home economics students take part in the general social work courses on alcoholism, alcoholism-specified content to their professions is integrated into modular components of their courses.

The School of Social Welfare's MSW curriculum is designed to develop strong methodological skills and problem analytic skills within the context of a particular problem area. These skills are further developed and reinforced by extensive field work coupled with field instruction and a monthly field seminar which integrates the field instruction and experiences with methods instruction. Field work with field instruction must be taken concurrently with a methods course and a field-methods integrative seminar.

All candidates for an alcoholism specialty take three semesters of field placement. The placements (16 hour/week in the first semester and 20 hour/week in the second and third semesters) are with local agencies providing alcoholism services (or planning and organising these services) and selected to match the candidates methodology specialisation.

Negotiations are currently underway to include this programme as a part of a new doctorate in human services to be offered by the School. This move is encouraging, for it would permit broadening to more professional disciplines and a clearer multi-discipline approach.

PITFALLS AND IMPLICATIONS

This is a multi-discipline approach to training and herein lies its strengths and weaknesses. Tremendous amounts of time go to administration and co-ordination. In addition, this approach places a premium on multi-discipline teams across need levels for continuum of care. Many agencies would need to re-allocate staff and co-operate with agencies with similar goals - an organisationally difficult task.

Professionals specialising in a family focused approach to the management of the problems of alcoholics and their families would definitely need to take coursework above and beyond that normally reserved for their specialisation alone. This approach contends that a generic base in human relations training must permeate all professional specialisations' curricula. In addition, a broader range of knowledge to include framework on "neighbouring" professions is essential. This is an argument for broader base training with implications of at least a baccalaureate base level for all. The "paraprofessionals" would also need similar broad training. This would be a departure from the stands of many a technically-oriented paraprofessional programme.

If alcoholism is to be handled as the major problem that it is, professional helping disciplines must turn their attention to an integrated approach, providing a continuum of care. The responsibility of case management and intra-organisational ties of systems of service systems will need to proceed far more rapidly than is the current case.

The family focused approach points out to agencies and to professionals alike that all too often their focus has paid attention to the individual alcoholic and not to his family. Agencies would need to turn to helping the family first; most agencies will need to structure their intake system to accomplish this task. Undoubtedly, it will be pointed out that at the time of intake many alcoholics did not or do not live with their nuclear family. The family focused approach immediately suggests the possible roles of family substitute systems, such as halfway houses and group homes. In addition, it points to the need to create a counter system which will serve to motivate the alcoholic to want to join it. This would create similar goal producing behaviour. If it does nothing else, the family focused approach helps one to understand the difficulty of helping an isolated alcoholic person.

REFERENCES

ALBERTSON, C.S. and VAGLUM, P. (1973). "The Alcoholic's Wife and Her Conflicting Roles". Scandinavian Journal of Social Medicine, 1. 7-12.

ALCOHOL AND ALCOHOLISM: PROBLEMS, PROGRAMMES AND PROGRESS. (1972). National Institute of Mental Health, National Institute on Alochol Abuse and Alcoholism.

BAILEY, M.B. (1961). "Alcoholism and Marriage: A Review of Research and Professional Literature." Quarterly Journal of Studies on Alcohol, 22. 81-97.

BAILEY, M.B. (1963). "The Family Agency's Role in Treating the Wife of an Alcoholic". Social Casework. 273-279.

BLANE, H., OVERTON, W.F. and CHAFETZ, M.E. (1963). "Social Factors in the Diagnosis of Alcoholism". Quarterly Journal of Studies on Alcohol, 24. 640-663.

BOSMA, W.G.A. (1975). "Alcoholism and Teenagers". Maryland State Medical Journal, 24. 62-68.

BOOZ, ALLEN and HAMILTON (1974). An Assessment of the Needs and Resources for Children of Alcoholic Parents. Final Report Submitted to National Institute on Alcohol Abuse and Alcoholism.

CHAFETZ, M.E. (1971). "Children of Alcoholics". Quarterly Journal of Studies on Alcohol, 32. 687-698.

CORRIGAN, E. (1974). Problem Drinkers Seeking Treatment. Quarterly Journal of Studies on Alcohol.

FINE, E., YUDIN, L.W., HOLMES, J. and HEINEMANN, S. (1975). "Behaviour Disorders in Children with Parental Alcoholism". Annual Meeting of the National Council on Alcoholism, Milwaukee, Wisconsin.

FOX, R. (1972). The Effect of Alcoholism on Children. National Council on Alcoholism, New York.

FUTTERMAN, S. (1953). "Personality Trends in Wives of Alcoholism". Journal of Psychiatric Social Work, 23. 37-41.

GAGE, M.G. (1975). "Economic Roles of Wives and Family Economic Development". Journal of Marriage and the Family, 121-128.

GLOBETTI, G. (1973). "Alcohol: A Family Affair". National Congress of Parents and Teachers, St. Louis, Missouri.

GORAD, S.L. (1971). "The Alcoholic and His Wife: Their Personal Styles of Communication and Interaction". Ph.D. Dissertation Boston University.

GUZE, S.M. (1972). "Criminality and Psychiatric Illness: The Role of Alcoholism". Joint Conference on Alcohol Abuse and Alcoholism.

HILL, R. (1965) "Generic Features of Families Under Stress". Crisis Intervention and Selected Readings, 32-52. H.J. Parad, ed. Family Service Association of America, New York.

HINDMAN, M. (1975). "Children of Alcoholic Parents". Alcohol Health and Research World.

HIRSCH, J. (1967). "The Disease Concept of Alcoholism". Opportunities and Limitations in the Treatment of Alcoholics. 1-20. J.Hirsch, ed. Charles C. Thomas, Springfield, Illinois.

JACKSON, J.K. (1954). "The Adjustment of the Family to the Crisis of Alcoholism". Quarterly Journal of Studies on Alcohol, 15. 562-586.

JACKSON, J.K. (1958). "Alcoholism and the Family". Understanding Alcoholism. 90-98 Seldon Bacon, ed. American Academy of Political and Social Sciences, Philadelphia.

JELLINEK, E.M. (1960). The Disease Concept of Alcoholism. Hill House Press, New Haven, Connecticut.

LEMERT, E.M. (1960). "The Occurrence and Sequence of Events in the Adjustment of Families to Alcoholism". Quarterly Journal of Studies on Alcohol, 21. 679-697.

MASLOW, A.H. (1954). Motivation and Personality. Harper and Row, New York.

MEEKS, D.E. and KELLY, C. (1970). "Family Therapy with the Families of Recovering Alcoholics". Quarterly Journal of Studies on Alcohol, 31. 414-423.

NATIONAL SAFETY COUNCIL. (1966). Accident Facts. Chicago.

PERROW, C. (1967). "A Framework for the Comparative Analysis of Organisations". American Sociology Review. 32. 194-208.

PLAUT, T.F.A. (1967). "Alcoholism and Community Caretakers: Programmes and Policies". Social Work, 12. 51-63.

PRICE, G.M. (1945). "A Study of the Wives of Twenty Alcoholics". Quarterly Journal of Studies on Alcohol, 5. 620-627.

RAPAPPORT, L. (1970). "Crisis Intervention as a Mode of Treatment". Theories of Social Casework. 304-305. R.Roberts and R.Nee, eds. Columbia University Press, New York.

SCHUCKITT, M.A. (1973). "Family History and Half Sibling Research in Alcoholism". Nature and Nurture in Alcoholism. 121-125 Annals of the New York Academy of Sciences.

SEELEY, J.R. (1960). "Alcoholism Prevalence: An Alternative Environmental Stimulation Method". Quarterly Journal of Studies on Alcohol, 21. 500.

STEINER, C. (1971). Games Alcoholics Play. Grove Press Inc. New York.

THOMPSON, G.N. (1956). Alcoholism Charles C. Thomas, Springfield, Illinois.

U.S. NATIONAL CENTER FOR HEALTH STATISTICS (1966). Vital Statistics of the United States. U.S. Public Health Service, Washington, D.C.

WHALEN, T. (1953). "Wives of Alcoholics: Four Types Observed in a Family Service Agency". Quarterly Journal of Studies on Alcohol, 14. 632-641.

WILLIAMS, R.J. (1951). Nutrition and Alcoholism. Norman : University of Oklahoma Press.

WISEMAN, J.Q. (1968). "Making the Loop: The Interactional Cycle of Alcoholism Rehabilitation". Ph.D. Dissertation. University of California at Berkeley.

WISEMAN, JACQUELINE (1975). "An Alternative Role for the Wife of an Alcoholic in Finland". Journal of Marriage and the Family. 172-179.

WOODS, R. (1955). "Nutrition and Alcoholism". Management of Addictions. 75-101. E. Rodalsky, ed. Philosophical Library, New York.

THE YOUNG ALCOHOLIC - APPROACHES TO TREATMENT

P.D.V. GWINNER

ROYAL VICTORIA HOSPITAL, NETLEY, SOUTHAMPTON

The title of the paper - "The Young Alcoholic - Approaches to Treatment" - was chosen by the conference organisers. The wording of the title, however, appears semantically appropriate; for a relative lack of clinical experience with the young alcoholic - not singular to the author, but currently shared by many workers in the field of alcoholism - renders any more definitive wording of the title unjustified.

It is considered essential, however, to define the term young alcoholic rigidly if a succinct account of treatment approaches to this particular group is to be presented. But definition has frequently been the rock on which alcoholism has recurrently appeared to founder. Mindful of such dangers, a conservative course has been steered and elected to define the young alcoholic, as did Rosenberg (1969) in an earlier paper in terms of the World Health Organisation definition of alcoholism. To that established, often criticised but broad definition is added a necessary chronological rider which simply states that to satisfy the diagnostic criteria for inclusion in the young alcoholic group, an individual must be aged 25 years or under. So the strict definition of the young alcoholic is that he or she is an excessive drinker 25 years or under whose dependence upon alcohol has attained such a degree that they show mental disturbance, or an interference with their bodily and mental health, their interpersonal relations and their smooth social and economic functioning, or who show the prodromal signs of such development they therefore require treatment.

All treatment approaches - that is the consideration of treatment aims, the development of therapeutic strategies and the establishment of treatment facilities - evolve following the recognition of a specific patient group and the definition of its particular needs. Recognition of alcoholism in the young and the acceptance of such individuals as a specific group are comparatively recent phenomena. The explanation for this novel awareness of the young alcoholic is probably associated with these patients' increasingly intrusive numerical significance as reflected in statistics emerging from a variety of treatment and other agencies.

This apparent increase of drink related problems in the youthful can be attributed to various factors which must include the affluence of the group, the particular vulnerability of the young and inexperienced to the relaxation of social and moral restrictions in our increasingly liberal and relaxed culture, the concomitant loosening of traditional family ties, the ready availability of alcohol and the apparent bias towards the young consumer evident in much advertising of alcoholic beverages.

Recognition, however, does not merely depend upon the numerical significance of the group under study, a heightened awareness among those who are employed in treatment and other agencies where the young alcoholic is likely to present is also necessary. It appears unfortunately that there still remains in a minority of such workers a considerable reluctance to define the young excessive drinker in terms of alcoholism and that alternative diagnostic euphemisms or the nebulous rubric of personality disorder are descriptively utilised This manoeuvre undoubtedly avoids stigmatisation of the individual but it also often engenders a paralysing therapeutic negativism. A parallel situation existed in the past with an often acknowledged reluctance to make a diagnosis of alcoholism in the female patient or client. The explanation of these inhibitions, which may significantly undermine treatment approaches to the young alcoholic, are possibly associated with puritanical influences, inherent in the culture in which so many were nurtured, which seemingly equate youth and femininity with innocence, and middle age and masculinity with the loss of that original state of unblemished purity.

The paper continues with a brief account of a pilot study conducted in the summer of 1974 in the alcohol treatment unit at the Royal Victoria Hospital, Netley. This hospital, where the author is employed as a consultant psychiatrist in charge of the Alcohol Treatment Unit, is a biservice psychiatric hospital providing inpatient facilities for serving members of the Royal Navy and British Army. Throughout 1974 it became increasingly apparent that the treatment facilities provided within the alcohol treatment unit satisfied the needs of the middle aged alcoholic, but failed to provide a relevant treatment programme for the increasing number of young patients who were presenting with alcohol related problems.

It was therefore decided to conduct a small pilot study to evaluate the characteristics of these younger patients, to compare them on a variety of parameters with their older colleagues, and to assess whether they were truly alcoholic and conformed to diagnostic criteria inherent in the World Health Organisation definition. The study was further intended as a pragmatic assessment upon the results of which would depend the type of treatment facility to be established. The study could in itself therefore be considered as an approach to treatment.

The study consisted of a comparative assessment of two numerically equivalent groups of patients, each group totalling 45 individuals. All patients included in the study were male and all were serving members of either the Royal Navy or the British Army. Consecutive admissions to the alcohol treatment unit at Netley were allocated to their appropriate group according to their age on the day of admission - those patients aged 25 and under being allocated to the young Group, the remainder to the euphemistically titled middle aged group. The results of this study and their relevance in considering approaches to treatment will now be described with reference to a short series of tables.

TABLE I: PILOT STUDY ALCOHOL TREATMENT UNIT RVH NETLEY 1974

YOUNG (MA = 22.1 years)	DRINKING MILESTONES (N = 45) AGE OF ONSET	MIDDLE AGED (MA = 35.9 years)
17.2	DAILY DRINKING	19.2
20.0	SUBJECTIVE AWARENESS OF LOC.	29.1
20.3	EARLY MORNING DRINKING	30.5
21.0	EARLY MORNING TREMULOUSNESS	30.5
	POST ALCOHOLIC AMNESIA	

Table I illustrates the "drinking milestones". It shows statistics related to the temporal quality of the two groups different drinking patterns. This information was acquired by extrapolation from a questionnaire which formed part of a larger survey of patients being made by a senior member of the nursing staff. The Table shows significant differences between the two groups in terms of their drinking pattern illustrating the accelerated rate of onset of loss of control and other signs of dependency exhibited by the young drinkers. It appears that these younger drinkers had rarely achieved a posture of social drinking at all, whereas the middle aged drinkers had enjoyed some ten years social or non disruptive drinking prior to exhibiting alcoholic stigmata. This finding, if accepted, is of considerable impact in view of current arguments suggesting possible treatment goals of social drinking rather than total abstinence propounded by a variety of workers including Drewery (1974) and Orford (1974). The explanation for this accelerated loss of control is not entirely clear but may be understood in terms of the young drinkers gargantuan daily consumption of alcohol. It proved impossible to accurately quantify this rate of consumption but empirical observation leads to the belief that it is considerably in excess of that presented by the older group.

Table II is merely a pictorial representation of the findings related to rate of onset of dependency and is included to underline the importance of envisaging the two groups of drinkers as separate rather than as part of a continuum, a theme which will be returned to later.

TABLE II: PILOT STUDY ALCOHOL TREATMENT UNIT RVH NETLEY 1974

1 YEAR	ESTABLISHED ALCOHOLISM YOUNG
10 YEARS	ESTABLISHED ALCOHOLISM MIDDLE AGED

Table III is included to illustrate the broad spectrum of damage suffered by these young drinkers, which is significantly similar and on occasion in excess of, the damage suffered by their senior counterparts.

TABLE III: PILOT STUDY ALCOHOL TREATMENT UNIT RVH NETLEY 1974

N = 45

YOUNG		MIDDLE AGED
20	WITHDRAWAL HALLUCINOSIS	7
20	EVIDENCE OF LIVER DYSFUNCTION ON HAEMATOLOGICAL INVESTIGATION	17

This Table illustrates two comparative parameters - alcohol induced hallucinatory states and haematological evidence of liver dysfunction. The presence of both these phenomena are widely accepted as indicative but not singularly pathognomic of established alcoholism and the inference drawn from this Table is that the young patients are a truly alcoholic population who satisfy the criteria of the World Health Organisation if it is applied to them. This inference is reinforced by investigation of the young group in terms of social damage, in which area they exhibit a high incidence of military and civil alcohol induced offences and of civil charges related to drunken driving.

It is intended to leave consideration of this pilot study, which was admittedly of simplistic design, and to describe some clinical experiences with the young alcoholic. Initially an individual approach to these patients was favoured, surmising that they were possibly too disturbed to effectively participate in a group setting. It was discovered however, that their response in the one to one situations which characterise individual therapy was disappointing. Within such a relationship they maintained postures of passivity, abrogating all responsibility to the therapist whom they frequently attempted to endow with near magical powers, which, if only he would utilise properly, would lead to the cessation of their drinking without significant effort on their part.

In a group setting, however, they exhibited considerable group dependency which often could be mobilised by the therapist to reinforce positive therapeutic responses. Early experiences with these patients in either resident or outpatient group settings taught the pragmatic importance of separating them for purposes of formal therapy from middle aged and elderly alcoholics (Gwinner 1976). If this separation is not achieved, mutual failure of identification between the two age groups occurs, with frequent associated acting out behaviour being exhibited by the younger members. The young overtly resent time spent by the middle aged in discussion of problems germaine to their life situation - marriage, career, prospects, and finance, and frequently identify the middle aged patients with their own fathers who were significantly often habitual excessive drinkers. The middle aged for their part display a paternalistic but often ambivalent attitude towards the young, doubting the validity of the latter group's alcoholism but simultaneously expressing resentment that they themselves were not afforded treatment earlier. In a resident milieu social diffusion between the two age groups outside formal group meetings is mutually well tolerated and appears to lead to no disruption. It was insisted initially, that sleeping accommodation for the two groups should be separate but they are currently sharing a dormitory ward with little difficulty but often one suspects mutual denial of each others' presence. This dichotomy between the two groups is further reflected in the young alcoholics' distrust of Alcoholics Anonymous, an agency which in a paper by Edwards, et al (1967) has been described as catering primarily for the middle aged.

Thus the therapeutic separation of the young from the middle aged appears to diminish the hostility of the former group, a hostility reported by both Rosenberg (1969) and Hassall & Foulds (1968) and envisaged by some workers as a gross impediment to the therapeutic responsiveness of the young alcoholic. This hostility can be further controlled by the careful selection of medical and nursing staff and by restricting the size of the resident or non resident group to no more than eight patient members. Staff selection is of considerable importance in establishing treatment facilities for young alcoholics. Empirically it appears that staff attitudes should not be permitted to polarise as the young alcoholic responds negatively to heavily authoritarian or totally permissive therapeutic regimes. The age and status of staff have proven important - extremely youthful staff however enthusiastic and proficient are poorly tolerated by the young alcoholic as are more mature staff members who hold crystallised and rigid views about the young. The anti-authoritarian propensities of the young alcoholic are further reflected in negative attitudes maintained towards those whom they envisage as of privileged or special status. This category undoubtedly embraces psychiatrists, other medical practitioners and psychologists. It is clearly neither desirable nor possible to exclude all such individuals from the treatment setting but more favourable responses are undoubtedly achieved in the course of therapy if clinical responsibility for the facilitation of the group is seen to be invested in an individual who like the mental nurse or social worker is not overtly identified with the medical establishment. Consistency of staff as in all group work is essential.

Treatment goals of total abstention have surprisingly been readily accepted by the majority of young alcoholics referred as a therapeutic necessity - who unlike the middle aged alcoholics are experientially fully aware of their limited capability in terms of achieving a posture of social drinking. It is significant that they are punitively intolerant of relapses among their peers in therapy. Relapses in the course of therapy occur more frequently than in older patients but almost immediate loss of control appears to exert a braking device and many relapses are ephemeral. Nor do relapses appear commonly to be denied, a surprising number of young out-patients have re-established immediate contact with the unit following such a relapse.

Patient selection has at no time been rigid although individuals whose presenting or past histories indicate the existence of significant psychopathy should be excluded. Since September 1974 only six patients selected for the young group following assessment had to be removed from the treatment setting because of the disruptive quality of their behaviour in the treatment milieu. It is important that such restive patients, representatives of the subgroup who have assuredly earned the young alcoholic the possibly undesired reputation of negative therapeutic motivation, be removed from the treatment setting instantly if the integrity of the group setting is to be maintained.

It is evident that many of the young alcoholics treated have experienced significant emotional deprivation in childhood. They appear to have developed few social skills, to be unduly sensitive to criticism and to be of limited stress tolerance. As one of the nursing staff succinctly expressed it they appear to have missed out, because of their seeking oblivion through drink, on essential maturing influences. An important approach to treatment with this group of patients therefore is that of education, not only in the narrow didactic sense of providing information about alcohol and alcoholism, but more importantly in understanding the treatment process as a learning situation in which skills related to the handling of interpersonal relations can be developed in a non-judgemental and supportive milieu. This is certainly most readily achieved in an inpatient setting with a fairly lengthy - four months - admission but it is certain that much can also be achieved with outpatient group counselling after a short admission to ensure initial detoxification and abstinence.

The essential approach to treatment of the young alcoholic is that former therapeutic pessimism should be discarded. The author believes that ultimately this group will do no better and no worse than others providing the treatment facilities are manipulated to satisfy their needs.

REFERENCES

DREWERY, J. (1974). "Social Drinking as a Therapeutic Goal in the Treatment of Alcoholism". Journal of Alcoholism, 9, No.2, 43.

EDWARDS, G., HENSMAN, C., HAWKER, A. and WILLIAMSON, V. (1967). "Alcoholics Anonymous the Anatomy of a Self Help Group". Social Psychiatry, 1. 195-204.

GWINNER, P.D.V. (1976). "The Treatment of Alcoholics in a Military Context". Journal of Alcoholism, 11, No.1, 24.

HASSALL, C. and FOULDS, G.A. (1968). "Hospital Among Young Alcoholics". British Journal of Psychiatry, 63, 203.

ORFORD, J. (1974). "Controlled Drinking in the Existing Behaviour Repertoires of Alcohol Dependent Man". Journal of Alcoholism, 9, No.2, 56.

ROSENBERG, C.M. (1969). "Young Alcoholics". British Journal of Psychiatry, 115, No.519, 181.

DETOXIFICATION - THE FIRST STEP

J. R. HAMILTON

UNIVERSITY DEPARTMENT OF PSYCHIATRY

ROYAL EDINBURGH HOSPITAL, EDINBURGH

There has been in recent years increasing realisation of the ineffectiveness of penal measures in dealing with public drunkenness offenders. A Home Office Working Party (1971) has recommended the establishment of detoxification centres and legislation has been enacted but not yet implemented (Section 91 of the Criminal Justice Act, 1967) to remove the penalty of imprisonment for those convicted of being drunk and disorderly. This paper describes the characteristics of drunken offenders and gives some of the results of the pilot detoxification centre in Edinburgh, the first step in decriminalisation of drunkenness.

The public has a stereotype of a drunken offender as being either a 'spree' drinker or a Skid Row alcoholic, but how many are vagrants, how many have homes, how many are first offenders, how many habituals, and how many are alcoholics ? By the term drunken offenders is meant those guilty of simple drunkenness or committing the offence of being drunk and disorderly. In Scotland the charges are usually for being drunk and incapable or committing a breach of the peace, but other offences can be included such as vagrancy, begging or being a nuisance. Many, perhaps most, of those who are drunk and incapable will not be apprehended by the police, but those who are alcoholics and those who are homeless have a probability that they will be arrested more often.

A group of 50 random drunken offenders were studied in Edinburgh Burgh Court, comprising 41 men and nine women in an age range of 17 to 75 with an average of 45 years. Only 20% were married, whilst 36% were single, 10% were widowed and 34% were living apart, separated or divorced. 26% lived in the Grassmarket area of Edinburgh and a further 14% in another socially deprived area, compared with 2% and

5% respectively of the Edinburgh population. 44% lived in a night shelter, common lodging house, Salvation Army hostel or had nowhere to live, including one woman who spent all her time travelling in trains without paying the fare apart from when she was arrested and put in prison. She said she had nowhere to live and she liked trains.

One third of the offenders were in social classes I, II and III, half as many as in the Edinburgh population, and half were unskilled labourers, five times that of the general population. Only a quarter had been in employment continuously in the last year. The offenders were asked a lot of questions about their drinking habits and on the basis of this and their symptoms of addiction to alcohol it was reckoned that certainly half, and more probably two-thirds, were alcoholics. This figure was the number of individuals with two or more symptoms of chemical dependence on alcohol - 70% had experienced amnesia, 56% shakes and 24% D.T.'s. One quarter drank every day and a similar number on most days of the week. They were asked why they were drinking when arrested and replies such as "drink every day", "celebrating birthday", "I'm a chronic alcoholic", "lost my pension book", "at football match", "depression", and "just a wee daft sort of notion I had", were given. The offenders seemed to be in three roughly equal groups, those who drank every day, those who were celebrating or at a party or football match, and those who mentioned "psychological" reasons such as tension or depression.

Only 20% were first offenders, whilst two-thirds had had three or more lifetime convictions for drunkenness offences. 42% had had three or more convictions in the last year and would thus meet the criteria used by the Home Office Working Party on Habitual Drunken Offenders for being an "habitual drunken offender". It is clear that the penal management of these people was quite ineffective in preventing further offences and in providing treatment for their alcoholism. Only 28% had ever had any treatment for alcoholism from any source.

Edwards et al (1974) found 37% of vagrant alcoholics in London were known to medical sources and less than one-third were in contact with psychiatric hospitals. He also found that the figures for resident male alcoholics were similar to the vagrant group with less than half known to a medical source. The characteristics described are in agreement with those found by Parr (1962), Ratcliff (1966), Edwards et al (1966) and Gath (1969) and if these figures are not convincing enough that the offenders are a group of individuals with a high degree of social and psychological pathology, it must be pointed out that about a dozen individuals known to be alcoholics and habitual offenders were excluded for purposes of statistical analysis and comparison from this survey. These men were those participating in a recently completed research project establishing a detoxification centre for alcoholic offenders and their characteristics are described elsewhere (Hamilton 1974). These hundred men were usually aged 40 - 60 and Scottish, though the Irish were over-represented. Most were single and of the others, most showed breakdown of marriage. The majority could be classed as homeless, though many men who had lived for many years in a common lodging house would dispute that this was not their home. By definition, all these men were alcoholics with a history of repeated convictions for drunkenness offences.

These men were randomly allocated to two groups, half of whom were given immunity from prosecution for one year and the police, instead brought them to the detoxification centre. The men could also refer themselves when intoxicated or showing alcohol withdrawal symptoms. The aim was to see if the management of the public drunken offenders could be transferred from the penal to medical and rehabilitative resources and this was successful. The number of convictions for drunkenness fell to almost zero in the 'detox' group and they also showed a large rise in the number of admissions to psychiatric hospitals for treatment for their alcoholism, and to rehabilitative hostels.

TABLE I : NUMBER OF COURT APPEARANCES FOR "DRUNK AND INCAPABLE"

	In the year before enrolment	In the year after enrolment	% change	
Probands: N = 47	189	20		$t = 4.124$; $df = 46$; $p < 0.001$
per 100 probands	402	43	- 89%	
Controls: N = 44	107	156		$t = 1.536$; $df = 43$; NS
per 100 controls	243	354	+ 46%	

$t = 4.127$; $df = 89$; $p < 0.001$

The establishment of this alternative system met many problems and hostile attitudes and the location was switched after one year from the Regional Poisoning Treatment Centre to a psychiatric hospital for these reasons. It was found that nurses and doctors with psychiatric skills were much more able to handle the men in a professional manner without letting their subjective feelings get the better of them. The full time workers on the detoxification project were a psychiatrist and a social worker and the author worked very closely with the Grassmarket Project in Edinburgh, a team receiving an Urban Aid Grant who ran a hostel (Thornybauk) specifically for male alcoholics.

85% (though seven had died) of the men were followed up after one year and it was found that the patients had improved compared with the control subjects in aspects of their life, such as health, accommodation and drinking habits.

TABLE II : NUMBER OF DAYS IN TREATMENT
Rehabilitative Hostel + Psychiatric Hospitals

	In the year before enrolment	In the year after enrolment	% change	
Probands: N = 47 per 100 probands	142 302	1620 3447	 + 1041%	t = 3.5023 df = 46 p < 0.005
Controls: N = 44 per 100 controls	500 1136	269 611	 - 46%	t = 0.6999 df = 43 NS

$t = 3.1146$; $df = 89$; $p < 0.005$

TABLE III : CHANGE IN ACCOMMODATION DURING YEAR OF ENROLMENT
TYPE OF PREMISES

	Probands (N = 42) %	Controls (N = 36) %
Improved	38	11
Same	48	67
Worse	14	22

$X^2 = 10.2$; $df = 2$; $p < 0.01$

TABLE IV : QUALITY OF LIFE DURING YEAR AFTER ENROLMENT

	Probands (N = 42) %	Controls (N = 36) %
Better	52	28
Same	17	42
Worse	26	25
Other	0	6
(Not known)	(5)	(0)

$X^2 = 11.2$; df = 2; $p < 0.01$

The patients also considered that the quality of their lives had improved to a significant extent.

Because of the high degree of medical morbidity encountered hospital-based detoxification centres are specifically recommended though it is realistic to acknowledge that other centres will be sited according to circumstances such as finance and the availability of enthusiastic staff. It is fully recognised that hostels have an essential place in the process of rehabilitation following detoxification and perhaps the traffic from detoxification centres to hostels should not be considered 'one way' but more of a 'roundabout', for relapses among some hostel residents are inevitable. As alcoholism is often a chronic condition, characterised by relapses and remissions, 'success' in treatment may be construed as establishment of an effective system of management rather than an outcome measured only in terms of sobriety.

Detoxification centres must be seen to be units providing primary care of drunkenness with management of withdrawal symptoms, assessment and referral for rehabilitation. Detoxification is but the first step.

REFERENCES

EDWARDS, G., KYLE, E., NICHOLLS, P. (1974). "Alcoholics Admitted to Four Hospitals in England: Social Class and the interaction of alcoholics with the treatment system". Quarterly Journal of Studies on Alcohol, 35, 499-522.

EDWARDS, G., WILLIAMSON, V., HAWKER, K., HENSMAN, C. (1966). "London's Skid Row", Lancet, 1, 249-252.

GATH, D. (1969). "The Male Drunk in Court". The Drunkenness Offence Cook, T. (ed.) 9-26, Pergamon Press, Oxford.

HAMILTON, J.R. (1974). "Detoxification of Habitual Drunken Offenders - A first report". 20th International Institute on the Prevention and Treatment of Alcoholism, International Council on Alcohol and Addictions, Lausanne.

HOME OFFICE (1971), "Habitual Drunken Offenders" Report of the Working Party, H.M.S.O., London.

PARR, C. (1962) "Offences of Drunkenness in the London Area: A pilot study". British Journal of Criminology, 2, 272-277.

RATCLIFF, R.A.W. (1966) "Characteristics of Those Imprisoned in Scotland in 1965 on Convictions for Primarily Alcoholic Offences". Health Bulletin, 24, 68-70.

THE TREATMENT OF DRUG DEPENDENCE - A TAXONOMY OF APPROACHES

IAN HINDMARCH

DEPARTMENT OF PSYCHOLOGY

UNIVERSITY OF LEEDS

"When they haven't their own axes to grind, they've got their theories; a theory's a dangerous thing. Now my theory is . . . "

John Galsworthy

Each of the professional disciplines concerned with drug dependency has its unique approach to treatment. The focus on a particular treatment modality is in accord with the theoretical explanation adopted by that profession to explain the aetiology of drug dependency. It can be argued that such a specific concentration on limited aspects of the complex area of drug abuse leads to a 'blinkered' approach to the treatment of drug dependence. Several authorities (Harms 1964) have argued for a multidisciplinary approach to the treatment of drug dependence, but in practice this entails a collection of professions each with their own theories and approaches to treatment with little integration of commonly held principles. The net result of a multidisciplinary approach is a fragmentation of effort and it is proposed to argue that the most effective approach to treatment is interdisciplinary.

First, some of the approaches to treatment adopted by the different disciplines with an intimate concern in the treatment of drug dependency will be reviewed. The list is not exhaustive nor are the distinctions made in any way mutually exclusive since some formal disciplines subsume others. It is also outside the scope of this paper to examine and evaluate specific therapies. The merit, or lack of merit, of behaviour, aversion, drug and group therapies are beyond this presentation but they are not to be discounted or seen as the exclusive preserve of a particular discipline
in discussing the practical approach to an integrative treatment of

drug dependence. Since the approach to treatment is determined by the theoretical tenets of the adopted aetiology then it is impossible to omit a discussion of both the putative cause of drug dependency and the resulting approach to treatment in reviewing the various treatment models.

THE MEDICAL APPROACH

The physician or psychiatrist holds the assumption that all drugs are toxic and it is the drug's toxicity in the host organism that produces the problems associated with drug abuse. Classic psychiatric texts (Henderson and Gillespie 1964) have classified alcoholism and narcotic abuse under organic psychoses: firmly asserting that the chronic exogenous administration of toxins is the prime cause of the disorder. Drugs of dependence are characterised :

(a) by producing a developing tolerance in individuals taking them; tolerance is a state of progressively decreasing responsiveness to the drug whereby the tolerant individual needs an increasingly larger dose; and

(b) by producing a state of dependence, as manifest in individuals exhibiting withdrawal symptoms upon cessation of drug administration. These withdrawal symptoms can usually be terminated by readministration of the drug.

The medical approach to treatment thus embodies these two aspects viz. withdrawal of the toxin i.e. detoxification and administration of a drug to allay the effects of toxin withdrawal or abstinence.

Traditionally there are four approaches to withdrawal (Nyswander 1956) :

1. Total abstinence and total withdrawal of all drugs, the so called 'cold turkey'.
2. Total withdrawal of all the addictive drugs, but some medication for treatment of abstinence effects e.g. night time sedation for insomnia.
3. A gradual reduction in the amount of addictive substance used i.e. a reducing maintenance therapy.
4. Substitution of the addictive drug by another, non-addictive drug.

However, "withdrawal and detoxification" as an approach by itself has certain drawbacks. The addict usually has a total lack of motivation to abstain since he feels no anxiety about his drug use and is probably well supported by a sub-cultural group of friends that do not see his use of drugs as a problem that should give rise to concern. Total withdrawal and abstinence is often envisaged as an end in itself i.e. as a 'cure', but this is a difficult position to maintain if the only alternative to drug dependence that is provided is no drug dependence. Cohen (1972) makes this most crucial point in a short interview sequence from his book 'Journey beyond Trips' :

Psychiatrist:	Why do you use drugs ?
Addict:	Why not !
Psychiatrist:	How can I convince you to stop ?
Addict:	By showing me something better.

Many investigators (Alknse 1959, Kooyman 1975, Hunt and Odoroff 1962) have shown that relapse rates following 'withdrawal and detoxification' without any form of rehabilitation are too high to consider withdrawal per se as a fruitful approach to treatment.

History also illustrates the futility of substitutive withdrawal, alone, as an approach to treatment. The American Civil War produced the first generation of morphine addicts as a result of excessive use of the drug as a pain killer in field surgeries. The approach to this 'problem' was to produce a morphine substitute by 'diacetylizing' the drug and then administering the substitute to wean the addict off his dependency producing drug. The history of diacetyl morphine, i.e. heroin, is also well known and the attempts to substitute physeptone or methadone for heroin are proving that history has a habit of repeating itself. Even the administration of minor tranquillizers to help allay the withdrawal symptoms are simply a substitution of one psychoactive substance with dependency producing properties by another with equally potent addictive tendencies.

THE ADMINISTRATIVE APPROACH

Administrators in government and health authorities accept the multidisciplinary approach as a necessary condition for the effective treatment of drug dependency. The following figures illustrate possible approaches (Erie County Department of Mental Health 1970, Louis 1971) to the treatment of drug dependency.

The effective treatment of drug addicts is naturally dependent upon economic and logistic resources being made available to the psychiatrists, social workers, psychologists etc. involved in the implementation of treatment. The task of the administrator is to organise the appropriate dispersal of funds and physical amenities

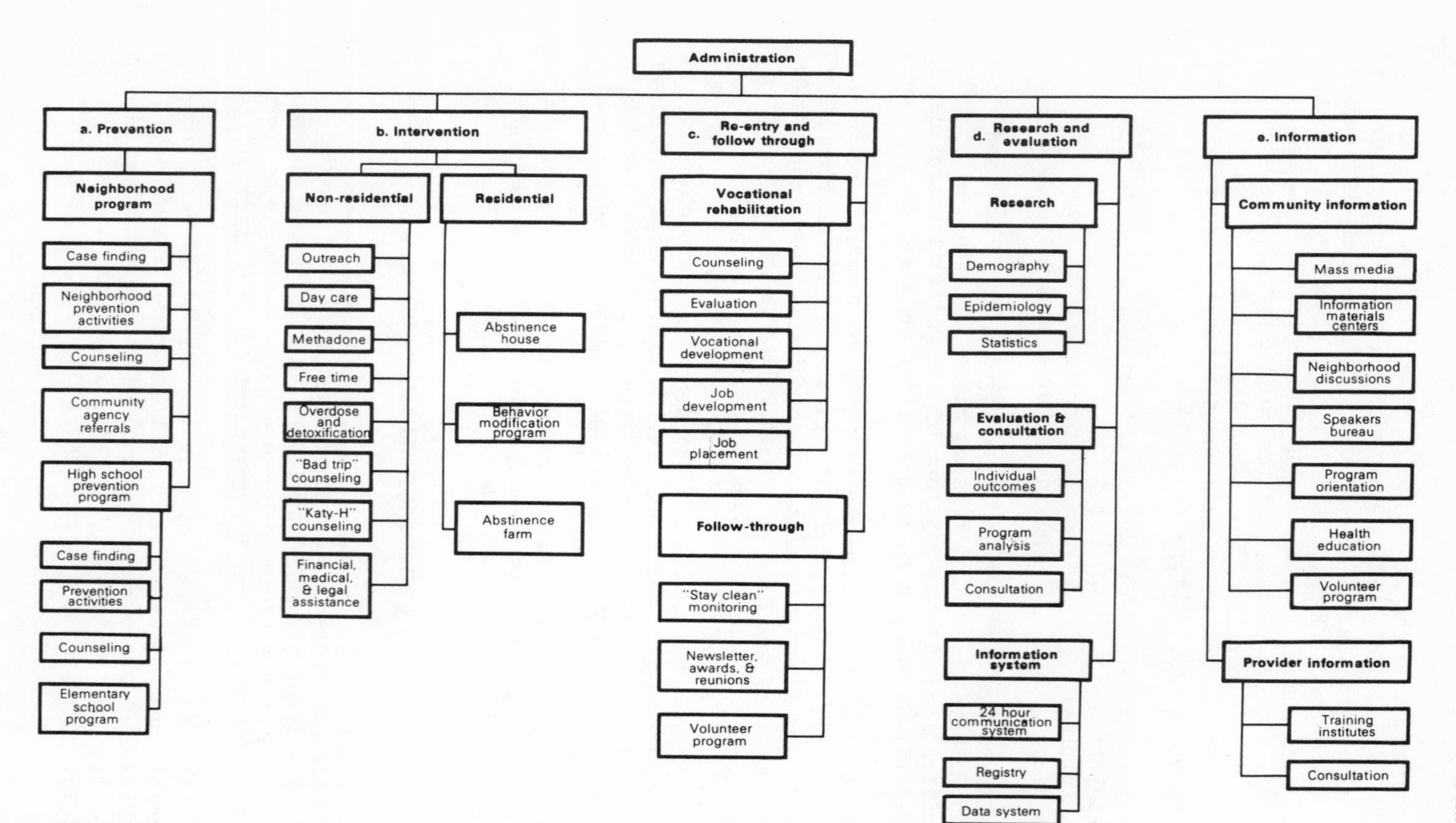

FIGURE 1. A Model for the Development of Services to Control and Treat Drug Dependency Within the Community.

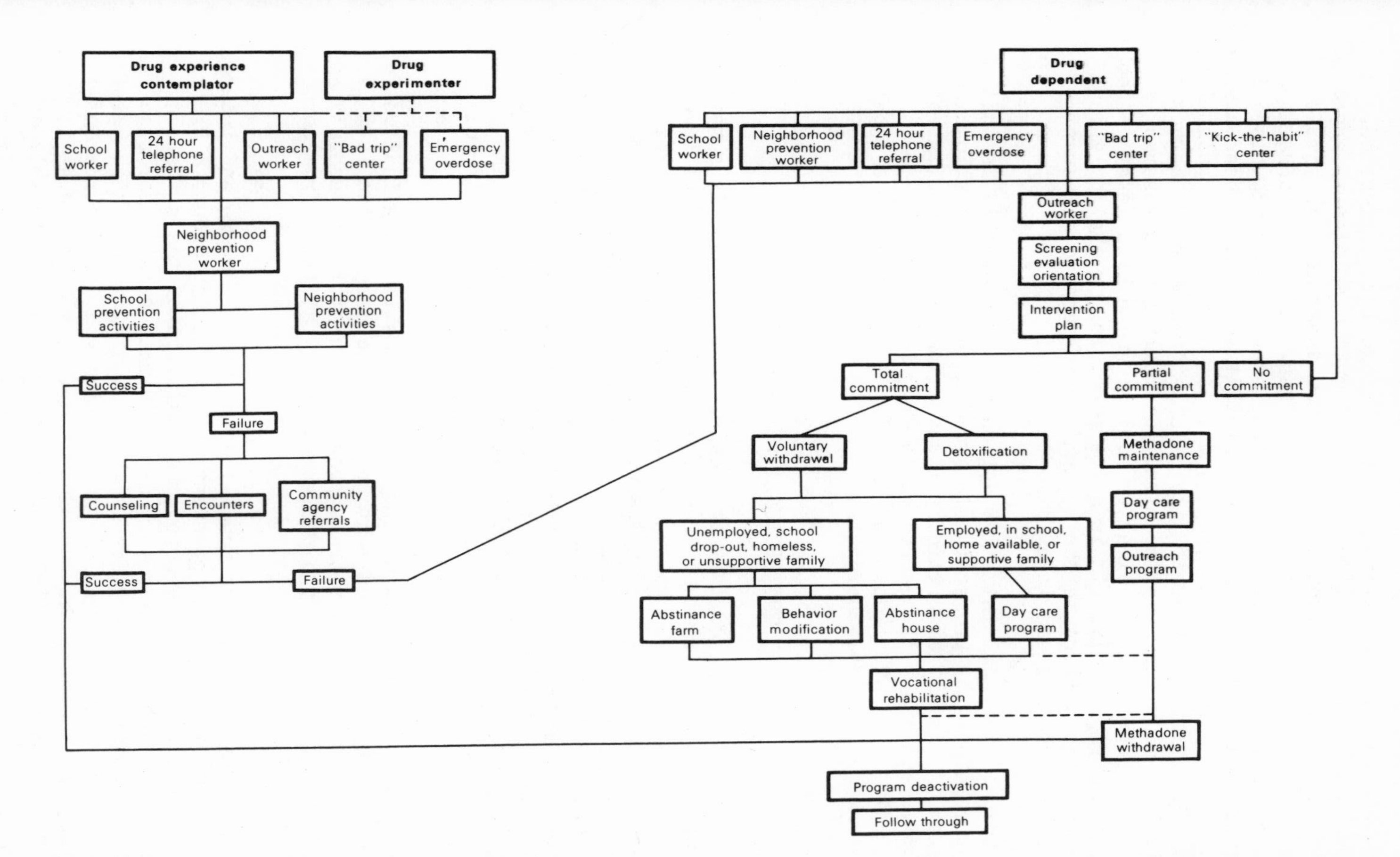

FIGURE 2. A Flow Diagram for Three Examplar Drugs Users (the Contemplator, Experimenter, and Dependent) Through the Community Based Services Available for Treatment

to ensure the maximal utilisation of the facilities provided. However, most administrators accept the multidisciplinary approach to treatment, as is evidenced in the flow diagrams with little or no regard to an integration of disciplines. The 'flow diagram approach' keeps the various professions separate and organised into a hierarchy with the clinic or hospital at the apex. These channels of communication open only above and below a particular point in the chain without any possibility of a lateral movement across the disciplines. Furthermore, such multidisciplinary approaches require widespread administrative control and a high level of book-keeping and paper work. It is quite possible for the individual addict to become 'lost in the system' or subject to treatment by bureaucracy and mislaid in a forgotten file.

Some of the administrative approaches are sensible in that they do not separate the professionals from one another and allow the integration of multidisciplinary ideas. The other essential of an effective administrative approach to drug dependence is to include and involve the educational and training media within the total framework of available services.

THE LEGISLATIVE APPROACH

It could be argued that the legislative approach is a means of prevention not treatment of drug dependency. The present level of legislation (Pharmacy & Poisons Act 1933, Medicines Act 1968, Misuse of Drugs Act 1971) in the U.K. is, according to some (Bradshaw 1972) only a result of successive attempts to control the increasing abuse of pharmaceuticals and other non-medical substances. However, the presence of comprehensive laws relating to individuals convicted of offences against one of the various drug control acts has important implications for treatment. The provisions included in the Dangerous Drugs Legislation include those for the compulsory treatment of drug addicts.

Compulsory treatment usually implies institutionalisation in a prison, hospital or treatment centre. The drug addict is thus viewed as someone who has transgressed the statutes of society and his treatment must embody a punishment i.e. isolation from normal society.

For some time international bodies with a concern and interest in the treatment of drug dependents (U.N. Working Groups & Expert Committees 1972) have advocated the abolition of the so called medico-legal approach to treatment. The main reasons for such bodies expressing these views is the vast difference which exists between the environment in the institution and the environment which is pertinent to the everyday experiences of those at large in society. Most importantly the patient, or convict, is able only to play the role of patient or convict. His behaviour if and when it changes is reinforced by staff and not by his fellow peers as would be the case in society. There is little opportunity for developing adequate staff/patient relationships since such 'fraternisation' is discouraged. In short the institutional environment does not allow for the development of skills and behaviours necessary for adjustment to the real world.

The institutional approach will only become effective when realisation is made of the importance of group processes (Zucker 1961) and the worth of the individual is emphasized. In permitting staff/patient relationships to develop so the ultimate transition from institution to society will become smoother and there is more likely to be a lower relapse rate since the individual has been allowed to adjust and cope with his behaviour in a realistic and life-like manner.

THE SOCIOLOGICAL APPROACH

Sociologists would argue the importance of sociological factors (Plant 1975) in arriving at the aetiology of drug dependence and would strongly favour an approach to treatment which involves changing the social structure (Young 1971) of the society which 'produces' the drug using behaviour. At a more microscopic level the pragmatic approach of social workers bears examination. They would argue that drug dependency is but a facet of a total life style which naturally relates to the particular socio-economic and sub-cultural group to which an individual belongs. Since the genesis of drug dependent behaviour is due mainly to a breakdown of interpersonal relationships within groups or an alienation (Young 1971) from familial influences then the approach to treatment must involve a restoration of the feelings of personal worth and reduction of anomie via group centred therapy. Although a specific therapy and, therefore, outside the scope of this paper there are implications of the group centred approach to treatment which are relevant to this present discussion.

The aim of a group centred approach is to utilise the forces of peer group coercion, which were instrumental in producing the drug using behaviour, to help the individual gain insight into his own condition and to provide him with an interpersonal framework of relationships upon which to base his social behaviours. The group approach is invaluable since drug dependents, particularly the adolescent ones, have a history of gang membership (Einstein and Jones, 1964). They are particularly influenced by their interaction with their peers and since they are members of a definite sub-culture they possess the argot and jargonese which helps them to relate and identify with others. Furthermore, the necessary verbal skills for interaction in a one to one therapeutic relationship are not always present in drug dependents who prefer motor to verbal responses (Yablonsky 1959). An interesting practical application of the group approach has been in the 'seeding' of treatment groups with non-addict (Patrick 1964) adolescents in the hope that the non-addicts will help the addicts resolve their feelings of depersonalisation and alienation.

THE SYMPATHETIC APPROACH

This approach is adopted not by a clearly defined discipline or profession but rather by a range of helping organisations including the Salvation Army, Alcoholics Anonymous, Helping Hands and "telephone therapy" groups. The oldest aetiological explanation for drug addiction attributed the cause to a lack of moral fibre resulting in a voluntary development of abherrent tastes i.e. drug misuse. The common feature of all sympathetic approaches is the exhortation of the addict to 'kick the habit'. The first therapists using this modus operandi were the temperance movements of Victorian England and their subtle mixture of song, religion and group support is still effective. The thirteen steps (Patrick 1964) of the Narcotics Anonymous given below, illustrates the important ingredients of the sympathetic approach.

The Thirteen Steps

1. Admit the use of narcotics made my life seem more tolerable, but the drug had become an undesirable power over my life.

2. Come to realise that to face life without drugs I must develop an inner strength.

3. Make a decision to face the suffering of withdrawal.

4. Learn to accept my fears without drugs.

5. Find someone who has progressed this far and who is able to assist me.

6. Admit to him the nature and depth of my addiction.

7. Realise the seriousness of my shortcomings as I know them and accept responsibility of facing them.

8. Admit before a group of N. A. members these same shortcomings and explain how I am trying to overcome them.

9. List for my own understanding all the persons I have hurt.

10. Take a daily inventory of my actions and admit to myself those which are contrary to good conscience.

11. Realise that to maintain freedom from drugs I must share with others the experience from which I have benefited.

12. Determine a purpose in life and try with all the spiritual and physical power within me to move towards its fulfillment.

13. God Help Me! These three words summarise the entire spirit of the 12 preceding steps. Without God I am lost. To find myself I must submit to Him as the source of my hope and my strength.

In the N.A. approach to treatment the addict is first withdrawn from 'temptation' into the security of a group situation which provides not only physical but also psychological support. It is inherent in Maslow's (1962) notions of the development of self actualisation that personal growth cannot begin until more basic needs of security and physical well being are established. The emphasis of many 'community centred' approaches to detoxification (Healey 1976) exemplify the need to satisfy comfort and nutritional requirements of the individual drug dependent before attempting to change his drug using behaviour.

However, this approach is not simply 'withdrawal and detoxification' in a sympathetic milieu - it also provides an alternative to drug use, as in step 13 above. The provision of an alternative to drug use is an aspect of treatment spoken about later.

THE PSYCHOLOGICAL APPROACH

It is not without a certain prejudice that the psychological approach is presented. Previous papers (McKechnie 1976; Lowe 1976) have emphasized the importance of an individual's motives and attitudes as causative agents of drug misuse. We have also seen that one of the hurdles to effective drug withdrawal therapy is the contrary motives of many adolescent addicts.

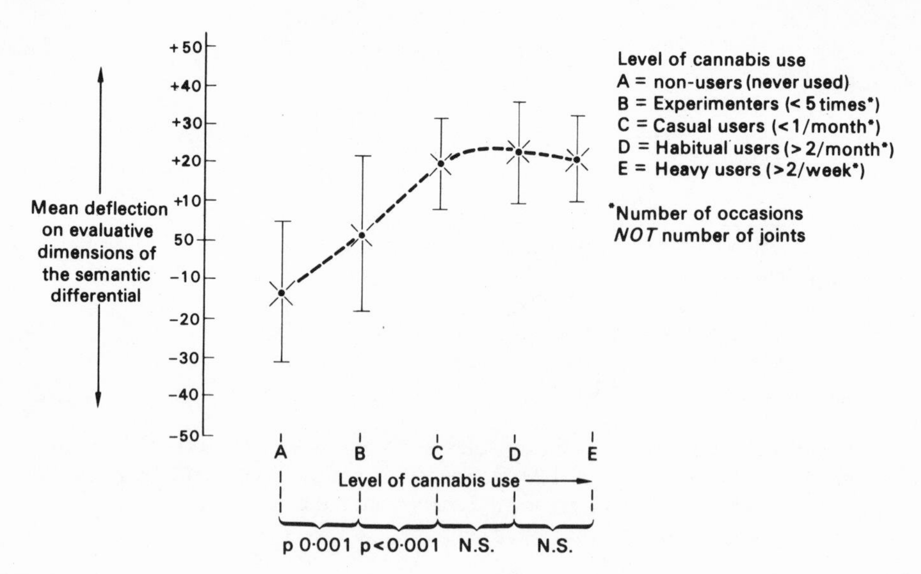

FIGURE 3. Levels of Cannabis Use and Mean Scores on Evaluative Dimensions of the Semantic Differential

As one would expect the more involved in drug using an individual becomes the stronger and more positive his attitudinal appraisal will become. Fig. 3 illustrates this for cannabis users (Hindmarch 1975) and shows that as the behavioural involvement with a drug increases (in this case a greater frequency of cannabis use) so the attitudinal component (as measured on the evaluative dimension of the semantic differential) intensifies. Although the present example relates to cannabis use it is suggested that similar findings would be found for any drug using behaviour and recent work (Martino and Truss 1973) has shown, on a variety of measures, that attitudes to drugs are positively related to the actual level of drug use and drug involvement. Also of importance, Fig.3 is the significant difference ($p < 0.001$) between "experimenters" and more frequent users of cannabis in the attitudes they hold towards cannabis. These differences tend to contradict notions of escalation of drug use based on developing pharmacological tolerance since, before an individual can progress towards a persistent chronic use of the drug he must overcome an "attitudinal barrier". A similar "attitudinal barrier" exists between the"non-user" and "experimenter". Attitudes are regarded as a product of the individual's personality but they also embrace the results of past, present and even future or anticipated behaviour. Such important processes must not be forgotten when speaking of an approach to the treatment of drug dependency, for at the hub of the treatment programme is an individual whose response to treatment is greatly determined by the stereotypes, beliefs, prejudices, motives and attitudes intrinsic to his cognitive system.

There is a basic tendency in human organisms to maintain a consistent view of themselves and their behaviour in a social environment. In some situations the internal belief (attitude) and the external (behavioural) state may be in conflict so generating dissonance (Festinger 1957). Dissonance produces tension within the cognitive system and so to lessen this tension an individual will restructure either his behaviour or his internal beliefs to restore equilibrium and consistency (Doob 1947; Campbell 1963; Hindmarch 1975).

This approach to the relationship between internal motives and beliefs and external behaviour in this instance drug use, gives rise to two important corollaries in approaching the treatment of drug dependence.

The first is the recognition that all behaviour is motivated and, therefore, all drug dependent behaviour is motivated.

The second is that a change in attitude will result in a change in behaviour.

THE INTERDISCIPLINARY APPROACH

Some of the approaches to treatment adopted by different disciplines connected with the treatment and rehabilitation of drug dependents have been reviewed. A multidisciplinary view has been taken, but individual approaches can be somewhat sterile if practiced in isolation, and, therefore, an integrative interdisciplinary approach is proposed.

The 'problem' has two aspects, the first is to find an alternative to drug use, as Cohen's addict said 'show me something better'. If it is accepted that drug dependent behaviour is motivated then the motives can be isolated and with a little imagination alternatives provided. These alternatives are not the exclusive domain of any one profession but require co-operation between physician, sociologist and psychologist and the logistical support of the administrators for their implementation. This is not a new concept and when it has been practised previously as in the Synanon style community a fair degree of success has been claimed.

Table I illustrates some of the possible 'motivators' together with alternative approaches to treatment.

The second aspect of the 'problem' of treatment relates to attitude change and how to accomplish this with drug dependents. By far the most important tools by which to change attitudes are education and information. Just as it is impossible to

TABLE I

Possible Motivators	Putative Needs	Alternative to Drug Use
PHYSICAL	Need for energy and activity	Athletics, hiking, joining Army, dancing.
SENSORY	Arousal and stimulation	Bio-feedback, saunas, body beautiful, sky diving.
EMOTIONAL	Relief from anxiety	Sensitivity training, encounter weekends.
INTERPERSONAL	Freedom from 'pain' of Psychological hangups	Computer dating, pen friends, marriage.
SOCIOCULTURAL	To promote social change	Help the aged, V.S.W., Friends of the Earth.
POLITICAL	To identify with anti-establishment.	Underground Press, non-party lobbying.
INTELLECTUAL	Cognitive Awareness	Chess, bridge, debating societies.
CREATIVE	Enjoyment of Imagery	Theatrical productions, amateur painting.
PHILOSOPHICAL	Meaning of Life, the world + death	Literature, Yoga, religious debate, evening classes.
SPIRITUAL	Transcendence, visions, insight	Orthodox and non-orthodox religion, 'Divine Light'.
"KICKS"	General need for risk, excitement, drama and sharpening of experience	Gocart racing, outwardbound, freefall parachuting, clay pigeon shooting, and A MEANINGFUL EMPLOYMENT.

Figure 4: Some of the possible motivators and needs of a drug dependent together with tentative suggestions as to how these may be met via alternatives not involving drugs.

consider approaches to treatment without recourse to the underlying aetiological theory it is as impossible to conceive of an approach to treatment that can succeed without due attention to the role of, and necessity for, education.

In short, the approach to treatment advocated would be to utilise all the skills of the various disciplines concerned with dependence, not as many different professionals each with their own "axes to grind and theories to develop", but as an interdisciplinary team approaching not so much drug dependence in isolation from other behaviour phenomenon but approaching treatment as an attempt to improve the quality of life and provide an effective satisfaying alternative to drug abuse.

REFERENCES

ALKSNE, H. (1959). "A Follow Up of Treated Adolescent Narcotic Users". Health Resources Board, Mimeograph.

BRADSHAW, S. (1972). Drug Misuse and the Law, MacMillan, Basingstoke.

CAMPBELL, D.T. (1963). "Social Attitudes and Other Acquired Behavioural Dispositions in Psychology". A Study of Science, 6, S. Koch (ed.) McGraw-Hill, New York.

COHEN, A.Y. (1972). "A Journey Beyond Trips". Heroin in Perspective. 186, Prentice-Hall, New Jersey.

DOOB, L. (1947). "Behaviour and Attitudes". Psychology Review, 54 135-56.

DRUGS ACT, MISUSE OF (1971).

EINSTEIN, S. and JONES, F. (1964). "Group Therapy with Adolescent Addicts". Drug Addiction in Youth, E. Harms (ed.), Pergamon, London.

ERIE, COUNTY DEPARTMENT OF MENTAL HEALTH (1970).

FESTINGER, L. (1957). Conflict and Cognitive Dissonance, Stanford University Press, Stanford.

HARMS, E. (1964). Drug Addiction in Youth, Pergamon, London.

HEALY, P. (1976). Third International Conference on Alcoholism and Drug Dependence, Liverpool

HENDERSON, D.K. and GILLESPIE, R.D. (1964). A Textbook of Psychiatry, 6th Edition.

HINDMARCH, I. (1972). "Patterns of Drug Abuse among School Children". Bulletin on Narcotics XXIV, 3, 23-26.

HINDMARCH, I. (1975). "The Psychology of Cannabis". Cannabis and Man, Connell and Dorn, Churchill-Livingston , London.

HINDMARCH, I., HUGHES and EINSTEIN (1975). Bulletin on Narcotics XXVII, 1, 27-36.

HUNT, G.H. and ODOROFF, M.E. (1962). "Follow up Study of Narcotic Drug Addicts after Hospitalisation". Public Health Reports, 77, 1. 41-54.

JULIEN, R.M. (1975). A Primer of Drug Action, Freeman, San Francisco.

KOOYMAN, M. (1975). "From Chaos to a Structured Therapeutic Community". Bulletin on Narcotics, XXVII, 1, 19-26.

LOUIS, N.B. (1971). "An Application of a General Model to the Development of an Operational Drug Plan in a Community". Second International Institute on the Prevention and Treatment of Drug Dependence. Baden.

LOWE, G. (1976). "Psychology of Alcohol". Third International Conference on Alcoholism and Drug Dependence, Liverpool.

McKECHNIE, R.J. (1976). "How Important is Alcohol in Alcoholism"? Third International Conference on Alcoholism and Drug Dependence, Liverpool.

MARTINO and TRUSS (1973). Journal on Counselling Psychology, 20, 2, 120-126.

MASLOW, A.H. (1962). Towards a Psychology of Being, Van Nostrand, New Jersey.

MEDICINES ACT (1968).

NYSWANDER, M. (1956). The Drug Addict as a Patient, Grune and Stratton, New York.

PATRICK, S.W. (1964). "Our Way of Life - A Short History of Narcotics Anonymous Inc.", Mimeograph.

PHARMACY & POISONS ACT (1933).

PLANT, M. (1975). Drug Takers in an English Town, Tavistock, London.

TENNANT, F.S. (1973). British Journal of Addiction, 68, 327-380.

U.N. WORKING GROUPS AND EXPERT COMMITTEES (1972). "Drugs in Modern Society : Community Reactions to Drug Use by Young People". United Nations, Geneva.

WEIL, A.T. (1972). Dealing with Drug Abuse, Praeger, New York.

YABLONSKY, L. (1959). "Group Psychotherapy and Psychodrama for Drug Addicts". Journal of the Probation and Parole Association, 5, 63-70.

YOUNG, J. (1971). The Drug Takers, McGibbon and Kee, London.

ZUCKER, A.H. (1961). "Group Psychotherapy and the Nature of Drug Addiction". International Journal on Group Psychotherapy, 11, 209-218.

AIMS OF TREATMENT

B. D. HORE

REGIONAL ALCOHOLISM UNIT

MANCHESTER

Traditionally, the aims of treatment of the alcoholic have been designed to produce total abstinence; it being felt that this will be accompanied by an improvement in other areas of life. This aim remains that of the majority of treatment agencies in the U.K. Recently, this view has been challenged and the possibility of aiming for a different pattern; that of "normal", "controlled" or social drinking has been suggested. Further, doubt has been expressed as to whether total abstinence does produce improvement in other areas, e.g. Gerrard et al (1962); also it is clear that on a wider issue that different treatment facilities in the same area, (Pattison 1973) may have different aims. It is the aim of this brief review to examine some strands in this continuing debate.

During the late 1930's and early 1940's, two major areas of treatment developed, i.e. aversion treatment and that of Alcoholics Anonymous. Both aimed for total abstinence but the latter aimed for more. Its aims appeared to be a total reform of the individual including moral and spiritual aspects. Alcoholics Anonymous came to the United Kingdom in 1948 (London) and has had a major influence on treatment philosophies. Indeed the fact that Alcoholics Anonymous had shown something could be done if alcoholics were grouped together, rather than left languishing on the wards of mental hospitals, stimulated the development of alcoholic treatment units, (Glatt 1961). This influence has persisted, and many such units today work intimately with Alcoholics Anonymous, and demand total abstinence as a major treatment criteria. Perhaps the one major difference is that such units also examine reasons behind drinking, e.g. anxiety neuroses, sexual conflicts, etc. and offer additional treatment for their conditions.

Alcoholics Anonymous appears to neglect this, being primarily concerned in encouraging identification amongst alcoholics irrespective of reasons which led to their drinking excessively in the first place.

Davies (1962) in a carefully prepared series of case histories showed that some alcohol addicts were able to return to a more normal pattern of drinking. This conflicted with the concept of Alcoholics Anonymous and also that of Jellinek (1960) (himself basing his data on Alcoholics Anonymous members) which suggested once physical addiction had occurred, drinking would rapidly return (because of the state of physical addiction existing in the person). Major doubts have been experienced (Keller 1972), on the validity of the immediate "loss of control" concept and as to whether it has a physical basis. Keller stresses that what differentiates the alcohol addict from the non addict is the inability to consistently control drinking, over prolonged periods, and in clinical practice this appears to fit better the observed facts. It is further suggested that "loss of control" can be explained in psychological terms. Glatt (1973) accepting these limitations still insists that most addicts will continue to drink in an uncontrolled way and that controlled drinking is most unlikely. What is being debated is not the fact that some people with physical addiction can return to some form of controlled (rarely normal) drinking but whether this is a frequent occurrence or a rarity. It is this pragmatic point which it is essential to clarify. Not all patients with typhoid or even with some malignancies will succumb without treatment, but as the majority do, we advise treatment. If the majority of physical addicts are not able to drink in a controlled manner, then we must be careful in our advice.

Experiments in "controlled" drinking have varied from complex procedures using highly skilled personnel based on aversion techniques to more simple attempts to train people in their manner of drinking. No one doubts the value of these techniques in an experimental manner, but it remains to be shown that adequate samples of physically addicted subjects followed up for a sufficiently long period (at least two years) can drink in a controlled manner throughout the period, and if so how they can be differentiated from those who can't. Patients who are not physically addicted fall into a different category and increasingly such patients are being seen at our community based centres. Some may not ever be psychologically dependent and educational techniques naming the danger of using alcohol "as a crutch" and aiding self awareness may enable drinking to continue.

Let us now examine two other areas already mentioned. Gerrard et al described in a follow up study how neurotic traits were evident in the abstinent alcoholic. This would seem hardly surprising, e.g. the alcoholic who drank to relieve tension may be more tense on abstinence. This does not seem an argument against advising a person to become abstinent. Abstinence can be sought and then application of non-pharmacological treatments to relieve reasons given. It is the corollary of the conclusion that seems important, i.e. continuous drinking is liable to cause difficulties and abstinence at least social improvement, (Emrick 1974). It is also forgotten that on a pragmatic level there is also a time scale. In clinical hospital practice when the alcoholic is first seen, he is often on a knife edge. For example, his marriage or his job is in the balance. The spouse or employer is not prepared to let him experiment for a further two years. If further difficulties arise, there will be no spouse or job left. Again therefore, we have to be careful regarding our advice. Some of the differences reported by different workers in their attitudes to treatment may reflect the type of client they see and the stage of dependency their clients are at (Pattison 1973), explained in one town how treatment aims could range from an attempt to understand in analytical terms conflicts behind drinking (private psychologically orientated clinics for alcoholics) to keeping people out of society's way (police farm). A current debate in our half-way house programmes is, should the aim be primarily "shelter" or "change" ? In terms of Skid Row hostels, it would be that the number "changing" is not likely to be high if by that is meant an independent socially useful life outside whether totally abstinent or not.

Treatment of alcoholics in any large numbers using specialised techniques is very recent in this country (apart from Alcoholics Anonymous). Many doctors still require convincing that alcohol addicts can reach sobriety in any way and the prognosis is regarded as hopeless. Much work will have to be undertaken before this occurs and that alcoholics deserve any sympathy. This fact raises anxiety about talking too loudly in relation to controlled drinking as people with little knowledge of problems involved in treatment believe this to be a well and successfully established aim of treatment. Certainly in the United Kingdom at present, only a small minority of treatment agencies (whether half-way houses, out-patient clinics, in-patient units) have "controlled" drinking as their prime treatment, the vast majority do not. This, of course, does not mean the latter are correct, but it will require convincing data to be produced to show in practical terms in relation to the physically addicted subject that the experience of the last thirty years should be abandoned and treatment aims altered. The onus is on the innovators to prove their case.

REFERENCES

DAVIES, D.L. (1962). "Normal Drinking in Recovered Alcohol Addicts" Quarterly Journal Studies on Alcohol, 23, 94.

EMRICK, C.D. (1974). "A Review of Psychologically Orientated Treatments of Alcoholism. I. The Use and Inter-relationships of Outcome Criteria and Drinking Behaviour following Treatment". Quarterly Journal Studies on Alcohol, 35, 523.

GERRARD, D.L., SANGER, G. and WILEY, R. (1962). "The Abstinent Alcoholic". Archives of General Psychiatry, 6, 83.

GLATT, M.M. (1961). "The Treatment Results in an English Mental Hospital". Acta Psychiatrica Scandinavica, 37, 143.

GLATT, M.M. (1973). "The Hazy Borderline of the Loss of Control". Presented at 30th International Conference on Alcoholism and Drug Dependence, Amsterdam.

JELLINEK, E.M. (1960). "The Disease Concept of Alcoholism". Hillhouse Press.

KELLER, M. (1972). "On the Loss of Control Phenomenon in Alcoholism" British Journal of Addiction, 67, 153.

PATTISON, E.M. (1973) "Drinking Outcomes of Alcoholism, Treatment, Abstinence, Social, Modified, Controlled and Normal Drinking in Alcoholism". A Medical Profile - B.Edsall & Co., London.

CONTROLLED DRINKING IN THE ALCOHOLIC

A SEARCH FOR COMMON FEATURES

TOBY LEVINSON

CLARKE INSTITUTE OF PSYCHIATRY

TORONTO

Total abstinence versus controlled drinking as a treatment goal for the alcoholic is a controversial issue which is currently receiving a great deal of attention. Two opposing camps seem to have evolved over the question and each puts forward convincing arguments essentially representing divergent views on "what is and makes an alcoholic".

There seems little doubt that everyone working in the field recognizes the complexity of the problem and agrees that alcoholism is multifaceted in its symptomatology and etiology. The more traditional group representing many reputable alcohologists and certainly all recovered alcoholics stresses the physiological component in regarding alcoholism as a progressive irreversible disease in which biochemical changes cause an alcohol allergy such that a single drink by the dry alcoholic inevitably retriggers loss-of-control drinking. The opposing view held largely by an ever growing group of behavioural psychologists places greater stress on the psychological factors involved and believes that the alcoholic is made not born - his body is not biologically different. Behaviour is learned and this includes bad drinking habits which can also be unlearned.

The abstinence versus control issue is not altogether a new one. Over the years researchers who have undertaken follow-up studies in their attempts to evaluate various treatment programmes in which abstinence was the only treatment goal, have frequently reported small but significant percentages of subjects who were drinking "normally" or at least not "alcoholically".

Dr. D.L. Davies of the Maudsley Hospital in London was perhaps among the first to focus his attention on this group. In an article published in 1962 Dr. Davies reported that out of 93 patients treated for alcoholism by what he describes as a general measure including disulfiram, discussions and social help during a hospital stay of two to five months, seven were found on follow-up to have been drinking socially for seven to 11 years after discharge. None of them had been drunk in that time and all were socially better adjusted than they had been for the year prior to admission. While he could find nothing distinctive about this group with regard to sex, age, length of hospital stay and length and type of drinking history, he did note that significant social changes e.g. occupational and marital did seem to play a part. But he concluded that such cases are more common than have hitherto been recognized and that the generally accepted view that no alcohol addict can ever again drink normally should be modified although he states "all patients should be advised to aim at total abstinence". This seemingly modest suggestion evoked a swift and almost universally critical reaction on the part of some of the most prominent alcohologists in the United States and Canada (Davies 1963a, 1963b). A little over ten years later Drs. Linda and Mark Sobell (affiliated with Vanderbilt University in Nashville, Tennessee) have ammassed a list of over 70 scientific papers that refer to normal drinking in recovered alcoholics.

Starting back in the '60s too a number of researchers sceptical of Dr. Jellinek's disease concept of alcoholism (Jellinek 1960) tested the assumption with surprising and controversial results. Merry (1966) in one study and Engle and Williams (1973) in another, for example, examining the consumption of alcohol by alcoholic subjects in controlled settings found no evidence to support the theory that one or many drinks or even a bout of drunkenness produced a loss of control and thus question the so called physiological craving.

In more recent years a number of studies had been undertaken which attempt to demonstrate that certain selected alcoholics can be taught to become controlled social drinkers.

One of the most impressive of these is that of Drs. Linda and Mark Sobell, who introduced a technique of "Individualized Behaviour Therapy for Alcoholics" in an experimental study conducted at the Patton State Hospital in California (Sobell & Sobell 1973). After two years of exhaustive follow-up they report their results as follows: "Male gamma alcoholics treated by the method of individualized behaviour therapy were found to function significantly better after discharge than respective control subjects treated by conventional techniques". Men who by anyone's definition had been alcoholic had learned to practice what the Sobells called "controlled drinking".

A number of such experiments and indeed treatment practices had been enthusiastically reported in recent years. Some, like North Carolina psychiatrist Dr. John Ewing after a five-year trial period express scepticism as to the efficacy of the approach on any long-term basis.

Others are more adamant: "the mischievous suggestion that alcoholics might return to social drinking will soon be rejected" predicted Dr. Andrew Malcolm (1975) to the tenth Annual Meeting of the Canadian Foundation on Alcohol and Drug Dependence in 1975. "We do not have the ability to resolve the problem of compulsiveness in any significant number of people and until we do, the exciting concept of social drinking as a goal, will have to be regarded as premature - to think otherwise is to be fundamentally unserious and lacking in common sense and humility."

Both camps are vehement. "I have been aware of a certain amount of uneasiness among patients in various stages of their recovery as they try to evaluate public reports which indicate that some people can drink again socially" states Dr. John D. Armstrong (1963), formerly of the Addiction Research Foundation of Ontario. It seems very difficult for a person whose whole future depends on what he does about alcohol to weigh realistically the odds for him in attempting to emulate the special subjects of Dr. Davies' paper. The opposing argument expressed by Addiction Research Foundation psychologist Dr. Sanchez-Craig is that the advantage of teaching control drinking is the possibility of helping larger numbers of problem drinkers who have refused to seek treatment because they are unwilling to become teetotallers for life.

It is into this kind of a dilemma that researchers such as myself have been thrown. Most alcohologists who have devoted a number of years to the treatment of the alcoholic patient admit to the baffling nature of the task and to the frustration of its limited success. Discouraging too is the rate of relapse which reinforces the thesis that "once an alcoholic". By the same token though one is motivated to seeking out any innovative treatment approaches, even to considering the possibility of "controlled drinking for the alcoholic techniques". It is in this spirit that the present investigation was undertaken. It addresses itself to the question raised by Dr. Peter Nathan (1970) of Rutgers who declares "what we don't know are the kinds of people for whom abstinence and the kinds of people for whom controlled drinking as therapeutic goals are most useful".

This study is an extension of a five-year follow-up programme (Levinson 1975) completed at the end of 1974, the goal of which was to evaluate the effectivenenss of the treatment programme offered to the alcoholic at the Donwood Institute in Toronto, Canada. Like so many of the other follow-ups studies it too provides some empirical evidence that certain alcoholic patients seem to be able to establish a regime of moderate drinking with concomitant improvements in life style. The purpose is to examine these subjects in some depth in order to determine whether any common characteristics distinguish them from the rest of the group.

The original study is summarized briefly in order to provide a clearer picture of the subjects under consideration and of the nature of the treatment programme which they experienced.

The Donwood Institute is a special public hospital supported by the Government of Ontario for the treatment and the rehabilitation of adult patients afflicted with an uncontrolled dependence on alcohol and drugs. The underlying philosophy and theoretical considerations represented in the multi-disciplinary, community oriented treatment programme are perhaps in modern day terms rather traditional. Addiction is considered an illness in which physical, psychological and social components are equally significant with repair needed in all of these areas. Sobriety alone is not enough; the patient must recognize that his alcohol free life is more rewarding and comfortable than his dependent one, if he is to be expected to give up his chemical comfort. The alcoholic patient is seen as a potentially responsible human being and attention is focused on his resources rather than his pathology. His loneliness, loss of identity and insecurity highlight the need to help him establish closer interpersonal relationships and to break down the firmly established barriers to communication.

The hospital itself, fondly referred to as the Donwood Hilton, does resemble a luxurious hotel, but more important than the abundant facilities is the unmistakable atmosphere of warmth, acceptance and mutual respect which prevails. The programme is divided into three phases. Phases I & II take place during a four week residential period; Phase III goes on for at least one or two years after discharge. Phase I consists of detoxification when necessary and the examination and treatment of acute physical disorders. It is during this first week too that a comprehensive assessment of psychological and social status is carried out. The main function of Phase II, a combined educational and therapy programme, is to promote the understanding of the illness and to improve insight and motivation. The didactic instruction by way of closed circuit T.V. lectures and discussions, and an intensive 26 hour experience in group therapy, along with physiotherapy, physical education and relaxation training makes for a rather rigorous and well structured three weeks for the patient. Close relatives, employers and friends are also invited to participate in a special family day. The spouse is included too in Phase III which involves the patient in weekly semi-educational, semi-social group sessions with his fellow patients and some staff - also in social functions generally arranged by a very active Donwood alumni association. Sustained contact which is considered crucial to good treatment is maintained with out-of-town patients also on a weekly basis by correspondence, by telephone or occasional return visits. Responsibility for maintaining this on-going relationship is undertaken by the clinical secretary, a semi-volunteer from the community who acts as a kind of liaison officer between the patient, the hospital, the staff and fellow patients. The use of the protective drugs antabuse or temposil is promoted for all patients during the recovery phase and membership in Alcoholics Anonymous also urged for many.

The subjects for the follow-up study consisted of all those patients primarily addicted to alcohol who entered hospital to begin the three Phase treatment programme during the first four months of 1969. This random sample numbered 154 and represented one in three of all admissions for that year. The composition of the group was as follows: 120 were males, 34 females, their average age at intake was 46. They had completed an average of 12 years of formal education and were typically clerical, sales and technical workers or owners of small businesses. They had been drinking heavily, on an average for 11 years; 14% were binge drinkers, 69% steady drinkers and 17% combined binge and steady drinking. In terms of marital status, 62% were married 38% single, separated, divorced or widowed. Of those in the labour force 85% were employed and 15% unemployed - some of these retired.

These patients were all examined by the same psychologist - who also did all the follow-up interviewing, shortly after their admission to Hospital, to evaluate their status in the three areas toward which treatment was geared - physical, social and psychological. One questionnaire was drawn up to provide an estimate of the patient's alcohol intake and its affect on his health, another assessed his social adjustment in terms of marital and family relationships, living arrangements, occupational record, community involvement etc., and selected psychological tests were administered to determine his level of intellectual functioning with specific reference to possible impairment, and to appraise his general level of psychological integration and underlying pathology.

One year following discharge, in 1970, all of these patients (147 of 154) were contacted and the questionnaires and psychological tests readministered (Bell & Levinson 1971). During 1972 a telephone survey was conducted, in part as a further check on progress, but also as a safeguard in maintaining the ongoing contact (Levinson 1973). The final follow-up representing 5 years post treatment was completed in 1974 and 115 of the 154 subjects successfully contacted and reassessed once again on the same scales. Twenty-three subjects were known to be deceased (de Lint & Levinson 1975) so that 90% of the original sample is accounted for. A generous travel allowance which permitted the researcher to cover

over 25,000 miles through Canada and the United States to see subjects in their own homes, their place of employment in other hospitals or institutions, accounts for the relatively large numbers traced.

A numerical scoring system which provided objective ratings and comparisons of subjects' physical, social and psychological status and a combined estimate of general degree of improvement at each of the designated intervals was applied. Patients were accordingly categorized as recovered, significantly improved, moderately improved meant that the patient had been totally abstinent or had had up to three short-lived slips a year but had been abstinent for at least three months at the time of re-examination or alternatively that he had had a period of uncontrolled drinking within the five year interval but had not been drinking during the last one. His health was satisfactory and he was showing improvements in social adjustment i.e. family relationships, occupational record, community involvement etc., also in psychological state in that he was usually functioning at an increased intellectual level. Moderately improved referred to those individuals who had reduced their alcohol consumption - indeed claimed "controlled drinking" and were showing similar gains in the other areas. The unchanged or deteriorated category is self explanatory and includes the 23 deceased subjects.

Results of the five year follow-up were as follows: 56 subjects or 41% were recovered or significantly improved, 29 subjects or 21% were moderately improved, and 53 subjects or 38% were unchanged or deteriorated. Interestingly there was a considerable degree of shifting between categories among individual patients at the different time intervals of this survey - and it would seem that, for the most part, it was in a negative direction, 39 of the subjects who had done well in the first year subsequently resumed their old drinking habits and life style. On the other hand 17 subjects who had done poorly at the outset were later able to give up their "habit" and improve their life situations, frequently after further treatment either at the Donwood or some other facility.

The group with which the present study is concerned represents the "moderately improved" category. It was, as one might have speculated the most unstable one. Of the 29 subjects rates this way at the one year mark, only 11 representing 8% of the patient population managed to maintain a regime of reduced drinking and improvement in psychological functioning, and social adjustment. This then defines the "controlled drinkers" of 5 years duration - the subjects of this study. The group is not entirely homogeneous in that they did not all adopt this pattern from the outset and sustain it for the entire period. Four subjects were controlled drinkers from the time they resumed drinking; after initial periods of total abstinence lasting up to six months they subsequently drank moderately. Three subjects experienced occasional bad bouts up to twice a year of 2 to 5 days duration, but managed to recover from the bouts on their own (only one required help) and then return to their newly established drinking habits. The final three subjects both experienced short-lived periods of excessive drinking and also relatively lengthy intervals of total abstinence. Generally speaking though these 11 patients were drinking in significantly reduced amounts when they were drinking.

The variables were considered and analyzed in an effort to determine whether there was anything distinctive about the aforementioned outcome groups. Intake demographic features such as sex, age, occupational level, education, marital status, employment record, drinking history prior to admission in terms of number of years of heavy drinking and of its pattern, binge, steady or combined - and estimated intellectual functioning based in Information and Block Design sub-tests of the Wechsler Adult Intelligence Scale. No overall differences were found among any of the recovery groups, so that the variables considered provided no prognostic indicators of the long-term capacity for patients to deal with alcohol.

The more crucial question as to whether there were any common characteristics that distinguish the 11 "controlled" drinkers from the rest of the total group was examined further. Marital status was the only significant intake variable. A greater percentage of the "controlled" drinkers were married when they came for treatment and during the recovery phase. An interesting trend too, though not statistically significant, was that all of the "controlled" drinkers in the working force were employed and at full-time jobs.

Apart from these factors - which have also been shown to be general predictors of success (Levinson & McLachlan 1973), the 11 controlled drinkers were essentially indistinguishable from the rest of the group. To describe them briefly, there were eight males and three females, ranging in age from 35 to 58, the mean being 48. Nine were married, two separated and living alone, two had resumed moderate drinking immediately after discharge from Donwood - one of these had nevertheless attended the Phase III programme regularly - the other lived a considerable distance from Toronto and maintained contact through correspondence. The remainder had had initial periods of total abstinence ranging in duration from three months to 24 months - with an average of 10 months. Six attended Phase III regularly - the other three sporadically. Thus all of these patients had maintained an ongoing contact with the Donwood for a year after discharge.

Fewer than half (five subjects) had sometime during the five year follow up period experienced major social changes in their lives, which Dr. Davies (1962) felt "may have tipped the scales" for his subjects: two had remarried and three had entered new occupations. Again nearly half (five subjects) had "built-in" controls such as, in one case, a retired husband who was at home every day and another, a wife who immediately upon the first sign of out-of-hand drinking simply packed up and moved herself and the children into a nearby motel (charging all bills to her husband), and another man, part of a first generation Italian immigrant family, who quite literally was surrounded by a watchful wife, a mother, sister-in-law and daughter. Allowing for overlap, social change and built-in controls apply to eight of the 11 subjects. Of the other three, one lived on his own, and two in looser marital situations, and each returned to his former employment.

Recently, in fact within the last few weeks, which now represents seven years since their Donwood experience, a further attempt was made to establish telephone contact with these people - with reasonably good success. Two had left their place of employment and were reported to have moved out of their communities one in Northern Ontario - one back to the U.S. - and could not be traced. Of the remaining nine, seven reported that they continue to sustain the regime of controlled drinking reported when they were last interviewed. One lady had gradually slipped back to her pre-Donwood drinking pattern and given up the part-time job she had undertaken, indeed she would now be reclassified as unchanged. Another man who had experienced occasional drinking bouts indicated that these had become more frequent and that his work efficiency had deteriorated. Still he did not consider that this situation was as bad as it had been prior to Donwood. Neither of these subjects could provide any explanation for relapse. By way of contrast, on the other hand, one "controlled drinker" related a very traumatic experience - a pending marriage to a 50 year old widow, who died very suddenly just days before the wedding - which naturally precipitated a grief reaction but without any inclination to return to drinking. Thus only seven, possibly nine of the 11 subjects continue to be "controlled drinkers" seven years later.

It is difficult to draw any firm conclusions from the material just presented. If the key word is selection of subjects for a controlled drinking programme as opposed to the total abstinence tradition, the results of this survey provide few, if any definitive guidelines. Beyond the fact that more of these subjects were married and steadily employed (which are good prognostic signs of success, be it total abstinence or controlled drinking), there is nothing distinctive about the small group which managed to achieve controlled drinking and improved life style. Significant social change and permanently built-in controls helped some - not all.

Generally speaking one might interpret the present survey to be discouraging as a source of support for the "controlled drinking - an alternative to abstinence" concept. The number of patients that finally qualify as controlled drinkers is so small as to be almost negligible and their self-selection more or less chance. However, it may well be that the very premise upon which this study is based is a faulty one insomuch as it presupposed that all alcoholics can be considered as a homogeneous group. All of these patients were, after all, identified alcoholics who had already entered a treatment programme for alcoholism, and as such, they may represent a fairly chronic group sufficiently damaged that for them, there is no clinically justifiable alternative to total abstinence.

Greater consideration must be given to the distinct likelihood that there are different kinds of alcoholics for whom a wide range of therapies in different treatment facilities are needed. "It is unrealistic to set a pre-determined course for everybody" states Dr. Gordon Bell, president and director of the Donwood Institute. He outlines five classes of help that he feels should be made available and concedes too that the philisophical tenet that rehabilitation from alcohol dependence necessarily involves total abstinence must be reconsidered.

In the author's practical experience, for example, a small number of patients considered to be "reactive alcoholics" have been encountered. They had experienced traumatic and disruptive life situations to which they had responded by withdrawal through alcohol. For them, support and psychotherapy was the treatment of choice and after their circumstances had been resolved these people were able to return to social drinking. Another group for whom empirical evidence suggests that this might apply is the heavy social drinker - those individuals whose ordinary life style involves frequent occasions throughout their waking hours in which appropriate alcohol consumption just naturally occurs and for many this "normal pattern" marks the beginning of problem drinking and then transition into heavy dependent drinking. Perhaps for this group earlier intervention by getting them to face up to their problem realistically might be facilitated by something other than total abstinence as a basis upon which to work.

In spite of the fact that the present survey does not lend support to the "controlled drinking" phenomenon, it is a notion that cannot be discarded and warrants further research and experimentation. Perhaps though, in our present stage of knowledge, it is most applicable in a preventive programme. As the Sobells (1974) point out, there are at present no separately identifiable services for individuals just developing drinking problems, and as a result they often are referred to programmes oriented towards treating chronic alcoholics. One might expect that early problem drinkers would find the availability of a treatment orientation of "controlled drinking" to be both appealing and acceptable. Such efforts would be likely to meet with more success than current alcoholism education programmes.

REFERENCES

ARMSTRONG, J.D. (1963). "Comments on the Article by D.L. Davies". Quarterly Journal of Studies on Alcohol, 24.

BELL, G. and LEVINSON, T. (1971), "An Evaluation of the Donwood Institute Treatment Programme". Ontario Medical Review.

DAVIES, D.L. (1962). "Normal Drinking in the Recovered Alcoholic". Quarterly Journal of Studies on Alcohol, 23

DAVIES, D.L. (1963a). Comments on the article by D. L. Davies. Quarterly Journal of Studies on Alcohol, 24.

DAVIES, D.L. (1963b). Comments on the articles by D.L. Davies. Quarterly Journal of Studies on Alcohol, 25.

ENGLE, K.B. and WILLIAMS, T.K. (1973). "Effect of an Ounce of Vodka on Alcoholics' Desire for Alcohol". Quarterly Journal of Studies on Alcohol, 33.

JELLINEK, E.M. (1960). The Disease Concept of Alcoholism.

LEVINSON, T. (1973). "Four Years Later - An Ongoing Follow-up Study of Treated Alcoholics". Unpublished Paper 19th International Institute on the Prevention and Treatment of Alcoholism, Belgrade, Yugoslavia.

LEVINSON, T. (1975). "The Donwood Institute - A Five Year Follow-up Study". Unpublished Paper presented 31st International Congress on Alcoholism and Drug Dependence, Bangkok, Thailand.

LEVINSON, T. and McLACHLAN, J.F. (1973). "Factors Relating to Outcome in the Treatment of Alcohol Addiction at the Donwood Institute". Toxicomanies, VI. 3.

de-LINT, J. and LEVINSON T. (1975). "Mortality among Patients Treated for Alcoholism, A Five Year Follow-Up". Canadian Medical Association Journal, 113.

MALCOLM, A.I. (1975). "Gazing into the Crystal Ball". Address to the 10th Annual Meeting of the Canadian Foundation on Alcohol and Drug Dependence.

MERRY, J. (1966). "The Loss of Control Myth". Lancet, 1.

NATHAN, P.E., TITTER, N.E., LOWENSTEIN, L.M. SOLOMAN, P. and and ROSSI A.M. (1970). "Behavioural Analysis of Chronic Alcoholism". Archives of General Psychiatry, 22.

SOBELL, M.B. and SOBELL, L.C. (1973). "Individualized Behaviour Therapy for Alcoholics". Behaviour Therapy, 4.

SOBELL, M.B. and SOBELL, L.C. (1974). "Alternative to Abstinence; Addictions". Winter

A PROGRAMME OF GROUP COUNSELLING FOR ALCOHOLICS

J.S. MADDEN

REGIONAL ADDICTION UNIT

CHESTER

All treatment services are capable of improvement; this study reports on extension of the therapeutic facilities for alcoholics on Merseyside and discusses the implications. Although the alcoholism services in this region were developed earlier and more extensively than in most parts of the United Kingdom there had been no provisions outside hospital wards for group counselling of alcoholics and their spouses. This gap was closed the Merseyside, Lancashire and Cheshire Council on Alcoholism (M.L.C.C.A.), which is a voluntary agency with educational and therapeutic aims, (Madden 1968).

The M.L.C.C.A. considers that the group situation offers advantages to its clients. In a group there is pooling of experience and advice, learning by identification, and alleviation of unhelpful guilt feelings. Peers cut through the psychological defence mechanism of denial, that is so strongly developed in alcoholics - and sometimes in their normally drinking spouses - with respect to excess drinking by the former. Social skills are gained that may have been lost or never acquired. Mutual support is provided, the example of group members who make steady progress encourages those who are struggling. Of less importance, some economies of time are provided for counsellors compared with individual counselling techniques.

The Alcoholism Council therefore began to organise a programme of group meetings. The M.L.C.C.A. selects the group members (alcoholics, and when appropriate their spouses) on the likelihood of their co-operation. Each group consists of 12 members; all are resident in the community though some of the alcoholics have been recently detoxified by doctors, at the request of the Alcoholism

Council, in a medical or psychiatric ward, or at home. A group meets for ten sessions twice weekly in the evenings, the meetings are free of charge and last one and a half hours. Within each group there is no anonymity. Many of the members, either on their own initiative or on that of the Alcoholism Council, proceed to a second course of group counselling. The groups for the more advanced course, again comprise a dozen members, and meet on ten occasions, but the meetings are held only once a week.

A leader is provided by the Alcoholism Council for every meeting. The leaders, who are not paid, possess a wide range of backgrounds; they have included consultant physicians, a psychiatrist and social worker from a regional addiction unit, a University lecturer in psychiatry, general practitioners, an occupational medical officer, probation officers, magistrates, a solicitor, M.L.C.C.A. staff, and recovered alcoholics. A topic for discussion is given for each session; at the start of a course the members are given a list of the topics, as set out in Tables I and II. (See next page)

Neither the discussion topics, nor the group sessions aim for profound investigation of emotions or for radical personality change. Group counselling centres on the alcoholism, on the inculation of insight about excessive drinking and on ways in which total abstinence, with some better adjustment in other psychosocial spheres, can be achieved. Certain members default after a few meetings, either because they feel they do not need, or like, group counselling, or because in the case of certain alcoholics they resume drinking.

METHODOLOGY

An intensive study was made of the first 98 alcoholics who received group counselling. Their drinking and its consequences delineated them as alcoholics within the scope of the World Health Organisation's definition (1952). They were referred from several sources (Table III). (See over). Thirty eight had not been previously treated by doctors for alcoholism; 42 of the rest had been detoxified by doctors at the request of the M.L.C.C.A. in order to prepare them for group counselling. Their other characteristics are in Table IV. (See over). The social stability of males was scored with the scale devised by Straus and Bacon (1951) for residential, marital and occupational steadiness; the highest scoring subjects are the most stable. The number of years of alcoholism was assessed from the subjects' accounts of the onset of withdrawal symptoms or the repeated use of alcohol as a drug.

1. Definitions of alcoholism; phases of the illness
   - a. Basic definition
   - b. W.H.O. Definition
   - c. Phases of alcoholism
   - d. Jellinek types
   - e. Complications of alcoholism

2. Effects of alcoholism
   - a. Personal
   - b. The family
   - c. Employment
   - d. The community

3. Attitudes to alcoholism
   - a. Personal
   - b. Public
   - c. Medical
     The general practitioner
     The psychiatrist
     The physician
   - d. Illness or moral weakness

4. Difficulties in maintaining abstinence
   - a. Stress
   - b. Relief drinking
   - c. Experimental drinking
   - d. Promotion of other interest

5. The family and alcoholism
   - a. Role changes
   - b. Co-operation

6. Avoiding a drink

   Group discussion based on personal experiences

7. Treatment
   - a. The general practitioner
   - b. Counselling and advisory service of alcoholism council
   - c. Hospital
     Psychiatric wards
     Other wards
     Special alcoholism unit
   - d. Out-patient clinic
   - e. Alcoholics Anonymous and other sources of help

8. Rehabilitation
   - a. In the family
   - b. In employment
   - c. In society

9. The community and the alcoholic

   To reveal or not to reveal the condition

   To relatives
   At work
   In general

10. Never to drink again

    Comments and conclusions

TABLE I "LIVING WITHOUT ALCOHOL"; TITLES FOR TEN GROUP MEETINGS

TABLE II TITLES FOR A FURTHER COURSE OF GROUP MEETINGS

1.	What it means to be an alcoholic.
2.	Rewards that alcohol offers and reasons for drinking.
3.	Reactions to threatening situations during early sobriety.
4.	a. Medical aspects. b. Why the alcoholic cannot drink again.
5.	Precipitation of an isolated relapse or of a full return to drinking.
6.	Family attitudes to recovery.
7.	Tolerance of others.
8.	Problems of being a non-drinker.
9.	The alcoholic at work.
10.	a. Further involvement in help to others. b. General discussion.

TABLE III - SOURCE OF REFERRAL TO THE ALCOHOLISM COUNCIL, AND TREATMENT BY DOCTORS FOR ALCOHOLISM AT ANY TIME BEFORE GROUP COUNSELLING

Referral Source	Self	42
	Spouse	15
	Other relative	4
	General Practitioner	7
	Psychiatrist	9
	Physician	3
	Occupational Medical Officer ..	5
	Personnel Officer	2
	Probation Officer	5
	Social Worker	1
	Alcoholics Anonymous	2
	Others	3
Prior treatment by doctors for alcoholism	No treatment	38
	General Practitioner Only ..	4
	Psychiatric wards	18
	Medical Wards	44
	(of whom six were also in psychiatric wards)	

TABLE IV - DETAILS OF THE ALCOHOLICS

Sex	Male	80
	Female	18
Marital Status	Married	77
	Separated	4
	Divorced	3
	Widowed	2
	Single	12
Social Class	I	10
	II	26
	III	38
	IV	18
	V	6
Accompanied by spouse	By wife	36
	By husband	4
Drinking pattern	Bouts	54
	Continuous	44
Age in years (range 21-62; mean 41)	21-30	13
	31-40	36
	41-50	35
	50 plus	14
Years of alcoholism (range 2-30; mean 11)	Under 16	80
	16 plus	18
Social stability score (males)	0	4
	1	2
	2	9
	3	13
	4	52
Offenders	Drinking and driving	12
	Drunkenness	8
	Prison for assault	3
	Prison for fraud	2

The minimum follow-up period was six months and the maximum 36 months. Progress was gauged by personal contact, using home visits if necessary. If a subject claimed abstinence since starting group counselling, or after not more than two drinking relapses to be currently abstinent for at least six months, the claim was checked from another source (close relative, friend, general practitioner, or employing company if the latter had referred the alcoholic).

Subjects who had developed either one or two drinking recurrences, but who had abstained for at least the final six months of their follow-up period, were classed as 'recovered' along with those who had no drinking episodes. Cessation of group counselling after five or less sessions was defined as defaulting.

RESULTS

Six three (64%) of the subjects recovered, of whom 48 (49%) were continuously abstinent (Table V). The outcome of six individuals could not be accurately assessed. There were two deaths, in males, respectively from coronary thrombosis, and from a traffic accident as an inebriated pedestrian.

TABLE V - OUTCOME

	Recovered	Not recovered	Died	Unknown
Continuously abstinent 48 (49%)	63	27	2	6
1 relapse, abstinent for last 6 months 9				
2 relapses, abstinent for last 6 months.... 6				
Total recovered.... 63 (64%)				

Outcome was not associated with sex, marital status, social class, drinking pattern and age, nor for married subjects whether their spouses came with them for group counselling (Table VI). (See next page). Recovery was related to an alcoholism duration of less than 16 years, and to an absence of convictions; all the subjects with an alcoholic history of this length, or with convictions, were male. Recovery was also associated among males with a high social stability score. Unsatisfactory progress was

TABLE VI: OUTCOME RELATED TO DETAILS OF THE ALCOHOLICS

		Recovered	Not recovered
Sex	Male	48	32
	Female	15	3
Marital Status	Married	50	27
	Separated	4	0
	Divorced	2	1
	Widowed	1	1
	Single	6	6
Social Class	I	7	3
	II	17	9
	III	22	16
	IV	13	5
	V	4	2
Accompanied by spouse (married subjects)	Accompanied	26	14
	Not Accompanied	24	13
Drinking pattern	Bout	34	20
	Continuous	29	15
Age	21-30	8	5
	31-40	20	16
	41-50	26	9
	50 plus	9	5
Years of alcoholism*	Under 16	56	24
	16 plus	7	11
Social stability score** (males)	0, 1 & 2	2	13
	3 & 4	46	19
Offenders *** (all males)	Offenders	11	14
	Non-Offenders	52	21
Type of Offence +	Drinking and driving	9	3
	Other offences	2	11
Medical ++ Detoxification	Required before group counselling	21	21
	Not required	42	14
Medical +++ Detoxification (males)	Required before group counselling	16	20
	Not required	32	12

* (with Yates' correction) = 4.91; 1d.f.; $P < 0.05$.
** (with Yates' correction) = 14.4; 1d.f.; $P < 0.001$.
*** (with Yates' correction) = 4.89; 1d.f.; $P < 0.05$.
\+ $P \leq 0.02$ (Fisher Exact Probability Test.
++ (with Yates' correction) = 5.48; 1d.f.; $P < 0.02$.
+++ (with Yates' correction) = 5.47; 1d.f.; $P < 0.02$.

related, among the subjects with convictions, to offences not connected with drinking and driving and, also among males, to the need for detoxification prior to group counselling.

Persistent attendance at sessions was associated with recovery among the subjects as a whole and among males; in the male sex persistence was also related to high social stability scores (Table VII). (See next page). A higher proportion of females than males showed a satisfactory outcome and persistence in attendance, but the low number of three females who did not recover helps to explain the absence of statistically significant differences of variables in the female sex.

DISCUSSION

Outcome was judged by drinking behaviour. Preferably other criteria - e.g. psychological well being, interpersonal relationships, employment record - should also be used, (Pattison et al, 1969), but there is some correlation between a reduction in the drinking of alcoholics and their psychosocial adjustment, (Rathod et al. 1969, Malcolm and Madden 1973). Pokorney and co-workers have analysed a representative sample of studies from several countries, and reported that up to one third (maximum 37%) of the subjects were abstinent and about a half (maximum 59%) were improved; in the present study 49% of the alcoholics were abstinent and 64% recovered, defining these categories as outlined above. Their drinking progress was also more favourable than the results given in a list of previous United Kingdom studies, (Willelms et al. 1973).

Features positively related to a satisfactory outcome in some British studies have included male sex, (Pemberton 1967), older age, (Ritson 1968), higher social class, (Ritson, 1968; McCance and McCance 1969; Willems et al. 1973), intact marriage, (McCance and McCance 1969), bout rather than continuous drinking, (McCance and McCance 1969), higher Straus-Bacon social stability score in males, (Edwards 1966), and absence of convictions, (McCance and McCance 1969). The subjects of the current study possessed a similar distribution of favourable prognostic studies to those in many of the other United Kingdom reports; including some degree of motivation, sparse antisocial tendencies, and a relatively high proportion from social classes I and II (professional and managerial strata). Within the study the alcoholics who persisted with group counselling did not differ markedly from the defaulters, yet the latter showed less favourable responses.

But the deduction cannot be reached that the satisfactory responses were produced by the group sessions. Hill and Blane (1967) have drawn attention to the difficulties in evaluation of

TABLE VII: PERSISTENCE OR DEFAULTING FROM GROUP COUNSELLING

		Persistence	Defaulting
Outcome *	Recovered	55	8
	Not recovered	17	18
Outcome ** (males)	Recovered	42	6
	Not recovered	15	17
Sex	Male	57	23
	Female	15	3
Marital Status	Married	57	20
	Separated	3	1
	Divorced	2	1
	Widowed	2	0
	Single	8	4
Social Class	I	7	3
	II	20	6
	III	31	7
	IV	11	7
	V	3	3
Accompanied by spouse (married subjects)	Accompanied	30	10
	Not accompanied	27	10
Drinking Pattern	Bout	39	15
	Continuous	33	11
Age	21-30	9	4
	31-40	23	13
	41-50	27	8
	50 plus	13	1
Years of alcoholism	Under 16	58	22
	16 plus	14	4
Social Stability *** Score. (males)	0, 1 & 2	6	9
	3 & 4	51	14
Offenders (all males)	Offenders	19	6
	Non-Offenders	53	20
Type of Offence	Drinking and driving	10	2
	Other offences	9	4
Medical Detoxification	Required before group counselling	36	6
	Not required	39	17

* (with Yates' correction) = 15.9; 1d.f.; $P<0.001$
** (with Yates' correction) = 13.55; 1d.f.; $P<0.001$
*** (with Yates' correction) = 7.02; 1d.f.; $P<0.01$

psychological methods of treatment in alcoholism. Since the present subjects were selected, and those who continued with group sessions were further self-selected, it is feasible that selection processes chose alcoholics who were more likely to recover. Nor can the group counselling be isolated from other forms of care received by the subjects. Sixty had previously been treated by doctors, 42 of whom were detoxified immediately prior to starting group sessions. All had undergone individual counselling at M.L.C.C.A. before joining a group.

It is possible to assert that a considerable number of selected alcoholics reduced their drinking when they received, as their main type of care, group counselling made available outside the statutory services. The assertion is hopeful since the current treatment arrangements even if expanded, are not likely to cope with the number of alcoholics requiring help, (Editorial, British Medical Journal 1972, Edwards et al. 1973). It should also be stressed that the group discussions were at a relatively superficial psychological level, focussing on drinking and its cessation. While all the counsellors were experienced with alcoholism, not all were knowledgeable about formal group psychotherapy; indeed some considered that psychodynamic factors have little relevance to the causation and treatment of alcoholism. The inference can be drawn that in the promotion of alcoholism services there is ample scope for counsellors who have not received a lengthy professional training, but whose skills focus on practical advice to clients.

Finally it may be noted that the therapeutic programme was initiated and controlled by a voluntary organisation. Agencies of this kind, that have evolved within communities and are maintained by local interest and enthusiasm, have an important role in the development of caring resources.

REFERENCES

EDITORIAL (1972). British Medical Journal, 2. 479.

EDWARDS, G. (1966). "Hypnosis in Treatment of Alcohol Addiction". Quarterly Journal of Studies on Alcohol, 27. 221-241.

EDWARDS, G., HAWKER, A., HENSMAN, C. PETO, J. and WILLIAMSON, V. (1973). "Alcoholics Known or Unknown to Agencies: Epidemiological Studies in a London Suburb." British Journal of Psychiatry, 123. 169-183.

HILL, M.J. and BLANE, H.T. (1967). "Evaluation of Psychotherapy with Alcoholics: A Critical Review. Quarterly Journal of Studies on Alcohol, 28. 76-104.

McCANCE, C. and McCANCE, P.F. (1969). "Alcoholism in North-East Scotland: Its Treatment and Outcome". British Journal of Psychiatry, 115. 189-198.

MADDEN, J.S. (1968). "The Role of the Merseyside Council on Alcoholism". Journal of Alcoholism, 3. 57-63.

MALCOLM, M.T. and MADDEN, J.S. (1973). "The Use of Disulfiram Implantation in Alcoholism." British Journal of Psychiatry, 123. 41-45.

PATTISON, E.M., COE, R. and RHODES, R.J. (1969). "Evaluation of Alcoholism Treatment; A Comparison of Three Facilities." Archives of General Psychiatry, 20. 478-488.

POKORNY, A.D., MILLER, B.A. and CLEVELAND, S.E. (1968). "Response to Treatment of Alcoholism; A Follow-Up Study." Quarterly Journal of Studies on Alcohol, 29. 364-381.

RATHOD, N.H., GREGORY, E., BLOWS, D., and THOMAS, G.H. (1966). "A Two Year Follow-Up Study of Alcoholic Patients." British Journal of Psychiatry, 112. 683-692.

RITSON, B. (1968). "The Prognosis of Alcohol Addicts Treated by a Specialized Unit." British Journal of Psychiatry, 114. 1019-1029.

STRAUS, R. and BACON, S.D. (1951). "Alcoholism and Social Stability. A Study of Occupational Integration in 2,023 Male Clinic Patients." Quarterly Journal of Studies on Alcohol, 12. 231-260.

WILLEMS, P.J.A., LETEMENDIA, F.J.J. and ARROYAVE, F. (1973). "A Two Year Follow-Up Study Comparing Short with Long Stay In-Patient Treatment of Alcoholics." British Journal of Psychiatry, 122. 637-648.

WORLD HEALTH ORGANISATION (1952). Technical Report Series, No. 48.

THE ONTARIO DETOXICATION SYSTEM: AN EVALUATION OF ITS EFFECTIVENESS

REGINALD G. SMART

ADDICTION RESEARCH FOUNDATION

TORONTO

The past 10 years have seen a major re-orientation in North America in the handling of the public inebriate or chronic drunkenness offender. Since about 1968 many states and provinces have been establishing alternatives to the old "revolving door" - that is the cycle of intoxication, arrest, trial, short term jail term then new intoxication, etc. Under this system police typically arrested only those drunks creating a disturbance or homeless alcoholics. The most typical changes to the older revolving door have been to replace jails with voluntary detoxification centres and halfway houses. Although these changes might have been defended solely on humanitarian grounds usually they were not. It was usually expected that the new non-criminal system would be rehabilitative for the inebriate and that there would be a major reduction in drunkenness arrests. Very few evaluations of any detoxication facility have been made which indicate what rehabilitative effects they have. The present study is believed to be the only one for a detoxication system in a large area - the province of Ontario. The aims of the current evaluation were to, (i) identify the effects of the system on drunkenness arrests, (ii) the nature of the population served by the system, and (iii) the extent of rehabilitation provided for the systems' clients.

THE ONTARIO DETOXICATION SYSTEM

Before describing the evaluation it is necessary to briefly outline the features of the detoxication system and the older methods which it replaced.

Prior to 1969 the typical revolving door functioned in Ontario.

Offenders usually spent a night in jail and appeared in court the next day to get a fine of $30 or 10 days in jail. Few paid the fine. During 1966 there were 56,290 drunk arrests for an Ontario population of about 7 million; about 40% of the arrests were in Toronto. Nearly 40% of those arrested in Toronto were repeaters with 3 or more arrests per year. It was estimated that there were about 2,000 skid row alcoholics in Toronto who contributed to the arrest rates. However, signs that they were being rehabilitated were almost non-existent. Arrest rates increased every year and specialized facilities for public inebriates were uncommon. Many of the facilities which did exist were more supportive of the system than concerned with ending it or rehabilitating the men. However, it was argued that chronic incarceration created some stability in the lives of skid row inebriates. It forced men to stop drinking for several weeks or months in each year and provided basic medical care and nutrition, at least temporarily.

As a result of several task forces involving the Addiction Research Foundation (ARF), Ministry of Health and Justice officials an alternative to jail was devised. A short trial with a medically managed detoxification provision suggested that very few public inebriates required any extensive medical care. It was therefore decided to develop a system of non-medical detoxication facilities and halfway houses which would have close connections with several hospitals but not be part of them.

The objectives of the system first developed in 1971 were to provide "care and rehabilitation for the chronic liquor offender". That is, the aim was to provide facilities for skid row or chronic inebriates rather than for the vast numbers of citizens who may be drunk and require some place to sober up. The aims were to remove the public inebriate from the criminal-justice system to health and social care systems.

Currently, the system is operating in all large and small cities in Ontario. There are 13 detoxication centres (265 beds) and 17 halfway houses funded by the system, plus 13 others. The detoxication centres are the legal responsibility of general hospitals. Typically they are not part of the hospital but are located a few blocks away. No physicians or nurses are employed as staff. Non-professionals are selected and trained to work in the centres. They are expected to admit only those inebriates who have no serious medical problems and to refer all others to the emergency ward of the hospital. Their second task is to refer clients, after detoxication, to a variety of treatment and rehabilitative agencies. That is they were not to become another revolving door facility.

The halfway houses (about 20 beds each) were planned to work in concert with detox facilities. It was hoped that they would take most of their referrals from detoxification centres. The philosophy of treatment varies from one house to another but most seek to provide a sober environment, a homelike atmosphere, referrals to Alcoholics Anonymous and in some cases formal treatment and vocational counselling.

THE EVALUATION SYSTEM

Since the Addiction Research Foundation was responsible for the evaluation of the system, research monitoring instruments were developed for both types of facility. Dr Helen Annis (1973) established a record system to be used in all detoxification centres. Each client gets a unique code number to permit the monitoring of contacts with detoxification centres. There is a "Detox Census Book" which is a registry giving each clients' number and data of first admission. The "Detox Face Sheet" is filled out for every admission but only once per year and lists code numbers, demographic and drinking characteristics. The Detox Patient Record of Each Admission - records the features of each admission, length of stay, and place of referral. All detox records are computerized. A similar system has been developed by Dr Alan Ogborne and Sarah Weber (1975) for the halfway house, except that more information is obtained on treatment experiences and reasons for discharge.

A preliminary analysis has been made of the characteristics of 3652 males and 504 females at six detox facilities. It seems that only about 12% have three or more skid row characteristics. A number are still employed (65%) and not living in skid row facilities (63%). The halfway house population included more skid row persons and more who had spent time in jail in the past year (63% and 56%).

THE EFFECTS OF THE SYSTEM ON ARREST RATES

If the system is fulfilling one of its major objectives it ought to decrease the number of drunkenness arrests. With the change to a social-health care model it should be possible for the police to spend less time arresting public inebriates. Ideally, they should be rehabilitated or in the process of rehabilitation and hence not be exposed to arrest. It should be noted that the new system did not remove the crime of public intoxication. What it did was to allow police once having arrested inebriates to take them to a detoxification centre instead of to jail. It did not obligate them to do so although police policy was to use the detoxification facilities if possible.

Two studies have been made of the effects of detoxification centres on arrests. A study by Annis and Smart (1975 b) examined arrests in Toronto for six detox years 1966-71 and for three post-detox years 1972-74. Rates per 1,000 population were calculated. Over the six pre-detox years arrests were decreasing slightly (e.g. 24,039 in 1966 to 22,094 in 1971). However with the opening of three detoxes in Toronto (125 beds) this trend merely continued. There was no evidence that the decrease was any greater after the opening of the detoxes. A study in Hamilton by Annis and Liban (1975) also showed that the opening of the detox there had no effect on rates of drunk arrests. Perhaps not enough time had been given for the system to work effectively but this seems unlikely.

The arrest statistics for a large Toronto police division, which included skid row, was examined. Only about 9% of persons arrested for drunkenness actually went to a detoxification centre - about 70% were refused at detoxification centres mostly because they were full and the remainder were held for other charges. Most of those arrested then did not get to a detoxification centre but were detained overnight or for a few hours and released to appear in court later. It is not surprising that detoxes were not having much effect on arrests. The difficulty is that extremely large numbers of beds are required for large North American cities. The average length of stay is 3.9 days. To process 22,000 arrests a year about 210 beds are needed - just for the police arrests. That is 40% more than now exist. It would be possible to reduce the length of stay but detoxication and referral are difficult to do rapidly. Time is needed to arrange for full drying out and transfer to hospitals or outpatient care for long term treatment. Also the police did not have guaranteed access to all detox beds - only to one-third of them. Referrals from various other agencies and hospitals made up the rest of the referral sources. Given all of these problems it is not surprising that the impact of Ontario detoxes on the revolving door has been small so far.

EFFECTS OF THE DETOX SYSTEM ON THE POPULATION

As stated earlier, an important aim of the system was to involve detox clients in long term rehabilitation and to thereby decrease their social deviance. Several large scale follow-up studies have now been made using the detoxification and halfway house monitoring systems. They are, of course, ideally arranged for follow-ups and those done in Ontario are the only large scale record follow-ups on detox populations.

The first study was done by Annis and Smart (1975 a) and it included a follow-up of 522 first admissions to three Toronto detoxes. A search was made of all treatment and arrest records for men. All treatment agencies likely to have seen them were contacted including 25 residential treatment facilities, 19 outpatient departments, four detox centres and seven police divisions. For each contact type of referral, length of stay, number of visits and type of referral on discharge were gathered.

In general, the results did not indicate that the system was having a major impact of a rehabilitative nature. About 53% had one or more arrests in the six months, mostly for alcohol-related offences. The average number of drunkenness charges was 4.3 per man. Less than half of the men had no readmission to a detox during the follow-up period (48%). Only 9.6% of the detox admissions resulted in a referral to a health agency. Only confirmed referrals are included. That is, referrals for which the person actually arrived. Many detoxes claim referral rates of 40 to 50% but only about a quarter of those actually arrive (10% of total).

Length of rehabilitation for those referred is also disappointing. Only 16% of the total first admissions to detox spent a month or more in residential treatment. Outpatient contacts were also few. Only 23% had any at all and all but 7.7% had only one or two. It is most unlikely that any major recovery could be made given the short periods of treatment received by the detox population as a result of their admission. Only 22% had no contact with a treatment agency, detox, or police in the six months follow-up.

A later follow-up was attempted for these 115 men in 1975 or two years after their admission. By that time 37 were known to have detox admission or to be drinking heavily - only 23 were known to be sober or to have periods of sobriety. That would mean that about 5% of original detox admissions retained some sobriety over the longer follow-up although an additional 10% could not be found.

The picture obtained from this study is that detoxes are having a very limited effect on the lifestyles of their clientele. Few are going to detoxes when they are arrested. Only about 10% are passed on as confirmed referrals but most who do so have such short periods of treatment that recovery cannot be expected. It was found that readmissions do not result in more referrals - i.e. further contact with detoxes may not create much improvement.

An attempt has also been made to examine the effects of specific types of referrals to community agencies on later detox admissions and public drunkenness. A study by Smart et al (1976) reported the relative value of referrals to an outpatient department, a halfway house and a long stay farm for detox patients. A comparison was made with detox clients who did not take any referral at all. Very few studies have been made comparing outcomes for inebriates in different types of facilities. Most of these studies have not used a standard follow-up period and objective records of recovery, (Ogborne and Smart 1974). Most such studies have used grapevine methods, guesswork on the part of agency staff or interviews with a large data loss at follow-up.

The facilities used in this study were probably typical of those for public inebriates in Ontario. The halfway house was exclusively for alcoholics and took about 75% of its clients from detox facilities. The farm was Bon Accord which takes only skid row persons, mainly from detoxes. The outpatient department was the ARF facility in Toronto connected with a large hospital.

Originally it was hoped to randomly assign detox clients who volunteered for a referral to one of these facilities, plus allowing another group called "self-referral" to choose their own. In fact, it was difficult to randomly assign clients and most of those who went did because they wished to do so. Hence another group called "refusals" assembled itself - they refused the place selected for them and typically went nowhere at all. In total 114 patients were included - with 42 refusals, 24 at the outpatient department, 22 at the halfway house and 26 at the Bon Accord Farm.

Again, the results of the study gave no optimism about post-detox recovery. Only about 60% actually arrived - even when firm commitments were made. However, those arriving and not arriving for treatment showed no pre-treatment differences in demographic characteristics or detox admissions. Nor were there any differences among the groups going to the various facilities. Pre- and post-treatment comparisons showed no post-treatment decrease in detox admissions, nor drunk arrests for the halfway house, farm or outpatient groups. However, the "refusal" group had significantly <u>more</u> detox admissions whereas the treated group did not. Surprisingly there was no difference between the treated and untreated groups. Perhaps motivation for treatment is a stronger variable in maintaining stability than is the treatment itself. Refusing treatment at a detox is a sign of poor motivation and deteriorating drinking. Perhaps detoxes should attempt to get larger numbers of referrals without worrying too much whether they will be accepted or not.

Another finding from this study was that treatment was of short duration. Those who went to the farm and the halfway house stayed only 50 days whereas three to six months is expected. Those who went to outpatients had only one or two visits with an average of 1.4. Referral agencies obviously have substantial problems in both attracting and holding detox clients.

FUTURE RESEARCH ON DETOX AND HALFWAY HOUSES

Much more needs to be done to understand the functioning and effectiveness of the detox system. Evaluation is not yet complete, especially that involving the halfway houses. Research is underway at ARF to examine the following :

(i) whether the system has resulted in a deterioration in

health among skid row inebriates. The earlier jail system allowed a certain amount of stability and health care which the current one does not provide;

(ii) More needs to be known about how successful halfway houses are and what types of programme are most successful. Studies of halfway houses treatment ideology related to outcome are now underway;

(iii) An experiment is in progress with an intensive counselling programme for skid row men. This programme does not wait for the men to appear but a trained counsellor approaches them on skid row and actively attempts to help. There is a control group of similar men not getting any help;

(iv) An attempt is being made to develop behavioural monitoring studies at Bon Accord which will allow a more personalized behavioural modification and therapy programme.

THE CURRENT POSITION OF THE DETOX SYSTEM ?

It is obvious from what has gone before that the Ontario Detox-Halfway House system has many limitations. It is unlikely that it has more limitations than any other but only that they are made more obvious by the extensive evaluation employed. No other system has had the degree of research evaluation which is currently being undertaken in Ontario.

The major limitations are :

(i) the system does not seem to affect drunk arrests in large cities - it does not replace the "revolving door". To do so it would have to be about twice as large as it now is and limit itself to police referrals;

(ii) the system does not provide enough impetus for long-term referrals. Only about 10% of those admitted take any sort of long-term care;

(iii) the system so far does not appear to have effective treatment agencies for the detox population. Even among those who are referred failure rates are very high and the rate is not better for those who arrive than for those who don't;

(iv) it looks as if the detox-halfway house system may deteriorate into a new "revolving door". Perhaps it would turn more quickly than the jail door but have fewer people passing through it.

Clearly there is a need to consider some alternatives and perhaps some of these could include :

(i) long stays at detoxes for public inebriates and contracts for them to engage in treatment;

(ii) a restriction of admissions to police referrals or public inebriates only;

(iii) a mandatory period of stay - e.g. until a referral had been arranged and accepted.

All this, of course, pre-supposes that we know how to successfully treat public inebriates and all that is needed is more opportunity to do so. However, this is almost certainly not the case and what is needed are more effective treatments, more efficaciously applied.

REFERENCES

ANNIS, H.M. (1973). "An evaluation system for Ontario Detoxication Centres". Addiction Research Foundation.

ANNIS, H.M. and LIBAN, C. (1975). "Drunkenness arrests and detoxication in the Hamilton area". Addiction Research Foundation, 717.

ANNIS, H.M. and SMART, R.G. (1975 a) "A follow-up of men referred from detoxication facilities: arrests, readmissions and treatment involvement". Unpublished manuscript, Addiction Research Foundation.

ANNIS, H.M. and SMART, R.G. (1975 b). "The Ontario Detoxication System: influence on drunkenness arrests in Toronto". Ontario Psychologist, 7, 19-24.

OGBORNE, A.C. and SMART, R.G. (1974). "Halfway houses for skid row alcoholics: a search for empirical evaluation". Addiction Research Foundation, 632. In Press, Addictive Behaviours.

OGBORNE, A.C. and WEBER, S. (1975) A data collection system for halfway houses". Addiction Research Foundation, 662.

SMART, R.G., FINLEY, J. and FUNSTON, R. (1976). "The effectiveness of post-detoxification referrals; effects on later detox admissions, drunkenness and criminality". In Press, International Journal of Social Psychiatry.

Section IV

Rehabilitation

THE ROLE OF THE PROBATION SERVICE IN THE TREATMENT OF ALCOHOLISM

T. CROLLEY

PROBATION AND AFTER-CARE SERVICE

LIVERPOOL

This paper begins by discussing two apparently unconnected ideas. The first concerns the nature of the Law; the second is to do with the nature of maturity, i.e. maturity as it affects normal thinking man. Hopefully, the connection between the two will become apparent as the argument unfolds.

The science of law has been facing the negative effects of alcoholism for a long time, and has been trying to prevent and fight alcoholism as a major cause of criminality. Alcoholism and criminality are inter-connected, but the precise relationship between the two is only now beginning to be examined and requires a strict psychiatric definition. It may also require a stricter legal definition. The problem for Probation Officers and others was highlighted in The Times, when the possible effects of drunkenness upon a defendant's Mens Rea was considered by the Court of Appeal. What the trial jury had to consider was whether the appellants, through drunkenness, were capable of forming the intention to do a criminal act. Presumably the jury felt they could, for they were found guilty and sentenced to imprisonment. On appeal, although the convictions stood, the sentence was reduced, reflecting perhaps Judicial uncertainty about this difficult problem. Thus there is a complex relationship between the law, criminal behaviour and alcoholism. Yet despite this complexity, the law is used to curb the effects of alcoholism and to define legal boundaries to alcohol-affected behaviour. The individual who crosses these boundaries comes within our responsibility.

This point should be kept in mind whilst we turn to an even greater problem, viz. the Nature of Maturity.

Man in his infinite variety is capable of change. It is realised change can mean decay, but it can also imply development leading to maturity. But what is maturity ?

Maturity is the ability to learn from experience and relate it to a given situation. Maturity is having authority and possessing the ability to use it in the correct way. Supremely, however, maturity is the assertion and affirmation of the personality without hostility (Storr). Maturity therefore validates one's being. It authenticates one's individuality and endorses one's self-worth. Maturity gives meaning and quality to the nature of man. Shakespeare's Hamlet makes the same point in another way, and in poetic language. "O what a piece of work is man! How noble in reason! How infinite in faculty! In form, in moving, how express and admirable! in action ... in apprehension ... the beauty of the world".

Now this is all very well and good! But if you feel that in this rarified literate and philosophical atmosphere the introduction of the word "alcoholic" defiles this picture of the mature man, then you have fastened on to the most profound yet the most fundamental consequences of alcoholism. For this is precisely the function of alcohol. It mars the face of maturity. To change the metaphor, it crumbles the very foundations of the will and disintegrates the superstructure of the personality. The drinker whose intake has led to an increased toleration for alcohol, soon finds he is indulging in secret and surreptitious drinking, which leads to the onset of guilt and remorse. Instead of validating one's being, alcohol undermines it. The endorsement of self-worth is substituted with loss of respect and self-doubt. In place of the assertion and affirmation of the personality, he withdraws into himself, and family and friends are avoided. "Noble reason" gives way to impaired thinking. In action and moving - interpreted as co-ordination - in the words of an ancient writer,the alcoholic "staggers to and fro like a drunken man". Apprehension is replaced with distorted perception. Work and money troubles follow. Amnesia sets in. Court appearances begin with being "simple drunk" and progress through to Drunk and Disorderly, Driving with excess alcohol, Stealing to get money for drink. Grandiose behaviour and unreasonable resentments erupt into violence and assaults on the Police, and, if the Chief Probation Officer of Bedfordshire is right, also an increase in violence against Probation Officers - a relatively new phenomena. The Court does its best, Fines, Probation and ultimately, Prison.

Now the question to be asked is this - have we painted a picture of an offender who, with all his other problems, is also an alcoholic ? Or have we described an alcoholic who, by reason of his alcoholism, has also become an offender ? The distinction is important and real. Uclesic, a Yugoslav, has written "Alcohol as a causative agent is present in all criminal deeds". This cannot be correct. But alcohol as a causative agent to criminal behaviour may be far greater than is realised. There is therefore, a need for further research and a greater understanding of alcohol-motivated offenders, which in turn has implications for the prevention and treatment of delinquency.

The connection between breaking of the Law and the de-maturation of the personality through alcoholism thus becomes evident. The connection is for all of us. For it is in this setting that the Police, the Prison Service and the Probation Service does its work.

This article cannot speak for the Police or the Prison Service. The brief is Probation. Hopefully what is said may have some relevance to whatever discipline one belongs. What is our role and contribution to the community within the framework of law-breaking alcoholics ?

First - identify the problem. Now this sounds obvious, but detecting alcoholism, particularly in the early stages, is a difficult problem. Moreover, there is some evidence to show that despite its history in dealing with "drunks", the Probation Service is not very good at discovering alcoholics in the social-scientific way described. One Officer who supervised a case for three years, variously describing the offender as "Depressive", "Psychotic", "Schizophrenic", yet in the final analysis when the alcoholism was revealed and dealt with, the criminality ceased. In a recent survey where Probation Officers were asked to identify those cases in which drinking had been assessed as a problem, only three cases out of over 200 were so identified.

There are however, more alcohol-related offenders than is first recognised. In an article entitled "The hidden influence of alcohol on crime", The Times of 26th March 1976 says, "Four times as many crimes are committed under the influence of alcohol as is apparent from the usual official statistics". This was the conclusion of a survey initiated by The Chief Constable of Bedforshire who said "The influence of drink on offences was so common that it was rarely brought to the attention of the Magistrates".
Drink is so often used as a plea of mitigation in Court that we fail to understand the true significance of what is being said. Less attention is given than is deserved. Further the Probation Service draws a substantial amount of its work from penal establishments. John Hanson, Deputy Governor of Ranby Prison, has just completed a survey of 217 inmates received into his Prison. 24.8% had a history of drinking offences and 48% admitted that drink had played a part in the offence for which they were then in Prison. For those who are supervising Parole and After-Care cases, how many in your case-load can you identify that might not only come within such a sample but have taken up drinking in a serious

way since release ? Smith-Moorhouse has estimated from a survey of three Midland Prisons that 30% of the Prison population is alcoholic. The need therefore, for Probation Officers to identify the problem becomes self-evident. We must beware labelling and over-identifying alcoholism, but the seriousness of underestimating the place of alcohol in the life of the offender cannot be too heavily stressed. Identifying the problem and acquainting ourselves with symptoms and knowledge is the duty of us all.

The second function of the Probation Service is to become part of the multi-disciplinary approach to the treatment of alcoholism. The sacred cow of Probation is the one-to-one relationship. It has to be slaughtered. In a survey of 500 cases where there was a higher percentage of identifiable cases, 48% of Probation Officers tried to treat the alcoholic on their own. It is widely recognised, however, that alcoholics are best treated by the mobilization of the totality of community resources and that individualized therapy is not enough. Social Workers alone, Psychiatrists alone, G.P.'s alone, Alcoholic Advisory Services alone, Probation Officers alone cannot be as effective as each working in conjunction with the other. This is an established truth.

Finally, the Probation Service must establish its objectives in dealing with the alcoholic. Is it one of containment in reducing as many Court appearances as possible ? Is it attempting to maximize their periods of sobriety and minimize their periods of intoxification ? What is the nature of our personal intervention ? Is it to analyse the pressures that make an alcoholic vulnerable to drink and make responses to reduce that vulnerability ? Is the aim total abstinence from drink or a reconstruction of the maturity of the personality ? Should the Service be working in the community to work for greater provision of facilities and to stimulate the need for treatment and seek prevention of its growth ?

Each of us must work out our own personal objectives. The author believes that the Probation Service, at its most primitive level, is in business to rehabilitate the offender. The Oxford Dictionary defines rehabilitation as "to re-instate, to restore to one's former position". Does this suggest therefore that Probation Officers have a responsibility not only to help towards individuals' recovery from alcoholism, but to do a much more permanent work by helping to re-create the spirit of maturity within those alcoholics who transgress the Law ? Is it our function, as with all our clients, to help provide them with opportunities to assert and affirm their personality in an acceptable way ? If this is so, the Probation Service has a valid role to play within the community by acting as an important link between the Law, criminal behaviour and alcoholism.

AN ANALYSIS OF CLIENTS USING ALCOHOLIC AGENCIES WITHIN ONE COMMUNITY SERVICE

SHENAGH DELAHAYE

DEPARTMENT OF PSYCHIATRY

UNIVERSITY OF MANCHESTER

We know that between only 11 and 25% of problem drinkers in a community are in contact with an appropriate helping agency (Edwards et al 1973). Prevalence studies vary in their estimates but one study based on the central part of Manchester (Wilkins 1972) found that the number of problem drinkers in one predominantly working class G.P. practice of 12,000 people to be 18.2 per thousand. Greater Manchester has a population of 2.7 millions and we would therefore expect there to be a large population of alcoholics and problem drinkers. The present paper describes that small part of this population who present for treatment.

The community treatment service in Manchester comprises six specialist and many non-specialist agencies. The present work is unique in studying a comprehensive service by describing individual agencies within one area, within one period of time. Each agency is described in depth in terms of its population and its treatment; the whole sample represents a large group of people seeking help for drinking problems, but, further than this, the comparisons drawn between agencies give us greater insight to the following questions: How do the populations of agencies within a comprehensive service differ ? Do their treatments differ accordingly ? What is the extent of multiple use of agencies ?

DESCRIPTION OF AGENCIES

Table I summarises the agencies involved in the present study. The Social Services hostel and the Church Army Hostel resemble other agencies reported in the literature (Cook et al 1966, Grass Market 1974, Orford et al 1974). A report from Helping Hand (1972)

TABLE I : DESCRIPTION OF AGENCIES

Specialist	No.of places in residential establishment	Sex	Comments
Alcoholism Treatment Unit	14	M & F	Intensive group therapy. Medical Supervision Compulsory activities Abstinence aim
Social Services Hostel	14	M	Daily group therapy Compulsory activities Two full time staff Abstinence aim
Church Army Hostel	9 - 13	M	Alcoholic group within a mixed hostel. Group therapy three times a week. Resident warden Abstinence aim
Helping Hands Hostel (2 Hostels)	1. 11 2. 6	M & F	Mostly A.T.U. follow up Intensive weekly therapy. Long stay. Work encouraged. Abstinence
Information Centre Manchester Council on Alcoholism	-	M & F	Counselling, Education and advice and referral. Walk in centre. Two full time staff. No rigid aim of abstinence
Alcoholics Anonymous	-	M & F	29 Groups in Greater Manchester. One walk-in centre. Daily groups provided. Acceptance of alcoholism compulsory. Abstinence aim.
Mental Hospital	No limit on beds	M & F	No specific ward. Group therapy five times a week. No programme of activities Sobriety not compulsory
Psychiatric Unit in General Hospital	No limit on beds	M & F	Approximately 75% on one ward. No specific treatment plan for alcoholics. Sobriety not compulsory
Private Psychiatrists	-	M & F	A.A. encouraged. Abstinence aim.

and the Merseyside Information centre (1970) enable broader understanding of two of the other specialist agencies.

AIMS AND METHOD

Aims The aims of the research relevant to the present paper are:

1. To compare the agencies in terms of their populations.
2. To give a description of the attenders of A.A.
3. To estimate the extent of multiple use of agencies.

Method The research began during February 1975, following a pilot study, and data collection was completed eight months later in October 1975. All specialist treatment agencies were included in the research as well as selected non-specialist agencies (see Table I). Where it was necessary to isolate subjects for the study within these agencies the following definition was used :

> "A client is defined as someone who admits to drinking to such an extent that he creates problems either for himself or for his family. He may or may not have symptoms of alcoholism" (Wilkins 1972).

The main method of data collection was personal interview of clients attending these agencies. Where possible this method was used in all agencies with clients seen within six days of admission to all residential establishments. When a sample of 50 was reached in an agency, data collection ceased. In the case of Alcoholics Anonymous, the Information Centre and the private psychiatrist, it was not feasible to conduct personal interviews. The alternative methods of data collection produced less detailed results and therefore are not comparable on all dimensions, with the main sample of 295 (Table II). Thirty four of the 295 were clients attending any agency for the second or third time within the eight month period. Over all there were seven definite case losses.

The main areas of investigation covered on the questionnaire are demographic details e.g. social stability (Straus and Bacon 1951), social class, complications of drinking (Hore and Smith 1975 Edwards et al 1972), former treatment for alcoholism, source of referral, use of A.A., use of Manchester agencies, view of drinking and reasons for entering treatment including choice of agency.

TABLE II : AGENCIES, METHOD OF DATA COLLECTION, SAMPLE SIZE

Agencies	Method of data collection	Sampling Method	Sample Size
Specialist A.T.U. Out-patients	Personal interview with a semistructured questionnaire	Alternate new Out-patients	33
" A.T.U. In-patient	"	Consecutive admissions	46
" Helping Hands Hostels	"	"	9
" Social Services Hostel	"	"	40
" Church Army Hostel	"	"	27
Non-Specialist General Hospital Psychiatric Unit	"	"	58
" Mental Hospital	"	"	61
			295
Private Psychiatric Clients	Search of case notes	Consecutive Clients over $1\frac{1}{2}$ years	68
Manchester Council of Alcoholism Information Centre	Questionnaires administered by centre staff	Consecutive new clients	89
A.A.	Self-administered questionnaire	Those attending A.A. groups over 2 week period in October 1975	172

For the A.A. study, all 29 groups were circulated with questionnaires which all those attending were requested to complete. Three groups did not reply at all due to administrative problems but apart from four of the groups, the response rate was good.

RESULTS

The results can be grouped under five headings. The results presented in this paper relate only to characteristics of the clients in individual agencies within the Manchester services.

TABLE III : CHARACTERISTICS OF THE SAMPLES

	5 specialist agencies 2 non-specialist Main Sample N = 295	A.A. N = 172	Private Patients N = 64	Information Centre N = 89
Mean age (yrs)	41.0	39.9	43.0	34.6
% Female	13	29	42	19
% Social Class I - II	9.1	22	54	10.2
% Married	25	52	75	28.4
% Suffering Shakes	85	84	75	74
% Marital Breakdown	33	26	6	-
% Scoring 4 on: Straus and Bacon Social Stability Scale	24	35	64	-

COMPARISON OF AGENCIES

The results in this section refer to the main sample of 296 subjects with reference to the Information centre clients where possible.

Age, Sex, Marital status, Class, Social Stability

The Social Services Hostel and the Church Army Hostel have no women at all; A.T.U. Out-patients and the General Hospital Psychiatric Unit contain most women but these only amount to 21% and 26% respectively.

The Helping Hands Hostel have 75% under 40 years of age, the A.T.U. 74%, whereas only 42.5% of the Social Services are under 40. Only 2.3% of the social services hostel are married, 35% single, whereas 41.3% of A.T.U. In-patients are married with 19.5% single. The mental hospital is closest to the Social Services Hostel in this aspect.

TABLE IV : SOCIAL STABILITY (STRAUS AND BACON)

	% Scoring 0	% Scoring 4
Social Services Hostel	62.5	0
Church Army Hostel	70.4	0
Helping Hands Hostels	50	0
General Psychiatric Unit	3.4	17.2
Mental Hospital	22.6	4.8
A.T.U. (In-patient)	10.9	17.0
A.T.U. (Out-patient)	7.5	21.7

TABLE V : CLASS

	Social Serv. Hostel	Church Army Hostel	Helping Hands Hostels	General Psych. Unit	Mental Hosp.	A.T.U. I.P.	A.T.U. O.P.	Info. Centre
% Class 1-11	2.5	0	0	18.9	14.9	17.3	9.4	10.1
% Class 5	45	25.9	0	8.6	21	6.5	4.3	29

View of Drinking

Denial of alcoholism is greatest in the A.T.U. out-patients where 50.9% deny it. None of the residents in the Church Army Hostel and only 9.5% of those in the Social Services Hostel do so.

Choice of Agency

Clients are asked why they chose the particular agency they were now in rather than other agencies. The Social Services Hostel had significantly more people coming because they felt it was the best treatment for them compared with all other agencies ($p < .008$). This was also true for all the hostels grouped together compared with other agencies ($p < .0001$) 69% of A.T.U. out-patients came through professional connections. Clients were also asked whether they would have come into treatment without pressure. 43% of the patients in the mental hospital would not have done so, which is significantly higher than all other agencies ($p < 0001$) particularly than the other non-specialist agency.

TABLE VI : FORMER TREATMENT

	Social Serv. Hostel	Church Army Hostel	Helping Hands Hostels	General Psych. Unit	Mental Hosp.	A.T.U. I.P.	A.T.U. O.P.
% no treatment before	4.8	0	0	24.1	18	8.7	37.7
% no In-patient treatment before	25.8	37	16.7	43	24.6	45.7	62.3

Source of Referral

A.A. refer only to the Social Services Hostel, the Church Army Hostel and the Information Centre. The Information Centre refers mainly to general psychiatric unit and the A.T.U. out-patients. Only the Social Services Hostel has people referred from other alcoholics. 29.6% of the Church Army Hostel are referred by probation officers which is significantly more than everywhere else ($p < .0001$).

TABLE VII

Agencies	% Referred by G.P.	% Referred by by Probation Officer or Social Worker	% Referred Psychiatrist
Social Services Hostel	7	24	7
Church Army Hostel	3.7	55.5	11
Helping Hands Hostels	0	100	0
General Psychiatric Unit	39	0	20.6
Mental Hospital	27.8	10	26
A.T.U. (I.P.)	19.5	4.6	63
A.T.U. (O.P.)	45.2	4	28.3
Information Centre	9	17.9	4.4

TABLE VIII : SIGNIFICANT DIFFERENCES BETWEEN A.T.U., A.A. AND THE INFORMATION CENTRE IN MANCHESTER (I.P.)

	significance
More attenders of A.A. are married than clients of the Information Centre	$p < .003$
More clients of the Information Centre are of Social class V than attenders of A.A.	$p = .0007$
More clients of the Information Centre are unemployed than attenders of A.A.	$p < .005$
More attenders of A.A. have suffered from morning shakes than clients of the Information Centre	$p < .05$
Mean age of all three agencies differ significantly from each other	

COMPARISONS OF GROUPED AGENCIES

TABLE IX : MAJOR SIGNIFICANT DIFFERENCES BETWEEN MEDICAL AND NON-MEDICAL AGENCIES

Conditions	Medical agencies N = 129 % fulfilling condition	Non-Medical agencies N = 76 % fulfilling condition	Significance
Acceptance of alcoholism	54	92	$p < .0001$
Over 25 complications of excessive drinking	33	72	$p < .0001$
Scoring 0 on social stability	11	61	$p < .00075$
Have had in-patient treatment for alcoholism	56	73	$p < .01$
Little self-determination in choice of agency	52	27	$p < .0007$
Referral by G.P.	33	4	$p < .01$

TABLE X : MAJOR SIGNIFICANT DIFFERENCES BETWEEN SPECIALIST AND NON-SPECIALIST AGENCIES

Conditions	Specialist agencies N = 175 % fulfilling condition	Non-specialist agencies N = 120 % fulfilling condition	Significance
Have not attended A.A.	20	35	$p < .01$
Have attended with pressure - do not see themselves as alcoholics	19	35	$p < .004$
Scoring 0 on social stability	32	13	$p < .00075$
Little self-determination in choice of agency	51	38	$p < .01$
Referral by general practitioner	20	33	$p < .01$

ALCOHOLICS ANONYMOUS

Out of a large study of A.A. the results presented here related only to the population of Alcoholics Anonymous as it compares and relates to other agencies in Manchester.

Age, sex, class, marital status, social stability, employment

Of the sample of 172 members, 79% are male. The mean age is 39.9 years. 52% are married, 16% single, 25% divorced or separated. 5.8% of the total sample fall into social class 1, 30.2% into social class 3 and 11% into social class 5. Only 19.7% are unemployed, 8% scored 0 on the Straus and Bacon social stability scale whereas 34% scored 4.

Complications of excessive drinking

Suffered shakes	84%
Drunk meths	15%
Arrested for a drunkenness offence	75%
Broken marriage through drink	26%
Lost job through drink	47%

Treatment in Manchester

29% of the 172 had attended a psychiatric ward for treatment for alcoholism. 20% had been in the A.T.U. as an in-patient and 19.7% had had treatment from a private psychiatrist in Manchester.

31% of the sample of 295 have rejected A.A., 26.7% have never attended and 41.3% attend at present. Out of the 216 who have ever attended A.A. 43% have rejected it; this is a large portion which will be analysed further and reported on at a later date.

MULTIPLE USE OF AGENCIES

Within one month of attendance at the present agency 58% had used only the agency that they were at present attending.

Within this same time period 2.6% had used four or more agencies.

Within the whole period of treatment in Manchester, 31% had used only one agency in Manchester.

Within this same time period 69% are multiple users of which 8% had used five or more agencies.

DISCUSSION

An attempt is made to enlarge upon the differences shown in the results section and to make initial comments upon the findings.

Characteristics of the samples

Private patients are older, of significantly higher class ($p < .01$) and significantly more are married ($p < .001$) than all other agencies. The low rate of marital breakdown recorded here is reported also in Carney and Lawes (1967). The lack of difference on physical complications supports other conclusions that there is less variation between classes on physical criteria. As private patients are significantly more socially stable than all other clients ($p < .001$) it would appear that the advantages of money and class protect against social implications of excessive drinking. The high number of women seeing private psychiatrists suggests either that there are more women alcoholics in social Class 1 and 2, or that our treatment agencies in general are not geared enough for women, particularly of lower social status. The latter conclusion is supported by the fact that only 13% of the whole sample in treatment are women, whereas Moss and Beresford Davies (1967) found the ratio of men to women alcoholics to be 4:1.

The Information Centre receives people younger than any other agency. There are also significantly fewer people suffering from shakes ($p < .02$) and fewer people with morning drinking and memory blackouts. This would support their own perception that they are receiving more early problem drinkers than other agencies. Perhaps the removal of the psychiatric stigma or the need to tell the general practitioner (referrals from general practitioners are very low) enables people to request help when they are beginning to worry about excessive drinking.

Alcoholics Anonymous in Manchester, as in London (Edwards et al 1967b, 1966a) helps people of a higher class than other agencies in Manchester excluding private psychiatrists ($p < .0001$), and the Alcoholics Anonymous population approximates more closely to that of private psychiatrists than any other type of agency.

Comparisons of agencies within the main sample

The emphasis in the present discussion is on comparisons within Manchester in order to describe the workings of a community service. The parallels with services all over the country will form the basis of a separate discussion in the future.

Age, sex, marital status, class, social stability

The main contrasts lie between the two hostels (run by the social services and the Church Army) and the A.T.U. in-patients. All the hostels see the more intractable problems with people who have less social support and therefore for whom rehabilitation depends primarily on the agency. The younger, higher class clients are concentrated in the A.T.U. providing a more fertile ground for intensive therapy and change. The likely explanation for the high number of social class V clients in the Information Centre is their links with the social services and their function as a referral agency for detoxification.

The mental hospital approximates more closely to the Social Services Hostel as the general hospital psychiatric unit does to the A.T.U. This reflects the movement from non-specialist to specialist agencies.

View of drinking, choice of agency and complications of drinking

The hostels contain people who tend not to deny their alcoholism, but in spite of a high level of suffering cannot find a way of maintaining sobriety. This differs from the A.T.U. out-patients where there is a majority who deny alcoholism and indeed who suffer less of the physical complications than clients of other agencies. Both non-specialist agencies and the Information Centre share this feature to a lesser extent. The policy of referral to A.A. for these people, who on account of denial or the lack of severe complications, cannot identify with A.A., is questionable. So far it is only in the Information Centre that an attempt is made to give to this group intensive treatment that is both educative and non-threatening. The Information Centre, however, does indeed have fewer people suffering from shakes ($p < .01$) than all other agencies put together and therefore probably has an easily definable population; but the A.T.U. out-patients still contain more people without signs of physical addiction.

Former Treatment: The hostels tend to admit people who have had treatment somewhere else before. In many cases there is a history of failure in treatment. Perhaps planners of services have to accept the view of the staff in these hostels, that there is a need for permanent sheltered care for a number of these chronic alcoholics who cannot maintain sobriety in the "outside world". But in addition to this, the Social Services hostel, particularly is getting people from Skid Row requesting help for the first time (Edwards et al 1966b).

Source of Referral Medical practitioners appear to refer within traditional lines rather than deciding upon appropriate care. There are more negative than positive feelings towards alcoholics within the general psychiatric service and this is not helped by the uncertainty over the role of psychiatric units and mental hospitals. Are they referral agents, detoxification centres or treatment agencies in their own right ? It is partly the nature of referrals to them that determines their role. It is notable that a similar link exists between the non-medical hostels and social referral agencies. The two hostels that are not connected to the A.T.U. seem to be reaching a different population most of whom would not be accepted in the A.T.U. But the Church Army Hostel is sometimes used as follow up service from the A.T.U.

Specific comparison of the A.T.U. in-patients, A.A. and the Information Centre

Hore and Smith (1975) found the populations in A.T.U.'s in England to be similar to that of Information Centres (Edwards et al 1967a) and A.A. (1966a). This does not appear to be true for Manchester.

The main differences in Manchester are on account of the Information Centre reaching a less socially integrated population than would be expected. The Information Centre meets a need that A.A. cannot do by seeing problem drinkers who are not yet addicted. The class structure in the A.T.U. and the Information Centre is dissimilar as there are so many social class V clients in the Information Centre. But on most dimensions there is a continuum whereby the A.T.U. falls in the middle in both physical and social terms.

COMPARISONS OF GROUPED AGENCIES

Medical and non-medical agencies

Alcoholics in medical facilities are more socially stable with less social deterioration, more often married, of higher social class and suffering less of the complications that accompany heavy drinking. There is no greater frequency of health problems amongst the clients of medical facilities. The agencies run by social workers deal with a population socially and physically deteriorated for whom alcoholism is acknowledged but invincible. But the medical facilities have different problems: only 4% of the sample in medical agencies are not addicted and yet 35% do not see themselves as alcoholics. In the light of this, psychiatric hospitals which are inevitably involved with alcoholics could evolve treatment programmes aimed at dealing with this ambivalence. It is perhaps appropriate that the alcoholics with most social difficulties are seen in agencies run by social workers. When social problems are severe, it is likely that physical addiction is advanced and personality inadequacies well entrenched. Thus the non-medical agencies have more than just the social problems to deal with. It is harder to see the relevance of medical care for the population in the hospitals.

Specialist and non-specialist agencies

The main distinguishing difference is the lack of motivation of the clients in the non-specialist agencies, not the level of suffering. There may well be a negative value in continuing to admit these people to give them a rest from drinking without developing some techniques to alter this motivational element. Either specialist agencies should be developed to give more intensive treatment to those reluctant to receive help or non-specialist agencies themselves could rethink their own programmes.

ALCOHOLICS ANONYMOUS

The similarities between the A.A. sample from Manchester and that from London (Edwards et al 1967b, 1966a) far outweight the differences. As previously stated, the A.A. population is more stable and of higher social status than the population in other agencies. A.A. does however see a proportion of the alcoholic that is more typical in the other agencies on account of its walk-in centre where almost daily groups are held and homeless and rootless alcoholics are given a welcome. Sixty-six percent of the sample have had no other treatment since joining A.A.; this shows the large body of people within A.A. who depend on it as the only external support. Thirty-nine percent had had no treatment in Manchester at all; this expresses the extent to which A.A. is an autonomous treatment agency in Manchester seeing people who do not reach A.A. through statutory treatment agencies.

MULTIPLE USE OF AGENCIES

The single users over both time periods are mainly in the A.T.U. out-patients, the mental hospital and the psychiatric unit. There are many first time users in the A.T.U. out-patients, and the non-specialist agencies tend to have repeat admissions from people who do not move on to specialist agencies.

A frequent combination is the Church Army Hostel with A.A.; both non-specialist agencies also have strong links with A.A. It was common to be in a psychiatric unit as a prelude to the A.T.U.

The extent of multiple usage makes it clear that evaluation of individual agencies in terms of their "success rates" is not straightforward as it is impossible to judge which element of treatment has had the greatest effect. We should be moving towards evaluating a community service as a whole in terms of its ability to reach and help all types of alcoholic problem and be less concerned about the comparative effectiveness of individual agencies.

CONCLUSION

The study demonstrates the wide variation in the characteristics of the population seeking treatment for alcoholism within one area. This is reflected in the clear differences between the client groups at each separate agency. The contrasts between treatment approaches in the different agencies are less clearly defined. If the agencies in the study were to co-operate more, they could rationalise their policies to allow wider and clearer alternatives of treatment. This in turn would encourage more appropriate matching of client with facility by the referral agents.

REFERENCES

CARNEY, M.P. and LAWES, T.G.G. (1967). "The Aetiology of Alcoholism in English Upper Social Class". Quarterly Journal Studies on Alcohol, 28. 55-69.

COOK, T., MORGAN, H.G. and POLLACK, B. (1966). "The Rathcoole Experiment. First Year at a Hostel for Vagrant Alcoholics". British Medical Journal, Volume I, 240.

EDWARDS, G., CHANDLER, J. and HENSMAN, C. (1972). "Correlates of Normal Drinking in a London Suburb". Quarterly Journal Studies on Alcohol, 6, 69-93.

EDWARDS, G., FISHER, H.K., HAWKER, A. and HENSMAN, C. (1967). "Clients of Alcoholism Information Centres". British Medical Journal, 4, 346.

EDWARDS, G., HAWKER, A., HENSMAN, C., PETO, J. and WILLIAMSON, V. (1973). "Alcoholism Known or Unknown to Agencies. Epidemiological Studies in a London Suburb". British Journal of Psychiatry, 123, 169-83.

EDWARDS, G., HENSMAN, C., HAWKER, A. and WILLIAMSON, V. (1967b). "Alcoholics Anonymous - The Anatomy of a Self Help Group". Social Psychiatry, 1, 195.

EDWARDS, G., HENSMAN, C., HAWKER, A. and WILLIAMSON, V. (1966a). "Who goes to A.A.?" Lancet, 382-384.

EDWARDS, G., WILLIAMSON, V., HAWKER, A. and HENSMAN, C. (1966b). "London's Skid Row". Lancet, 1. 249.

GRASS MARKET URBAN AID PROJECT TEAM REPORT (1974). "1 Thornybank - The First Year".

HELPING HAND ORGANISATION (1972). Annual Report. H.H.O. 8 Stutton Ground, London SW1.

HORE, B.D. and SMITH, E. (1975). "Who goes to Alcoholic Units?" British Journal of Addiction, 70, 263.

JELLINEK, E.M. (1946). "Phases of Drinking - An Analysis of an A.A. Survey". Quarterly Journal Studies on Alcohol, 17, 1.

MERSEYSIDE LANCASHIRE & CHESHIRE COUNCIL ON ALCOHOLISM (1970). Annual Report.

MOSS, M.C. and BERESFORD DAVIES, E. (1967). "A Survey of Alcoholism in an English County".

ORFORD, J., HAWKER, A. and NICHOLLS, P. (1974). "An Investigation of an Alcoholism Rehabilitation Halfway House : Types of Client and Modes of Discharge". British Journal of Addiction, 69, 213-224.

STRAUS, K. and BACON, S.D. (1951). "Alcoholism and Social Stability". Quarterly Journal Studies on Alcohol, 12, 231.

WILKINS, R.H. (1972). "A Survey of Abnormal Drinkers in General Practice". Paper presented at the 30th International Conference on Alcoholism and Drug Addiction.

CO-ORDINATION AND CO-OPERATION

B. D. HORE

REGIONAL ALCOHOLISM UNIT

MANCHESTER

In the United Kingdom, treatment services for the alcoholic have grown up from a variety of voluntary and statutory bodies in an ad-hoc way. There has rarely been any planning in relation to need, although with the "primitive level of our services" and the tremendous need (less than 10% alcohol addicts and problem drinkers are in touch with treatment agencies), each facility has a clear contribution to make. Co-ordination of facilities implies a linking of services, co-operation of agencies, clarification of each different services role and aims, and an understanding of the type of client with whom they are dealing. Studies of alcoholics attending different agencies, (Zax et al 1967, Delahaye, S. 1976), suggest, most use more than one treatment agency, and it is rarely possible in quantitative terms to weight the contribution to success or failure by any agency. Results of treatment appear to depend on variables such as social stability, (Straus and Bacon 1951) and absence of psychopathic personality traits more than any particular method of treatment and range from circa 70% sustained sobriety (U.S. Industrial Company programmes) with highly socially stable clients to 10% (Skid Row Hostels). If co-operation and co-ordination is going to happen (and naturally the opposite is more likely to occur) it is important that different agencies are clear about the aims and methods of other agencies. The principal agencies in the order of their historical development in the United Kingdom should now be examined.

ALCOHOLICS ANONYMOUS

This organisation came to the United Kingdom in 1948 and has steadily grown since, although the level of its activity is variable in different parts of the country. It aims for total abstinence, removes shame from the alcoholic, but also puts the responsibility for future conduct firmly on the alcoholic's shoulder. It offers group therapeutic experience, and has all the elements for attitude and behavioural change, i.e. group identification, "confession" and removal of denial (so important in the alcoholic), emotional arousal and a simple clear cut treatment goal, i.e. total abstinence. It appears less concerned with reasons for drinking which predate dependence.

REGIONAL TREATMENT UNITS FOR ALCOHOLISM

These began in 1951, and have now spread nationally. Studies, (Hore and Smith 1975 , Edwards et al 1967 a & b) have found their intake closely resembles in demographic terms that of A.A. and community based Information Centres, although individual units can differ widely in selection,(Hore and Smith 1975). Units have been very influenced by A.A. The aim of treatment is usually total abstinence and group therapy the principal therapeutic tool. There has also been an emphasis on in-patient treatment. Perhaps the major difference with A.A. is an awareness of other psychological problems behind drinking and need for appropriate treatment for these.

COUNCILS ON ALCOHOLISM

Councils pioneered in Liverpool are said to have functions of giving advice and information to individual clients and for their families, education (of lay and professional groups) and of producing a central co-ordinating body. Few Councils have done all, most have emphasised one or the other. Recently Madden & Kenyon (1975) have shown success in treatment of highly socially stable clients with a brief form of out-patient treatment. The majority of Councils aim for total abstinence in the advice they give to clients, although in cases of early drinking problems (increasingly being seen in such centres) advice on modifying drinking patterns may be given.

HOSTELS FOR ALCOHOLICS

These are largely the produce of the last ten years, and are very unevenly spread. They are run as therapeutic communities with total abstinence the treatment aim. Drinking usually leads to instant dismissal from the hostel. There is often confusion as to whether they are primarily in the "Change" business or in the "Shelter" business. Particularly in those hostels dealing with clients of low social stability, few are going "to change" sufficiently to return to "normal society",(Cook 1974). "Shelter is what is required". Some alcoholics of high social stability including A.A. members find it difficult to accept that alcoholics may need a permanent shelter. Included are agencies primarily involved in treating alcoholics. Other agencies, e.g. police, general psychiatric units, individual social workers may see alcoholics and its management in different terms. In the United Kingdom however, the differences seem less than in some parts of the U.S.A.,(Pattison 1973) and aims and techniques of the four agencies above resemble each other more than they differ. Some might say the resemblance is too close, and that lack of alternatives of aims and methods is, indeed, depressing. Experiments towards "normal drinking", re-evaluation of behavioural techniques, e.g. aversion are, therefore, to be welcomed providing it is stressed they are experimental and re-evaluative.

Co-operation and co-ordination are not the same. The former involves a working together, an appreciation of the contribution of the other, a "getting on" of individuals who run different agencies and accepting that clients will use different agencies which are most suited to his needs. Co-operation is not easy, however, and often does not exist. It involves a considerable amount of work and regular contact between agencies. Such contact can be provided by the vehicle of a Council on Alcoholism. Co-ordination involves more - ultimately it means a planning of service according to needs of clients and using agencies which are most suited to the task. Individual agencies might be asked to forego expansion in a particular direction in lieu of another agency - a very difficult thing to do. It is important in any city for a range of hostels or hospital units, or vice versa. The best services are those with such a range which should be balanced according to needs in these individual cities.

Co-operation and co-ordination are clearly helped by an understanding of the role and aims of other agencies and the clients they serve. They require regular "grass roots contact" but obviously involve emotional factors amongst the professionals, e.g. jealousy. Co-operation and co-ordination do not just happen; they are frequently difficult and take years. They can happen, however, if we accept that differing contributions are needed, and that there is no single road to sobriety.

REFERENCES

COOK, T. (1974) "Hostels for Alcoholics" - Paper presented at the 20th International Institute on the Prevention and Treatment of Alcoholism, Manchester.

DELAHAYE, S. (1976) "An Analysis of Clients using Alcoholic Agencies within one Community Service".

EDWARDS, G., HENSMAN, C., HAWKER, A. and WILLIAMSON, V. (1967) "Alcoholics Anonymous - Anatomy of a Self Help Group" - Social Psychiatry, I, 4, 195.

EDWARDS, G., FISHER, M., HAWKER, A. and HENSMAN, C. (1967) "Clients of Alcoholism Information Centres" - British Medical Journal, 4, 346.

HORE, B.D. and SMITH, E. (1975) "Who Goes to Alcoholic Units" - British Journal of Addiction, 70, 263.

MADDEN, J.S. and KENYON, W.H. (1975) "Group Counselling of Alcoholics by a Voluntary Agency" - British Journal of Psychiatry, 126, 289.

PATTISON, E.M. (1973) "Drinking Outcomes of Alcoholism Treatment, Abstinence, Social, Modified, Controlled and Normal Drinking" - Alcoholism a Medical Profile, B. Edsall & Co., London.

STRAUS, K. and BACON, S.D. (1951) "Alcoholism and Social Stability" - Quarterly Journal Studies on Alcohol, 12, 231-60.

ZAX, M., GARDNER, E.A., and HART, N. (1967) "A Survey of the Prevalence of Alcoholism in Monroe County, New York" - Quarterly Journal Studies on Alcohol, 28, 316.

THE PROBLEM - ITS MAGNITUDE AND A SUGGESTED COMMUNITY BASED ANSWER TO ALCOHOLISM

W.H. KENYON

MERSEYSIDE LANCASHIRE AND CHESHIRE COUNCIL ON ALCOHOLISM

LIVERPOOL

On the 5th April 1973, at the Second International Conference on Alcholism and Drug Dependence in Liverpool, the author read a paper entitled Community Involvement - Alcoholism, and the opening statement in that paper was as follows :-

"The great physicist and scientist Lord Kelvin once said 'When you cannot measure, then your knowledge is of a meagre and unsatisfactory kind. Therefore, it seems to follow that if your knowledge is meagre and unsatisfactory, the last thing in the world you should do is to make measurements.' There is little doubt that so far as this country is concerned knowledge about alcoholism is meagre, interest or concern, except amongst a few, is practically nil and in general, treatment and rehabilitative facilities, again except in isolated areas, are grossly unsatisfactory. Therefore if I accept Lord Kelvin's premise then with our meagre knowledge coupled with an undoubted dearth of satisfactory or scientific measurements I shall be on dangerous ground. Being neither a scientist, psychiatrist nor one with fixed ideas about alcoholism, I am going to venture onto this dangerous ground in an endeavour to establish some measurements."

During those three years there has been a remarkable change in some areas with regard to knowledge about alcoholism and approaches to the problem. However, it is still maintained that the action taken on a national, regional and local scale is grossly inadequate. We are in a position today to say to Lord Kelvin - we can measure, we are in possession of more information, and before it is suggested what steps should be taken and a proven community model examined, some of the facts should be looked at.

There are three basic indicators with regard to the prevalence of rates of alcoholism within any community - alcohol consumption; convictions for drunkenness, both pedestrian and motorist; and admissions to hospital of patients with either a primary or secondary diagnosis of alcoholic psychosis or alcoholism. The following statistics constitute a review of consumption, convictions and hospital admissions over the ten year period 1964 - 1974.

CONSUMPTION. In order to make the figures with regard to our national consumption of alcoholic beverages more meaningful, and more readily understandable, the three basic types of alcoholic beverages have been reduced to the common denominator of absolute alcohol.

UNITED KINGDOM

ANNUAL CONSUMPTION OF ALL ALCOHOLIC BEVERAGES IN TERMS OF ABSOLUTE ALCOHOL SHOWN IN MILLION GALLONS
1964/65 - 1974/75

(N.B. It has been assumed that all beers average 3% absolute alcohol, all wines, including light and heavy wines, average 12% absolute alcohol, and spirits average 57% absolute alcohol.)

Year	Beer Mill.Galls.	Wines Mill. Galls.	Spirits Mill.Galls.	Total Mill.Galls.	Increase Mill.Galls.
1964/65	32.3	4.4	10.9	47.6	
1969/70	36.1	5.1	10.2	51.4	3.8
1974/75	42.2	9.5	18.5	70.2	18.8

Over the ten-year period consumption has increased by 22.6 million gallons - a 47% increase. However, alarming as this increase might be, if the two five-year periods are compared an even more alarming picture is revealed:

Between 1964/65 and 1969/70 the increase was 3.8 million gallons or 8%.

Between 1969/70 and 1974/75 the increase was 18.8 million gallons or 37%.

Comparison of the first five-year period with the second five-year period reveals an alarming acceleration in the trend and the rate of increase was 363%.

During the same ten-year period the population increased by approximately 1.8 million. Between 1964/65 and 1969/70 the increase in population was 1.2 million or 2.2% and for the second five-year period - 1969/70 to 1974/75 - the population increased by only 0.6 million or 1.1%; a deceleration in population growth.

Basically we are still a nation of beer drinkers, approximately two-thirds of our present annual national intake of alcohol is in the form of beer. However, we are rapidly becoming a nation of wine drinkers. The figures reveal that over the ten-year period, by volume, our consumption of wine has increase by 116%. So far as spirits are concerned, over the same ten-year period there has been an increase of almost 70%, and the increase in the consumption of beer was 3.1%.

ENGLAND AND WALES

ANNUAL CONVICTIONS FOR DRUNKENNESS AND PROCEEDINGS AGAINST DRUNKENNESS AMONG MOTORISTS IN THOUSANDS
1964-1974

Year	Offences Drunkenness	Increase	Proceedings Against Motorists	Increase	Combined Totals	Total Increase
1964	76.8		8.1		84.9	
1969	80.5	3.7	23.7	15.6	104.2	19.3
1974	103.2	22.7	66.7	43.0	169.9	65.7

DRUNKENNESS AMONG PEDESTRIANS Over the ten-year period the number of convictions has increased by 26.4 thousand or 34%. However, if the period is broken down into two five-year periods the same kind of alarming picture is revealed as with consumption :

Between 1964 and 1969 the increase was 3.7 thousand or 6%.

Between 1969 and 1974 the increase was 22.7 thousand or 28%.

Comparison of the first five-year period with the second five-year period reveals an acceleration in the trend and the rate of increase was 367%.

PROCEEDINGS AGAINST MOTORISTS. Over the ten-year period the number of proceedings against motorists has increased by 58.6 thousand or 723%. Comparison of the two five-year periods shows that :

Between 1964 and 1969 the increase was 15.6 thousand or 195%.

Between 1969 and 1974 the increase was 43 thousand or 181%.

It should, however, be particularly noted that there is a distortion in the figures produced by the implementation of the Road Safety Act 1967 - "The Breathalyser"- which came into force on the 9th October of that year.

It is an undoubted fact that more and more young people are drinking, and at a much earlier age. This is borne out by a study of offences of drunkenness by age groups. For example, in 1964 offences of drunkenness committed by those aged under twenty one totalled 12% of the convictions, by 1974 this had risen to 19.5%.

ENGLAND AND WALES

HOSPITAL ADMISSIONS
IN THOUSANDS
1964 - 1974

Admissions to mental illness hospitals and units of patients whose primary or secondary diagnosis was alcoholism or alcoholic psychosis.

Year of admission	Male	Increase	Female	Increase	Total	Total Increase
1964	5.2		1.4		6.6	
1969	6.3	1.1	1.7	.03	8.0	1.4
1974	9.7	3.4	3.5	1.8	13.2	5.2

HOSPITAL ADMISSIONS. Over the ten-year period hospital admissions, both male and female, have increased by 6.6 thousand or 100%. Comparison of the two five-year periods shows that :

Between 1964 and 1969 the total increase was 1.4 thousand or 21%.

Between 1969 and 1974 the total increase was 5.2 thousand or 65%.

Comparison of the first five-year period with the second five-year period reveals an acceleration in the trend and the rate of increase was 210%.

Alcoholics are also admitted to non-psychiatric hospitals. The DH&SS produced figures regarding such admissions in the shape of a survey based on a 10% sample of NHS discharges and deaths from non-psychiatric hospitals in England and Wales. A study of these figures reveals that there were 3.2 thousand admissions for alcoholism in 1964, increasing to 4.8 thousand in 1974. Little import is given to such figures as many medical authorities agree that in the general hospital situation, although alcoholism could very well be a major causal factor for the patient's admission, it is simply not diagnosed. In fact, some experts would say that a correct diagnosis where alcoholism is present is made only in one case in ten.

It must also be borne in mind that so far as admissions to mental illness hospitals in England and Wales are concerned, many patients who suffer from alcoholism are simply not diagnosed as being alcoholic. In fact, in our experience over the years it is doubtful if more than one in four of alcoholics who are admitted to mental illness hospitals are correctly diagnosed.

The foregoing figures concerning consumption, convictions and hospital admissions, when looked at together reveal that the overall trends in the second five years of the ten year survey have in general terms increased by over 300% when compared with the first five year period.

In May of 1973 the Department of Health and Social Security (DHSS) produced a memorandum concerning community services for alcoholics, and a great deal of credit goes to Sir Keith Joseph, the then Secretary of State of Health and Social Services, in that he was prepared to look alcoholism squarely in the face and at least do something about it, or make certain recommendations. In that document the Government of this country for the very first time recognised the problem and put up an estimate with regard to numbers stating that there were probably between two hundred thousand and four hundred thousand alcoholics in England and Wales. Even at that time this was a gross under estimate. Governments are notorious under-estimators, and the figures they suggested would undoubtedly be based on information available for 1969/1970. In May 1970 Dr. Wilkins and his colleagues at the Department of General Practice, Manchester University, carried out a survey amongst three thousand patients in a particular group practice. The survey revealed between 32 and 52 alcoholics in that patient practice - a much higher rate than ever discovered before. Extrapolating Wilkins' findings to the rest of England and Wales produced a mean figure of 15 alcoholics per thousand head of population between the ages of 15 and 65. Therefore, in 1970 there would have been at least seven hundred thousand alcoholics in England and Wales, almost double the highest number suggested by the DHSS three years later.

In a recent survey - a Community Response to Problems of Alcoholism - a report produced for the DHSS by Shaw, Cartwright and Spratley, it was found that 2.1% of the population surveyed were excessive drinkers and/or alcohol dependent, and 3.5% of that population had experienced social and personal problems due to their alcohol intake during the previous twelve months. The survey was carried out in one of the London Boroughs. Why then today is the official estimate only four hundred thousand alcoholics in England and Wales?

The British have an international reputation for being the masters of understatement. Alcoholism is undoubtedly a disease of denial. Alcoholics are certainly guilty of denial. Denial is not the sole prerogative of the alcoholic. The Government, Health Authorities and society in general deny the extent of the problem as it exists today, and moreover it is reasonable to assume that few if any of those who have a responsibility for planning services to treat alcoholics have yet awakened to the stark reality of the situation as it is likely to be by 1980. If the indicators show the same kind of increase over the five years, 1975 to 1980, the trend will be maintained and any statistician worth his salt would predict that by 1980 the author's estimate two years ago of one million alcoholics by 1980 is unrealistic and there could well be a million and a half alcoholics by the turn of the current decade.

On the 26th February, the House of Commons, considered a Private Member's Bill, Licensing Amendment Bill No. 2. Just before the Bill was debated the Government Advisory Committee on Alcoholism, whose Chairman is none other than Professor Neil Kessell, in opposing the proposed changes in licensing laws in a press release stated that "admissions to mental hospitals for alcoholic psychosis have risen at an accelerating rate around 5% a year before 1969, an average of 10.5% per year since 1969, and now 13% in the latest years. If this continues, alcoholism will account for nearly a quarter of all mental hospital patients by 1985. If Professor Kessal is right, how much is that going to cost in hospital bed useage, consultant's time, etc.

The first part of the paper has been devoted to stating the case to leave no one in doubt as to what is considered to be not only the greatest single community problem that the nation faces, but the community problem that produces least concern, least action, and least money to alleviate. What is going to be done about it?

America's alcoholism problem is approximately four times greater than the United Kingdom's. The Government of the USA is currently spending - and this is Government money - some three thousand million dollars per year in supporting national and regional programmes to alleviate alcoholism, and remember they have no national health service. How much is this country spending? It is doubtful if the National Council on Alcoholism and all the regional councils put together are receiving Government funding, either direct from the DHSS, through regional health authorities, area health authorities, or social service departments, in excess of £250,000 currently. What a contrast.

When the question is asked, what is alcoholism, who are alcoholics, one enters a world of confusion, mystery, mythology, misconception and downright prejudice. Someone once said "evil is in the eye of the beholder". Alcoholism certainly is in the eye of the beholder. Unless one is dealing with a so-called chronic alcoholic, and by that is meant an individual who is so sick in body and mind, and so sick within society, that even the most uninformed individual would recognise him, we are up against individual perspectives and professional bias.

There is no doubt that the caring professions and organisations who deal with alcoholics see special samples of those who suffer. The psychiatrist, because he is a psychiatrist, will tend to see only those individuals whom it is considered need his professional care and expertise. On the other hand the Warden running a hostel for vagrant alcoholics sees an entirely different patient population.

Last year the MLCCA conducted a survey of some 3,000 individual cases seeking basic information only. The sample took in approximately a five year case load. It is believed that this is the only survey of its kind and size ever conducted in the United Kingdom. Again it is believed that this survey represents a truly comprehensive cross section of all those people within a patient population who suffer from alcoholism. Some important things were revealed by the survey which is reported fully in the Council's last report entitled "Alcoholism and its Variations".

<u>Firstly</u> there were no less than 452 individual professions or occupations and it was possible to identify certain professions or occupations that have a high risk factor.

Secondly there was no preponderance in any one of the five social classes. Dockers and doctors would appear to be just as much at risk as executives in social class 1, as too are housewives throughout the social spectrum. There appears to be a high incidence for example amongst seafarers, drivers of motor vehicles and those who are self-employed. High on the list come labourers and clerks, and surprisingly those who have retired. Equally surprising, nurses, school teachers and solicitors.

Thirdly 80% of the total sample were married or had been married and we found rates of separation and divorce to be four times higher amongst alcoholics than in a normal population.

Fourthly age groupings. Alcoholism was once considered to be a disease of middle life. That is no longer true. 44% were aged 40 and below. There were some 14 cases under the age of 20 and 25 cases aged 71 and over. There were approximately 2,400 in the sample who were married, separated, divorced or widowed, and there were almost 5,500 children involved.

Fifthly, 75% of the sample were still in employment.

Sixthly, the ratio of male to female was almost exactly three to one and it is interesting to note that in 1963 and 1964, the first two years of the Council's work, the ratio of male to female was almost ten to one. Last year, that is 1975, the ratio had dropped even further to about two and a half males to one female.

Lastly, of those needing medical care, 70% were either seen in or detoxified in the medical ward of general hospitals. 10% were detoxified at home under the care of the patient's general practitioner, 15% were either seen in or admitted to psychiatric hospitals, and 5% were referred to and treated in the alcoholism treatment unit. In every case, if the patient had not been referred by his general practitioner, the practitioner was informed of the Council's involvement.

The average alcoholic is likely to be a family man or woman with two or three children, in employment, reasonably socially stable and now under the age of 40, thus eradicating the stereotype image of the alcoholic. Only 2% of the 3,000 cases were the so-called typical male alcoholic, single, vagrant, and homeless. Only 5% of the women were the so-called typical female alcoholic - middle aged, middle class, menopausal, imbibing neat gin in the middle of the afternoon behind lace curtains in respectable suburbia.

We all take comfort from the fact that alcoholism never happens in our families, within our firm or organisation. We don't have alcoholics down our street. It is always in somebody else's family, in somebody else's firm and another neighbourhood. Alcoholism can happen to you and to yours, providing the circumstances are right or ripe.

There have not been many definitions mentioned during the Conference. Perhaps the simplest and most succinct of all and the basic definition the Council uses when assessing any individual patient is that proposed by Dr. D.L. Davies, Medical Director of the Alcohol Education Centre, London: "alcoholism can be defined as the intermittent or continual ingestion of alcohol leading to dependence or harm".

Dr. Davies points out that this definition cuts out drunkenness which is acute intoxication, it comprehends urge, compulsion, craving and the alternative, actual withdrawal symptoms such as tremor and also harm, whether physical, mental or social in the broadest terms whether to the individual or to others.

The aim of the definition is to state the least which all sufferers from alcoholism have in common. Any one of the three features, described, two of dependence and one of harm, is enough if it is a consequence of chronic ingestion. In his definition, Dr. Davies is asking two questions, whether the individual is dependent upon the drug alcohol, and whether harm results from his alcohol intake. The two points of dependence on the drug alcohol and the total harm from a medical and social sense that results therefrom constitute a logical basis for considering alcoholism.

The author suggests that alcoholism is a complex condition and there is no perceptible single factor responsible for it. There are a variety of causes and their individual preponderance differs from person to person. Alcoholism results from the interaction of the personality of the individual, the environment, both past and present, in which the individual functions, together with excessive drinking, this produces alcoholism.

Referring again to Dr. Davies, he has stated categorically that "the majority of alcoholics are diagnosed and treated outside medical channels and rightly so. Treatment of the alcoholic should not necessarily start with a doctor or in hospital". MLCCA work over the past thirteen years, during which time approximately 8,000 individual cases of alcoholism have been dealt with bears out the truth and reality of this statement.

If the Council's work was detailed in the areas of information, advice, referral, treatment, care and after-care of the alcoholic, and the service that is available to those who are directly affected by the alcoholic, whether it be a family member or an employer, it would require at least another paper of the same length.

Because of a terrific demand from doctors and social workers alike, and other Councils on Alcoholism, we have produced a booklet entitled "Treating Alcoholism - The Merseyside Experience", the booklet provides chapter and verse of MLCCA methods. Therefore, it is necessary only to summarise the main points with regard to what is considered to be a necessary comprehensive community based facility that is capable of providing for the needs of alcoholics and those who are affected thereby, a service which operates very much in the field of secondary prevention.

When the author took over as the Director of the M.L.C.C.A. in 1964, it was patently obvious that any service offered to the community must be of a highly professional nature in order to gain credibility. It had to be efficient and managed like any other business and above all it had to be attractive in that it was capable of dealing with any or all of the many needs of the alcoholic. It is simply not good enough for any medical practitioner or social worker, on coming into contact with the alcoholic, to say the answer to alcoholism is referral to the Fellowship of Alcoholics Anonymous (A.A.)

Because of A.A's success in helping many individuals to recover from alcoholism, there is a danger that too much may be expected of it. In fact many people who come into contact with alcoholics, through either a lack of knowledge or an unwillingness to be involved, are only too ready to pass the responsibility for treatment to A.A. when the fact of the matter is that many of the problems the alcoholic presents with are completely outside A.A's. terms of reference.

Each local group of the Fellowship is autonomous and there are no rules or conditions for membership other than an acceptance that one has become powerless over alcohol and one's life unmanageable. The group can only ever be as good as the members therein, they present the image and interpret the philosophy. In certain areas of the country the Fellowship is comparatively strong, particularly where the members are not only sober, but stable and are willing to co-operate with others to help sick people. In some parts the Fellowship is less flourishing and few groups exist.

Nevertheless, it is our experience that where groups of recovered alcoholics in A.A. are active, and sponsorship readily available, one can find no greater ally than the sober, stable member of the Fellowship who can be of tremendous help to those who are prepared to accept its principles and base their sobriety on its programme of recovery. It is of paramount importance that in the development of community resources, the role that can be played by A.A. should be appreciated in its proper perspective.

It must be appreciated that the A.A. approach is one of a great many. Inevitably it is self-selective and perhaps because of that, starts with a commonly accepted viewpoint which has therapeutic value for those who accept it, irrespective of the scientific validity of that basis. Alcoholics are so much in need of help, so many exist, and in so many varieties, that no form of aid is to be ignored if it appears to work. But claims that A.A. is the only, or the best, or the most effective agency should not be made without objective scientific scrutiny, and this of course applies to all such claims in this area from whatever source. Nor are the tenets of A.A. to be accepted as scientific facts without the obligation of submitting them to a similar scrutiny, though there is no objection to their propagation as articles of faith, as obtains in religious life.

The Council's experience has shown that a comprehensive service cannot be operated without the necessary highly qualified staff and the money to pay those staff, and above all there is a need for the confidence and co-operation of other agencies providing some of the elements of a comprehensive community service. For example, you cannot detoxify alcoholics without the co-operation of physicians and psychiatrists with bed availability. The help of social workers is needed where family problems are involved requiring their special skills and moreover it is of paramount importance that the general practitioner works with one in the interest of his patient.

There is provided within the community in Liverpool, St. Helens, Birkenhead and now in Sefton, and hopefully soon in Preston, a focal point to which alcoholics and those affected thereby can apply for help and treatment. In none of the situations where the counselling facility is provided is there about it either the awe of medicine nor the unfortunate connotations of a statutory social service agency, and of course one is assured of complete confidentiality. The service is free.

The author remembers in his very early days with the Council, Lord Cohen, who is the Council's President, telling him "Kenyon, the essence of good medicine is a correct diagnosis". The first requirement is to establish what type of alcoholism one is dealing with and what personal harm has been occasioned. Secondly, there must be an assessment of the individual's total situation. It is foolish to treat the patient's medical and/or psychological condition without having regard for the marital situation, the financial situation, and the work situation. The basic requirement for any individual's recovery is vested in a correct assessment of the whole of that individual's problems.

The majority of those seeking our help have a well established dependence on alcohol, many have a physiological craving, the majority are fearful, anxious and apprehensive. They are in fact saying to the alcoholism counsellor, "look this is my problem, I know that alcohol can temporarily dull my perceptions, shield me from reality and it can provide me with oblivion". He is looking for an offer of help. For far too long people have made offers to alcoholics that are grossly inadequate and only go part way to resolving the problem.

For example, it is simply not good enough to put the patient into a treatment situation and then forget him. Every patient hospitalised by the MLCCA is seen at least twice a week in hospital wherein, as soon as he or she is fit, the counselling, caring process starts. It is significant that the MLCCA is responsible for hospitalising approximately one hundred patients per year and is being called upon by both physicians and psychiatrists to see almost double that number of those patients whose alcoholism has been diagnosed or become evident whilst in hospital.

About one third of our total case load are now medical referrals of one type or another. Once the patient is fit and discharged from hospital, individual counselling follows at a frequency and over whatever period of time is necessary in any individual case. This is followed by group counselling and in the majority of cases this takes place with the non-alcoholic husband or wife. As soon as the individual is fit enough he is back at work. 10% of referrals are now from employers and the employee is given time off work to attend for regular counselling sessions. Obviously the same procedure applies where the alcoholic does not need any form of in-patient treatment. In all cases where another helping agency is involved, the aim is to co-operate and co-ordinate.

Initially, the responsibility of taking care of any or all of the problems that the alcoholic is unable at that point in time to cope with is undertaken by the counsellor. The alcoholic is supported and sustained, and as recovery progresses, so responsibility is returned to him or her. By the time treatment is completed, and this usually takes on average a period of six months, the alcoholic can be returned to society, sober, competent and capable of managing his own life and his own affairs without having to resort to alcohol to help him.

In the majority of cases the goal on offer is that of total abstinence but as more experience is gained, the Council is experimenting in some cases with certain types of alcoholics, the feasibility of returning to controlled social drinking, but it is too early yet to say how successful or how applicable this goal might be.

What has become known as the Merseyside model is not only viable but it has an enormous potential. Following on the re-organisation of the National Health Service and Local Government, it had to be decided to which authority - health or social service - M.L.C.C.A. should apply for the major portion of statutory funding. The Council opted to apply to the Health Service and went to the top tier to the Mersey Regional Health Authority and the following is a quote from that letter received in April 1973, when a grant of £10,000 was agreed by the Mersey Regional Health Authority.

"The Authority accepted that there was a need for substantial financial aid to your organisation from Health and Local Authorities in the Mersey Region and that the major contribution should come from the new Health Authorities". The Council now receives support from individual Area Health Authorities within the Mersey region and Social Service Departments.

When talking about alcoholics,please at all times remember that alcoholics are people; people with problems. They could include your wife or husband, father or mother, brother or sister, friend or employee.

Alcoholics unlike rats or mice, or pavlovian type dogs, are human beings with feelings who will respond to sympathy, understanding and care. Obviously they have a capacity to learn and assimilate knowledge. It is up to all of us at all times to remember that the alcoholic is worth treating. Given the right co-ordinating body and the necessary comprehensive facility within a community, Merseyside has demonstrated that at least two-thirds of those who have come under notice for treatment, and undertake treatment, recover and become once again useful members of our society, responsible parents and productive employees. The majority of the work is truly very much in the field of secondary prevention.

It is encouraging to note that strides have been made in the advancement of treatment, research and rehabilitation, but there is a long way to go yet. It is the duty of all informed individuals in this particular area of medical, para/medical, social help, to not only assimilate knowledge to help in this work, but also to make society in general aware that people who suffer from alcoholism can be returned to a normal life.

METHODOLOGICAL PROBLEMS IN EVALUATING DRUG MISUSE INTERVENTION PROGRAMMES

MARVIN A. LAVENHAR

DIVISION OF BIOSTATISTICS

NEW JERSEY MEDICAL SCHOOL, NEW JERSEY

The great tragedy of science - the slaying of a beautiful hypothesis by an ugly fact.

Thomas Henry Huxley

In response to the problem of widespread illicit drug use in the late 1960's, numerous intervention programmes were instituted in the absence of careful planning and tested hypotheses. A great amount of resources in time, energy, and money has been expended in preventive efforts. As the costs of these programmes have spiraled upward, with few apparent successes, the trend has been toward critical assessment of the efficacy of these efforts.

There is currently little empirical evidence available to demonstrate that intervention programmes have indeed been effective in reducing the problem of drug dependency. Most of the programme evaluations have not been based on objective scientific principles. In those limited cases where scientific methodology has been employed, for the most part the "beautiful" hypotheses of the 1960's have not been supported by the critical findings of the 1970's.

The intent of this paper is to provide an overview of some of the major difficulties and research shortcomings encountered in the evaluation of drug misuse intervention programmes, and to suggest directions and guidelines for future research. The following discussion will focus upon research methodologies for evaluation community drug education and treatment programmes.

Evaluation of Treatment Programmes

In the late 1960's, the increasing incidence of drug

dependence and the substantial recidivism among drug offenders demonstrated the ineffectiveness of the predominantly legal approach to drug abuse and triggered the proliferation of newly established drug treatment facilities in an attempt to find a therapeutic solution to the problem. Although the medical profession has been studying and treating drug dependence for more than a century, there is still a great deal of uncertainty and confusion about the causes and nature of drug dependence and about the rationale behind treatment. Although a wide range of treatment approaches has been employed, no method of treatment has been proven generally effective.

Evaluation in the past relied almost entirely on naturalistic observation; systematic data collection and feed-back have been, in general, nonexistent or extremely primitive. Judgements as to programme effectiveness have been based generally upon minimally documented evidence.

The most pressing problems encountered in evaluative studies have generally been lack of funds for in-depth evaluation, a lack of expertise, and a lack of objectivity. There have also been, and still are, many specific obstacles to effective evaluation which are extremely difficult to overcome. Some of the major difficulties encountered in treatment evaluation include:

1. <u>Limited generalizability of evaluative findings</u>

It is difficult to generalize from reported evaluation efforts because of the wide variation in populations studied, research methodology, sample size, time span of observation, criteria for admission, and criteria for classifying clients as cured or relapsed. Generalizations are also hazardous since the research subjects are usually the hospitalized or institutionalized drug addicts, not the unknown or invisible street addicts who avoid contact with official agencies. Therefore, evaluative findings may only be relevant to a particular population in a given treatment setting.

2. <u>Lack of agreement on aims, objectives, and criteria for effectiveness of treatment</u>

Treatments for drug dependence differ not only in method but also in rationale. Of course, all drug treatment programmes are dedicated to the elimination of drug dependence as their ultimate goal, but they may pursue this goal in different ways.

In the past, changes in drug involvement were used almost exclusively to assess treatment efficacy. More recently, it has been argued that no single criterion measure is sufficient to serve as evidence of treatment effectiveness. Sells (1973) categorized the goals and rationale of treatment into two major groups - intrapsychic changes and behaviour changes. Because of the difficulties in measuring intrapsychic changes in attitudes, personality organization, character, and values, these criteria have not been used extensively in evaluation studies. The most generally accepted criteria for behavioural change are those defining life style which include, in addition to drug use, such critical areas as employment, family relations, social adjustment, and participation in illegal activities.

3. Unavailability of appropriate outcome measures

In treatment evaluation we frequently measure what we can, not necessarily what we would like to measure. In other words, the selection of outcome measures is often dictated by the type of information that is available.

Two types of data may be generated in drug treatment programmes - management information and research and evaluation information. Management information tends to be cross-sectional and deals primarily with process outcomes such as rates of intake, rates of termination, readmission rates, and retention rates. Research and evaluation information tends to be longitudinal and focuses upon product outcomes such as substance use, criminal activities, and employment status.

Effective process analysis depends on the early implementation of a good management information system as an integral part of the day-to-day treatment routine. Large programmes generally require access to a computer. The major outcome measures of a product analysis are generally elicited by personally interviewing the client, and sometimes his friends, members of his family, or programme counsellors who are in close contact with him. This is a difficult, time consuming, and costly task which is frequently beyond the capabilities and resources of many treatment programmes.

4. Difficulties in obtaining objective and reliably measured data

Few evaluative studies of treatment effectiveness can withstand close scrutiny. Positive evaluations are readily accepted; negative evaluations are frequently challenged - usually on the basis of reliability and validity. The organization of reporting to assure reliable and valid data involves careful form design, thorough training of interviewers, and systematic built-in checks to minimize error.

Evaluations are frequently performed intramurally by individuals who are personally involved with the programmes. When budget cuts, staff reductions, or the very life of the programme is at stake, the data made available by programme administrators tend to be quite biased.

5. Practical difficulties in applying experimental designs

The degree of control of all factors relevant to treatment outcome that is needed to apply classical experimental designs is rarely, if ever, attained in a treatment setting.

In evaluating therapy, it is ethically indefensible to deny treatment to certain individuals in the name of research. Therefore, random assignment of drug-dependent individuals to treatment and non-treatment groups can hardly be justified. Frequently, a quasi-experimental approach is employed which utilizes a matched control group of individuals who never reach a significant treatment stage. However, addicts who do not receive significant treatment may be inherently different from those who do.

In most settings where more than one treatment is available, the choice among treatment alternatives is essentially haphazard, influenced by a number of factors such as the reputation of the programme, the type of facility, the proximity to the patient's place of residence, and the criteria for admission. Random assignment to alternative treatment regimens is difficult to implement.

6. Practical difficulties in patient follow-up

Most evaluation designs depend upon criterion measurement prior to treatment and at one or more points after the patient returns to the community. A major problem in implementing evaluation studies involves locating the graduates and drop-outs subsequent to treatment. Unless substantially all or a representative number of the subjects can be contacted, the results of evaluative studies may be seriously biased.

The New Jersey Medical School Treatment Evaluation Model

Since 1969, the Department of Preventive Medicine and Community Health of the New Jersey Medical School has administered a comprehensive narcotic addiction treatment and rehabilitation programme for the metropolitan Newark, New Jersey community in co-operation with six independent treatment agencies. A model for extra-mural treatment evaluation has been developed (Sheffet, et al., 1973) which is, at best, quasi-experimental in nature, but is nevertheless likely to represent a prototype of what is attainable in treatment evaluation.

The New Jersey Medical School evaluation model is based on a computerized management information system which includes a narcotics case register providing descriptive data on all drug dependent individuals who come in contact with any treatment component, and a patient monitoring system which follows each patient's progress in treatment. The model has evolved into three evaluation phases, each focusing upon different criteria for measuring treatment efficacy.

Phase I: Patient Retention Rates

Life table methods are used to calculate patient retention rates for each treatment modality, and for various homogeneous patient subgroups, based on the premise that the longer a patient remains in treatment, the greater the chances of successful rehabilitation.

Phase II: Psycho-Social Variables

To permit a more comprehensive analysis of treatment success, a detailed psycho-social questionnaire is administered to each patient upon admission to treatment and repeated at one-year intervals.

Phase III: Treatment-Based Variables

On-site assessment of treatment programmes is implemented to determine to what extent treatment success or failure is influenced by specific treatment regimens, the attitudes and/or skills of the programme staff, the interrelationships between programme staff and addicts, and other aspects of the treatment environment.

Future Directions in Treatment Evaluation

Evaluation research in the treatment of drug abuse has, in the past, been primarily concerned with the impact of the total programme. The trend has been towards focusing upon sub-components of the treatment process to determine which specific intervention procedures produce the most desirable changes in which patients under which conditions.

Since it has been clearly established that not all patients respond equally well to a given treatment, the focus of evaluation research has gone beyond assessment of efficacy to a search for those patient and programme characteristics which discriminate between successful and unsuccessful treatment rehabilitation. If this search is successful, it may be possible to measure prospective clients on one or more discriminating variables and predict their likelihood of success in a given treatment programme or modality.

Evaluation of Drug Education Programmes

Until the last decade, community drug abuse prevention programmes were identified exclusively with the law. Once it became apparent that the misuse of drugs could not be effectively controlled by harsh legal penalties, police surveillance, and efforts to curtail the influx of drugs, then communities turned to drug education as the best means of prevention. It was hypothesized that communication of information about the effects of substance abuse would dissuade individuals from misusing drugs or cause them to stop if they were already misusing them.

Despite the fact that a great quantity of resources in time, activity, and money has been expended in drug education efforts, several recently published reviews of evaluative research on these programmes (U.S. National Commission on Marihuana and Drug Abuse, 1973, Braucht, et. al., 1974; Goodstadt, 1974; Smart and Fejer, 1974) reached the same conclusion - there is little scientific evidence available to assess the effectiveness of drug education. Compared to the large number of implemented drug education programmes, it was recently reported (Smart and Fejer, 1974) that not more than a few dozen have undergone even a rudimentary evaluation, probably less than a dozen have employed the principles of scientific experimentation, no more than eight programmes have utilized a no-treatment control group, and fewer than eight programmes have studied their impact on drug use.

Not only has there been virtually no empirical evidence confirming the hypothesis that drug education can prevent or reduce drug use, but it has been suggested that under certain circumstances drug education may even encourage drug usage by allaying fears and by arousing curiosity (Bourne, 1972) and may also, in effect, prove to be a costly distraction from the more important moral and political issues underlying the drug problem (Halleck, 1971).

In view of the uncertainty as to the impact of drug education, the United States National Commission on Marihuana and Drug Abuse (1973) recommended a moratorium on all new school drug education programmes until operating programmes have been tested for effectiveness.

The major problems which have prevented the field of drug education from realizing its full potential and have complicated evaluative efforts include: 1) lack of clearly defined goals, 2) competition from mass media and other sources, 3) lack of well-trained, effective teachers, and 4) lack of a clear understanding of the individual's motivation to use drugs (Smart and Fejer, 1974). Most of the research limitations attributed to treatment programme evaluation are equally applicable to the vast majority of studies evaluating drug education programmes. The methodological requirements for effective evaluation are frequently difficult to attain in real life situations. In addition to methodological problems, the evaluator of school-based drug education programmes is frequently confronted with strong opposition by parents and school administrators. Parents may object to a programme which assigns their children to a control group which receives little or no drug education or may refuse to allow their children to answer questions on personal drug use, on the grounds that this constitutes invasion of privacy. School administrators may feel threatened by attempts to test the effectiveness of educational programmes. Many evaluation efforts have been unsuccessful because the evaluators neglected to educate, motivate, and involve the community.

Future Directions in Evaluation of Drug Education Programmes

In order to make sound decisions as to the future direction of drug education, it is not only essential to be able to assess the overall impact of programmes, but it is also critical to ascertain the relative levels of effectiveness of different approaches to drug education, with different target populations, in attaining different educational goals.

It is frequently difficult to obtain a consensus on what should be the ultimate of drug education. Goals range from an emphasis on abstinence to a stress on decision making and the formation of values (Globetti, 1974). In light of the widespread experimental use of drugs among adolescents and young adults, the trend has been towards compromising the objective of total abstinence and accepting the more realistic goals of lessening heavy drug use or confining use to the less dangerous drugs.

Recent studies have demonstrated that there is little relationship between drug-related knowledge and drug-using behaviour (Einstein, 1973; Swisher, 1974). It is becoming more and more apparent that, to affect changes in behaviour, a drug education programme must do more than provide information about drugs. Educators are looking to develop a more meaningful educational experience which will focus upon preparing individuals to face a variety of problems and decisions throughout life, including those pertaining to the use and misuse of drugs.

Clearly, more scientific research is needed to measure the effects of drug education programmes. However, it is also important to gain greater understanding of the underlying factors which determine the effectiveness of these efforts.

General Guidelines for Programme Evaluation

1. Evaluation mechanisms should be built into programmes at their inception and systematic data collection should be conceived as an integral part of the programme routine without imposing an unreasonable burden on the programme staff.

2. Any evaluation effort should start with an explicit statement of realistic programme objectives and of the criteria for effectiveness.

3. A variety of criteria should be considered for assessing the efficacy of a programme.

4. Evaluation of treatment efficacy should be based on measurement of criteria before and after programme exposure and should be of sufficiently long duration to permit the assessment of long-range effects.

5. Evaluation research should be objective and based upon sound, well-considered experimentation. If a true experimental design is not feasible, then some level of acceptable control can still be achieved through quasi-experimental designs.

6. All instruments, tests, and procedures used in the collection of evaluation data should be valid, reliable, and objective.

7. The internal validity of the evaluative findings should always take precedence over the external validity or generalizability of the results.

8. Evaluations should be done extra-murally by individuals who are not personally involved with the programmes and are unaffected by positive or negative appraisals.

9. Evaluations should focus upon the impact of specific components of the programmes as well as the impact of the total programme.

10. Evaluation should be viewed as a continuous, circular process providing timely and useful feedback to programme administrators and staff.

Summary and Conclusions

There has been a plethora of newly established community drug misuse intervention programmes during the past decade. While the costs of these programmes have increased dramatically, there is little empirical evidence to demonstrate their efficacy. The trend has been toward more critical assessment of intervention programmes to determine whether or not they merit continuing public and/or private support.

Some progress has been made in applying scientific principles to research and evaluation problems in the field of drug abuse, but much has yet to be done. An attempt has been made to identify some of the difficulties encountered in applying scientific methodology, and to suggest some directions and guidelines for future evaluation research.

Although the imposition of rigorous scientific control in community research may not often be possible, it is usually feasible to approximate the true experimental environment by means of quasi-experimental research designs which can yield useful information. If evaluators take the appropriate steps to maximize the reliability and validity of their data, and to exercise some degree of statistical control, then meaningful assessments of the efficacy of community intervention programmes are attainable.

It has been argued that scientific evaluation is too costly in terms of effort, time, and funding. However, community intervention programmes are also extremely costly and unless they are objectively evaluated, there is no way of assuring their effectiveness.

REFERENCES

BOURNE, P.G. (1973). "Is Drug Abuse a Fading Fad?" Journal of the American College of Health Association, 21. 198-200.

BRAUCHT, G.N., FOLLINGSTAD, D., BRAKARSH, D. and BERRY, K.L. (1973). "Drug Education. A Review of Goals, Approaches and Effectiveness, and a Paradigm for Evaluation". Quarterly Journal Studies on Alcohol, 34. 1279-92.

EINSTEIN, S. (1973). "Drug Abuse Prevention Education: Scope, Problems and Prospectives". Preventive Medicine, 2. 569-81.

GLOBETTI, G. (1974) "A Conceptual Analysis of the Effectiveness of Alcohol Education Programmes" Research on Methods and Programmes of Drug Education. Addiction Research Foundation, Ontario. 97-112

GOODSTADT, M. (1974). "Research on Methods and Programmes of Drug Education". Addiction Research Foundation, Ontario

HALLECK, S. (1971). "The Great Drug Education Hoax". Addictions, 18. 1-13.

SELLS, S.B. (1973). "Evaluation of Treatment for Drug Abuse". Fifth National Conference on Methadone Treatment, 2. 1362-68.

SHEFFET, A., HICKEY, R.F., LAVENHAR, M.A., WOLFSON, E.A., DUVAL, H., MILLMAN, D. and LOURIA, D.B. (1973). "A Model for Drug Abuse Treatment Programme Evaluation". Preventive Medicine, 2. 510-23.

SMART, R.G. and FEJER, D. (1974). "Drug Education: Current Issues, Future Directions". Addiction Research Foundation, Ontario.

SWISHER, J.D. (1974). "The Effectiveness of Drug Education: Conclusions Based on Experimental Evaluations". Research on Methods and Programmes of Drug Education. M. Goodstadt, Editor. 147-160. Addiction Research Foundation, Ontario.

U.S. NATIONAL COMMISSION ON MARIHUANA AND DRUG ABUSE. "Drug Use in America: Problem in Perspective", Second Report, U.S. Government Printing Office. (1973).

PLANNING FOR THE FUTURE - DEVELOPING A COMPREHENSIVE RESPONSE TO ALCOHOL ABUSE IN AN ENGLISH HEALTH DISTRICT

T. A. SPRATLEY, A. K. J. CARTWRIGHT, S. J. SHAW

MAUDSLEY ALCOHOL PILOT PROJECT

THE BETHLEM ROYAL & MAUDSLEY HOSPITAL, LONDON

This paper is based upon research done by the Maudsley Alcohol Pilot Project in a Health District in South London with a resident population of about 100,000 adults. The original brief of the research was to design a comprehensive community response to alcohol abuse for the resident population of that district. In order to do this it was felt necessary to concentrate upon three research questions :-

A. What was the size and nature of alcohol abuse in that district ?

B. What was the current response to alcohol abuse and what were its problems ?

C. How might the problems be overcome and a better comprehensive response developed ?

RESEARCH METHODOLOGY

To answer these questions a study of the general population of the district and a study of certain general community agents in the district were conducted.

GENERAL POPULATION STUDY

For the general population study, a representative sample of 286 residents aged 18 and over was taken, which represented a response rate of 80%. They were interviewed for one hour each about their drinking habits, their attitudes to drinking and their opinion of, and contacts with helpers, both lay and professional. The respondents also completed questionnaires which included screening for problems of alcohol abuse.

GENERAL COMMUNITY AGENTS STUDY

In the general community agents study, 34 general practitioners, 28 local authority social workers and 23 probation officers were randomly selected. Three general practitioners refused to be seen, leaving 31 who were interviewed. No social workers or probation officers refused to be interviewed. All the agents were interviewed for an hour each about how they felt and dealt with problem drinkers and alcoholics. They were also asked for their opinions of various other agencies, and completed an attitudinal questionnaire. They also discussed in detail with us the last case they had seen whom they felt had problems from drinking. This case is referred to as the "designated client".

It will not be possible to discuss in detail the data from all the studies here, but they are reported elsewhere. (Cartwright et al. 1975).

FINDINGS

A. What was the size and nature of the problem of alcohol abuse?

In the general population survey, estimates of the size of the target population were made in two ways. Firstly, people were asked directly whether they had had a period in their life during which they had problems from their drinking. One percent of the general population said this was the case currently, and a further 2% said they had had a problem period in the past. There was some indication, however, that of those who admitted only a past problem period, about half were probably still having significant problems at the time of interview.

The second way of measuring the size of the target population was to take measures of those three criteria for the diagnosis of alcoholism which are used by the World Health Organization (1952) namely, "excessive drinking", "problems from drinking" and "dependency". To this end, measures were made of the total alcohol consumption of the general population in the week prior to interview, their problems from alcohol in the year prior to interview and their experiences indicative of dependence in the year prior to interview.

These three measures were not able to pick out a group in the population whose drinking was fundamentally deviant and who could be called "alcoholics". Rather there were degrees of heaviness of drinking, of "problems from drinking" and of "dependency". Generally speaking, the more heavily people drank, the more problems they had. Any cut-off point had to be arbitrary. If one took excessive drinking to be 70 centilitres or more of absolute alcohol per week, then 2% of the general population were excessive drinkers. If one took a dependency score of reporting at least two indications of dependency (out of a possible three) then 2% of the general population were dependent upon alcohol. If one took a problem score of reporting at least five problems (out of a possible twelve) then 3½% of the general population were suffering from significant problems from drinking.

Nine years previously a similar study had been conducted in the same district, (Edwards et al 1972). Comparison of 1974 figures with those obtained in this earlier study revealed that in the nine years the average mean alcohol consumption had increased by over one third and the mean "problem scores" of the general population had doubled. Quite clearly the problem of alcohol abuse was getting worse.

What then is the nature and size of the target population ? It is believed there is no small discrete group of people in the community who are 'alcoholics'. Rather that there is in the community a large pool of problems from drinking affecting all areas of physical, social and psychological health in all degrees of severity. This does not simply constitute a disease, but a complex medico-social problem.

It is recommended that the district being researched, the target population should not be people called 'alcoholics' but any person who experiences social psychological or physical problems as a consequence of drinking.

Therefore, for the purposes of planning, the number of people in the general population with problems from drinking is probably the most relevant figure. The target population in the district is thus about 3½% of the adult population.

B. What was the present response to alcohol abuse and what were its problems ?

1. <u>Specialised Alcoholism Services</u>

The traditional response to alcohol abuse has been the specialist alcoholism services which comprise basically Alcoholics Anonymous and the psychiatric alcoholism services. Because of the problems of anonymity it was of course difficult to estimate the extent to which alcohol abusers in the district were using Alcoholics

Anonymous. Nonetheless, the only local group had an attendance of about 20 persons and there were approximately three or four new members per month.

During the study year 0.16% of the district's adult population was treated for alcoholism by a psychiatric service of any kind. Of those attending any psychiatric service with a diagnosis of alcoholism, only a minority attended the psychiatric alcoholism services. Although both Alcoholics Anonymous and the psychiatric alcoholism services probably dealt with a particularly severely damaged group of alcohol abusers, it would, however, appear quite clear that their impact upon the total pool of problems in the community was very small.

2. The General Community Agents and their Difficulties

The research on general community agents was restricted to general practitioners, local authority social workers and probation officers, but the problems they faced are almost certainly shared by most other doctors, nurses and social workers who work in other settings and who are not specialised in alcoholism.

Problem 1 : Non-Recognition

There is a failure by agents to recognise and diagnose the problem. Edwards et al (1973) in a study in the same district, "Alcoholics Known and Unknown to Agencies" had already shown this and studies confirmed it.

Problem 2 : A feeling of hopelessness

Many agents feel that the condition is hopeless to treat, - "I was interested (in them) ten years ago but I failed and gave up". The sense of hopelessness is due to many factors. Sadly, both in general practice and in social work situations the complainant is often not the drinker but his spouse. The agent is often frustrated in not being able to contact the drinker himself. Patients are only diagnosed, if at all, at a very late stage when treatment can be extremely difficult. There is a failure to get into a treatment relationship with the drinker where both agent and client agree upon the nature of the problem and what needs to be done. Few agents have a treatment programme which they can apply to drinkers themselves. As one agent said, "We are swimming around in the dark". An agent may attempt to refer the patient to a specialist alcoholism resource but may well find that the patient refuses to go, or if he does go, quickly defaults.

Problem 3 : Inter-agency difficulties

Referral between agencies depends partly on good relationships and upon a correct understanding and respect for what the other agency has to offer. Not altogether surprisingly the relationship between many general practitioners and their social work colleagues was often poor. Although they usually see that the other has something to offer people with alcohol problems they say that often, in practice, this help does not materialise. Neither were there good relationships between the general community agents and psychiatrists. General practitioners and social workers were prepared to refer patients to psychiatrists but they often admitted that they knew little of what psychiatrists did. Some were openly critical of psychiatry, pointing out the ineffectiveness and unreality of its treatment. One felt that sometimes alcoholic patients were referred to psychiatrists not for a real expectation of benefit but to get rid of the patient. Efficient referral between agents of course also depends upon both having a shared language to describe clients, but there was no concensus among agents as to what constituted "alcoholism". On this, there was marked intra- and inter-agency disagreement. Some agents were not prepared to label clients as alcoholics at all, as a matter of principle. Some agents gave definitions of alcoholism which are idiosyncratic, or impossible to apply practically, such as "He is not an alcoholic if he can give up (drink) for a week or two". "If he smells of alcohol in the surgery he is not an alcoholic".

Problem 4 : Problems of responsibility and roles

Although general practitioners state that alcoholism is a disease they do not see it as so clearly within their role as other physical disorders such as, for example, bronchitis. Several general practitioners saw alcohol abuse as being at least partly a moral problem.

The local authority social workers as is well known, have recently been re-organized and are in a state of some confusion. They have had difficulties fulfilling the roles for which they have a statutory obligation. There is no statutory obligation for social workers to treat alcoholics and consequently they sometimes avoid doing so. Furthermore, many social workers felt that they did not have the right to ask people about their drinking habits.

Problem 5 : The agents' lack of knowledge and skill

Many agents lack the necessary knowledge to pick up those of their clients likely to have drinking problems. Furthermore, they lack the skills necessary to elecit an appropriate drinking history, to come to a correct understanding of the situation, and to establish a treatment contract with the patient.

Problem 6 : The agents' fear of alcoholics

The general community agents were all asked whether they found alcoholics more dangerous, more demanding or more embarrassing than other patients. Many agents admitted that indeed this was the case.

3. Summary

The position in the district can be summarised thus. There is a large pool of alcohol related problems of differing kinds. People are quite prepared, when asked properly, to admit to these problems but they do not regard themselves as 'problem drinkers' or 'alcoholics'. The vast majority of those with problems receive no help for the underlying drinking, despite the fact that the problems from their drinking bring them frequently into contact with medical and social agencies. The significance of the drinking is not realised by the agent. Whilst the underlying drinking remains unchanged, then generally the problems persist to the disadvantage of the person and the frustration of the agent. In an unknown proportion of people, the situation escalates into a vicious circle of increasing drinking and increasing problems. Eventually, the damage and the drinking are so obvious that a diagnosis of alcoholism is made by some agent. The agent then sometimes tries to persuade the person of this but disagreement usually results with the person refusing help. More often, however, the agent feels incompetent to help the person himself. He then either has to refer the person to a psychiatrist or A.A. The patient may refuse such referral or if referred quickly rejects them. Failures result in more feelings of hopelessness by the agents who then tend to be even more reluctant to recognise and treat drinkers.

C. How might the situation be improved ?

There seems to be a choice of two solutions. The first possible solution is to massively increase the specialised psychiatric alcoholism services. This is not feasible in the present economic circumstances. Neither is it likely to be the best solution. Many alcohol abusers would refuse to attend and there is a doubt whether the psychiatric services are the best form of treatment for most of them.

The alternative approach to improving the situation would be to concentrate on improving the quality of care given by general practitioners, nurses and social workers already working in the community. These workers are already dealing with many alcohol related problems but failing to recognise and treat the underlying drinking. The advantage of this solution is that it would be possible to get very early diagnosis of the problem. People would not have to diagnose themselves as 'alcoholics' or 'mental patients' before receiving help. It is believed that earlier diagnosis would lead to easier treatment and better end results.

General practitioners, social workers and nurses working in the community will not be able to perform these tasks without a great deal of help. They will need much more education, training in skills and much more support before they will be able to effectively recognize and treat drinkers themselves. How can this help be given ?

It has been suggested that in the district being researched, a multi-disciplinary medico-social team be formed, called a Community Alcohol Team. It should consist of at least a part-time psychiatrist and full-time social worker with specialised knowledge and interest in alcohol problems. The functions of the Community Alcohol Team would be :-

(a) To provide for the educational, consultative and supportive needs to all general agents by visiting them regularly at their place of work. Case consultations should provide the basis for this work.

(b) To encourage and foster some general agents to take a specialised interest in alcohol problems in their own agencies.

(c) To act as a centre of the referral network. To provide a common language and to facilitate the relationship and understanding between agents by arranging joint case discussions and educational events together, etc.

(d) To provide an educational, consultative and supportive service to the local Council on Alcoholism and its counselling service. This counselling service could provide an appropriate alternative to a psychiatric alcoholism clinic providing it has consultant backing.

(e) To hopefully experiment with ways in which the psychiatric alcoholism services can change to provide a more useful service to the community, e.g. it might examine whether the needs of the community and the general agents would be better served by a rapid detoxification service and a day care setting rather than an in-patient alcoholism unit employing long term group therapy.

The operating model of the Community Alcohol Team would not be that 'alcoholism is a disease' but that excessive drinking leads to problems, and both the drinking and the problems need to be understood and responded to by the helpers, using the resources of the individual and his family.

CONCLUSION

It is not expected that the local changes we have recommended will solve all the local problems. It could solve some and create others. Furthermore, what happens locally will be greatly determined by what happens nationally. Three national processes in particular will affect every local district in this country in the next few years :-

(1) The relative price of alcohol.

(2) The quality of basic education and training in medicine, psychiatry, nursing and social work.

(3) The quality of the post-graduate training resources provided by the Alcohol Education Centre. The Centre will need to provide resources to augment the education and training efforts throughout the country.

REFERENCES

CARTWRIGHT, A.K.J., SHAW, S.J. and SPRATLEY, T.A. (1975). "Designing a Comprehensive Community Response to Problems of Alcohol Abuse". Report to the Department of Health and Social Security by the Maudsley Alcohol Pilot Project.

WORLD HEALTH ORGANISATION (1952). Technical Report Series, 48.

EDWARDS, G., CHANDLER, J and HENSMAN, C (1972). "Drinking in a London Suburb. I. Correlates of Normal Drinking". Quarterly Journal Studies on Alcohol 6, 69-93.

EDWARDS, G., HAWKER, A., HENSMAN, C., PETO, J. and WILLIAMSON, V. (1973). "Alcoholics Known and Unknown to Agencies: epidemiological studies in a London Suburb". British Journal of Psychiatry, 123, 169-183.

THE FEATHERSTONE LODGE PROJECT - PHOENIX HOUSE - ONE METHOD OF REHABILITATION

DAVID and SUZANNE WARREN-HOLLAND

PHOENIX HOUSE, LONDON

The Home Office Report on the Rehabilitation of Drug Addicts in 1968 stated : "... Rehabilitation principally involves persuasion to accept help, removal of physical dependence, and a longer process, removal of psychological dependence. Simultaneously with all these processes, attention must be paid to the requirements for the eventual social rehabilitation of the addict".

Phoenix House, a residential therapeutic community for the rehabilitation of long term ex drug abusers was started in late 1969. Its methods and concepts were based initially on the models provided by the therapeutic communities, Synanon, Daytop and Phoenix House, in America. Appreciating the differences between the legal, economic and social structures of the two countries, aspects of the American models have been changed to meet the needs of the British abusers, although the use of recovered abusers in the programme continues to be of central importance.

The methods used by Phoenix House are based on the following three premises: Firstly, an individual's reliance on chemical substances prevents him reaching solutions to his real problems, of which drug taking is merely a symptom. Therefore, total withdrawal from all intoxicating substances is essential, but is only a preliminary to tackling the problems themselves. Secondly, although there are many and varied facets of disturbed behaviour, the personality problem most commonly found among drug abusers, is that of emotional immaturity. This is characterized for instance, by irresponsible and demanding attitudes, avoidance of unpleasant reality and an inability to communicate effectively. Such manifestations may be countered by attempts to foster the abuser's emotional growth, to develop his sense of responsibility, his ability to communicate and to aid him to come to terms with his reality. Third, the assumption that abusers are helpless and incapable people, tends to deprive them of any opportunity to help themselves in their recovery and to assume responsibility for their lives. It is important, therefore, to oblige the abuser to meet the challenge of helping himself and others and bringing about his own recovery.

Thus the first step for any prospective resident is that he makes contact with the House himself, whether he is in prison or hospital, referred by a Social Worker, Probation Officer, or any other social work agency. The criteria for selection of prospective residents is not dependant on past history or initial strength of motivation, merely that he agrees to the conditions outlined to him; namely, that he give up all drugs before admission, and become aware that he needs help with the problems underlying his drug abuse.

The new resident walks into a well-thought out hierarchy structure within which everyone has a clearly defined function and status, and which provides both support and incentive to strive towards higher levels of responsibility. He will start work in one of the departments created in order to keep the house functioning, (the residents do all the cleaning, cooking, man the switchboard, etc. themselves). From the beginning he is made to understand that his progress is his responsibility, and if he wishes to have a better and more responsible job, then he has to earn it by becoming a more responsible person. In this he is helped by a form of role playing central to behaviour therapy. The method demands an immediate change in behaviour by asking each individual to act as if he is responsible and self-aware, and not allowing him to seek self-justification or excuses on the grounds of anything that has happened in the past, or the behaviour of others. The strict rules about behaviour have been designed to help residents give up functioning on an immediate gratification basis, and learn to function on a more constructive realistic principle. Therefore, a resident is not allowed to indulge in immature behaviour when under stress during the day's activities, but must learn to control this until given the opportunity in a constructive setting.

A typical day in the House starts at 7 a.m. when the residents are awakened. They wash, make their beds and clean their rooms thoroughly before breakfast - which is also prepared and served by residents at 8 a.m. At 8.30 a.m. the community gathers for the Morning Meeting and the day begins with a recitation of the Philosophy :

> "We are here because there is no refuge, finally from ourselves. Until a person confronts himself in the eyes and hearts of others, he is running. Until he suffers

them to share his secrets, he has no safety from them. Afraid to be known, he can know neither himself nor any other - he will be alone.

Where else but in our common grounds can we find such a mirror ? Here, together, a person can at last appear clearly to himself, not as the giant of his dreams or the dwarf of his fears, but as a man - part of a whole, with his share in its purpose. In this ground, we can each take root and grow, not alone any more, as in death, but alive to ourselves and to others".

This is followed by reading items of interest from a daily newspaper, and before the meeting ends each resident is encouraged to stand up and perform a short song, or tell a joke. This meeting should not be considered insignificant in light of the fact that most abusers start the day off with a fix. From 9 a.m. till noon, the residents do whatever jobs they have been assigned. Lunch is at 12 noon, and at 1 p.m. everyone assembles for a seminar, given by a resident or staff. All are encouraged to discuss not only the subject, but its presentation. Seminars are also used at times for a newcomer to tell his or her life story to the community. At 2 p.m. residents continue working until dinner which is at 5 p.m. The early part of the evening is usually spent cleaning the House. On Mondays, Wednesdays and Fridays, there are regular Encounter Groups which start at 7 p.m. On other days and week-ends there are various group activities in and out of the House. Residents are constantly encouraged to develop their initiative and talents within the structured framework of the programme, and to participate in the development of the community, since growth of individuals and the growth of the community are intimately related.

At this stage, no attempt is made to train anyone for any job or trade. The aim is rather to wean residents from their psychological dependence on drugs and to teach them self reliance and trust in other people.

The next few months become more intensive for the resident as he begins to seek more responsibility by progressing up the structure. Along with promotion comes the graded system of privileges to enable the resident to experience rewards for his efforts. Demotion can occur for a variety of reasons, and the loss of privileges encourages a resident to examine his attitude and behaviour, and to focus his energy on the aspects of himself that he desires to change or develop.

Whilst living and working together within such a small framework a resident's abilities and inadequacies become very apparent. Lack of social controls, social phobias etc., are everyday problems that residents can resolve with the help of their peer group and the community. Each resident writes a self-assessment once a week, and from this he can monitor his own progress or regressions and can also begin to see his development more objectively. Not all residents are so willing to see themselves realistically, but through the Encounter Groups and counselling sessions, each resident is given constant feedback on how others see him, and therefore he cannot avoid identifying the discrepancies between how he sees himself, in relation to how others see him.

The heirarchy structure of the House is balanced by the complete equality which is the rule in Encounters - a special form of group therapy. The apparently harsh and challenging way the group confronts an individual, stripping him of his rationalisations, illusions, self-defeating attitudes, and insisting that he sees himself realistically, is balanced by the frequently expressed and openly manifested support. In the Encounter the resident is encouraged to first express his feelings directly and with as much force and conviction as he is able, then is made to examine them, test their validity, and relate them to his general and habitual way of reacting and behaving. The resident is given much advice and guidance, support and concern which is continually being adapted to his particular needs and stage of development. He always has a choice and his view is always considered.

The principal aim in the Encounter and other therapeutic activities is to deter continued self-defeating behaviour. These methods may confuse an observer, sometimes being seen as destructive to self-respect, and therefore paradoxical. In fact, any confrontation is always used constructively, and provides a learning experience, not only for an individual but also for the rest of the community, who take part in its implementation. Thus the Encounter is perhaps the most important part of a carefully balanced therapeutic programme.

The total rehabilitation process is a continuous one without sharp divisions into stages. As far as the programme itself is concerned, however, the final stage of rehabilitation is taken to be the period, when the resident starts attending the re-entry group, and begins to spend a larger part of his time outside the House, being gradually weaned from his dependence on it. He may, after discussion with his peers and staff, return to his studies, undergo some form of training for a trade or profession, or look for employment. People are encouraged to be honest about their past when applying for jobs, though the decision is a personal one. Those that are honest,face the prospect of being turned down because of their past addiction. This has happened on a few occasions, but no one person has met such rejection repeatedly. Generally speaking, employers are encouragingly free from bias.

Once he has a job, the resident will move into the re-entry

flat in Phoenix. He will have considerable freedom to come and go as he pleases, the only demands being that he treat the community with consideration and attend the re-entry group. This group operates in three main areas. It can be supportive, enquiring or directive, depending on the needs at any one time. Its main function is to encourage the development of a cohesive re-entry peer group where mutual concern and support for each other is always present.

The re-entry phase is of prime importance as it is inevitably where the effectiveness of Phoenix methods of rehabilitation are really put to the test. On a more personal level, it is also a test for each individual of his development up to that point. Re-entry is where each person puts his increased awareness, increased social skills and abilities in the work situation into practice outside Phoenix itself.

For these reasons it is an extremely difficult stage in treatment where each individual is full of fears, insecurities and self-doubts. The aims during this stage are to be supportive without being protective. To ensure that each resident, before going into re-entry, is aware of the difficulties ahead, but more importantly, is aware of his own strengths and weaknesses, so that he will have realistic aspirations for the future, and will not place himself in situations which will demand too much of him.

There are considerable advantages in mixing pre re-entry residents with re-entry people in the re-entry group. Those already in re-entry help others to be aware of the type of problems they will be facing and ways of coping with these, while people still in treatment can often remind those in re-entry of parts of the Phoenix concept which they have overlooked and which can be very helpful in the struggle to clarify problems or difficulties.

In 1974 the Project approached the Lewisham Housing Committee and were able to get a semi-detached, short-life property for use in conjunction with the re-entry programme. Called 'Omega House' it consists of a basement flat (occupied by a staff member), plus six rooms with kitchen and bath, organised by those living there. The Omega Project is designed as another step away from the community support of Phoenix towards a more independent self-supporting life-style. At Omega, individuals can experience every day tasks such as buying and cooking food, cleaning, etc., whilst there is no obvious pressure to do so. As the house also has a communal kitchen and lounge, its occupants can learn how to integrate their life styles with one another.

Omega is the last stage before individuals break away from Phoenix to continue their lives with the minimum of support; as such it will inevitably become the final testing ground for the awareness, social skills and working abilities they have developed over the past eighteen months. For people who have been accustomed to a life, however miserable, where one can and does get 'high' with relatively little effort, this part of the re-entry stage can seem monotonous and frustrating. Few things change from day to day; small achievements, whether they be emotional, intellectual or material, require perseverance and consistancy. Many people in this stage, looking for excitement and a place where they can 'let themselves go', are tempted to visit old friends still associated with the drug scene. Some do this because they seek another place to 'belong', others because they want to see if they can resist drugs that are readily available. Alcohol in this context has always been a problem, and the staff of Phoenix have attempted to deal with this in the early part of re-entry. It is now accepted that people must be working for at least a month before they consume alcohol, and initial intake is limited. It is becoming noticeable that individuals are more able to apply their own brakes, especially when in periods of stress. It is important to bear in mind that successful residents are not people without problems - this is unrealistic; but people who are aware of their own personal limitations in certain areas, and have the ability to cope with problems as they arise.

It is becoming increasingly felt in the field of drug rehabilitation that less emphasis should be placed on addiction and more on the individual's underlying problems. This is entirely consistent with the policy which Phoenix House has pursued since its inception. However, the obvious contradiction which presents itself is: If the drug addict is not a special case, why should a special facility exist to cater for him ? In these circumstances, the rehabilitation process may well become a further stage in the mystification of drug abuse. Communities must of necessity, change in accordance with the needs of the client group. This, as represented by the current Phoenix House population, presents a very different picture from the stereotyped drug abuser of the late sixties, and reflects a significant change in cultural attitudes.

The current trend towards poly-drug use could be said to constitute one aspect of a broad spectrum of deviancy, with alcohol readily substituted for drugs of choice when these are not available. Clearly the individual with similar psychological problems which do not necessarily manifest themselves in drug abuse, would fit into this environment. Recidivists often experience similar problems to drug misusers, such as lack of confidence, emotional instability, and difficulty in developing mature relationships. Perhaps allowing both these groups to develop in a more comprehensive framework will produce a more constructive dynamic towards change.

Considerable discussion within the Phoenix House staff group and Management Committee, together with other interested parties, has resulted in seeking approval to widen the field of referrals. The London Boroughs Association Social Services Committee Working Party on Drug Addiction has now granted this, subject to the following conditions :

> "That the over-riding priority shall continue to be the admission of ex-drug addicts.
> That initially the scheme be for an experimental period of twelve months; that the number of 'non-drug' referrals be limited to not more than three at any one time; and that the places concerned would otherwise have remained unoccupied. Such referrals to be confined to persons in need of rehabilitation on discharge from prisons or referred through the Probation Service and who are likely to benefit from the particular type of therapeutic support provided by the Featherstone Lodge Project".

THE CO-ORDINATION OF CARE IN THE FIELD OF ALCOHOLISM

ERIC WILKES

DEPARTMENT OF COMMUNITY MEDICINE

UNIVERSITY OF SHEFFIELD

The last quarter century has seen very large scale benefits from improvements in medical knowledge. Many old people who would have formerly been living painful and restricted lives can, after their arthroplasty of hip, for example, live full, independent and pain-free lives. Almost all such advances have been consequent on specialised knowledge, specialised training and specialised experience and although alcoholism remains a difficult and indeed expanding problem, there have been in this area also, benefits from highly experienced doctors giving group therapy, acute detoxification units and day care to patients who in rather more primitive circumstances would have not been helped nearly so well.

It is however, possible that a managerial peak has been reached so that if specialisation increases then the undoubted benefits may well be overtaken by the inefficient fragmentation of care, already growing to a unique degree throughout the whole field of modern medicine. For example, a recent study funded by the King's Fund demonstrated that handicapped children needed help from many different organisations. Something like 15, or even 25, quite different organisations may be involved in varying ways to provide help for the child and the family, yet in spite of this tremendous richness and variety of help available it was proved that what really got help for the handicapped child was a determined and knowledgeable parent. This was far more helpful than a good general practitioner or any social work agency involved in the support of the family.

Because the acute infections are now a problem largely solved and more and more resources are being dedicated to the management of chronic disabilities, it is being learned again and again that what is most difficult to deliver and most expensive to maintain, rather looked down on by the professionals, and yet vitally important for the patients, is high-quality supportive care maintained over a very long period of time. The lack of this high-quality supportive care disfigures our management of diseases like multiple sclerosis, or certain cases of hemiplegia and rheumatoid arthritis. Alcoholism is a classic example of high-quality chronic supportive care being needed but not necessarily being made available.

In a small local study, about 100 psychiatric re-admissions were examined. These included 67 males and half the number of females. In this tiny sample none of the alcoholics were females and the 67 males were made up of 27 schizophrenics, 15 who were either recurrently or chronically depressed, and nine who were alcoholic. The other conditions - mental retardation, hypomania, psychopathic personalities, manic-depressive psychosis - provided only two or three cases each. The women included more depressives than schizophrenics, but otherwise similar diagnoses. These were youngish patients, mostly between the twenties and fifties and yet the general practitioner saw only a quarter of these patients regularly, slightly more than a quarter occasionally and the rest he apparently saw rarely or not at all. Liaison with the hospital services, when it was mentioned at all by these general practitioners was only mentioned critically - letters were late or never arrived at all. The patient was thought to be in the care of the psychiatrists even if he was being seen only by a junior doctor at fairly infrequent out-patient consultations, and psychiatric nursing or social work support reaching out to these patients in the community was in this particular area almost totally absent. Can it be surprising therefore that many of these patients had been admitted five, ten or even more times to hospital and that from this small series of 100 psychiatric re-admissions over a cenutry of in-patient care had been required ? And is it not apparent that neither practitioner, health visitor, psychiatrist, community nurse nor social worker are involved in continuing care for the majority of such patients ?

The setting up of a Council for Alcoholism in the city of Sheffield has been discussed recently. It would act as a liaison, co-ordinating and educational headquarters, and one is impressed, and to some extent also depressed, by the number of organisations, at any rate theoretically involved in alcoholism. As well as Alcoholics Anonymous we have Al-Anon, with its tremendous potential for supporting the family as the unit of care; we have the Social Services Department; we have the Area Health Authority; the District Management Teams; the Family Practitioner and the Local Medical Committees. We have the University Department of Psychiatry. We also have the local Association for the Care and Resettlement of Offenders while the Probation and After-Care Service also has played a vital, indeed a leading, role in trying to expand day care for these and similar long-term problems. Voluntary organisations also have to be included so the local Council for Social Services was obviously an important headquarters that must be involved. The local branch of the National Association for Mental Health, and for Battered Wives, the magistracy, and the police were also relevant and so on, and so on.

The whole situation reminded one of recent decisions made when the problem of discharging elderly patients home from hospital was discussed. The Age Concern report from Liverpool detailing the problems consequent on poor communication was interesting and impressive. Elderly patients were being discharged by harassed staff on the wards to ill-prepared and unheated houses without any adequate mobilisation of the Home Help Service, the Meals on Wheels, the ministrations of the District Nurse or bath attendant and without the rapid notification of the general practitioner. Such a lack of co-ordination of facilities allowed frail and vulnerable elderly patients to be discharged from the total support of the ward to the total neglect of the community. The answer to this problem put forward in the Home from Hospital report was to invent yet another new animal, and this animal was to be called the 'After-care Co-ordinator'. The after-care co-ordinator was going to check with the ward sister and perhaps the consultant or registrar, as to the physical and social needs of the patients and, when the discharge date had been agreed, to make sure that all the appropriate statutory and voluntary services and all the appropriate Health Service departments were alerted.

Service to the individual has moved a long way from the responsible ward sister and the involved practitioner for an enormous team of ward clerks, ward receptionists, primary health care teams involving secretaries and receptionists and district nurses and health visitors has been bred and when there is an after-care co-ordinator co-ordinating all these it may be that it will drift from a situation of occasional and unjustifiable neglect to one of profligate overuse of scarce resources. Many patients do not need an after-care co-ordinator but they are needed for a tiny minority. If there is a co-ordinator to co-ordinate, however, one must co-ordinate. If the co-ordinator is worth employing at all then the job will grow. The author is not terribly enthusiastic at the organisational future that is foreseen. The after-care co-ordinators will themselves have to be co-ordinated, first at hospital, then at District, then at Area, perhaps even at Regional level, and again and again money will be wasted on expensive administrative machinery not directly participating in patient care. This is seen at its most flagrant and distressing in the increased professional status of those nurses, who, as their salary increases, become less and less relevant to the nursing care of patients. The doctor, who is perhaps most suitable for an almost exclusively managerial, consultative, and planning role, insists on a personal, yet intermittent, involvement with individual patients, and this perhaps is a luxury that cannot be afforded for many more decades. But at this sort of stage, and before a headquarters for the co-ordinating of co-ordinators is embarked upon, the family with the handicapped child should be examined again.

The parents who demand it most often get the care their child needs. No multiplication of anonymous associations, no matter how well intentioned, can replace the parental role. It is suggested that the co-ordination of care for the alcoholic and for the family of the alcoholic should be very much the responsibility of the family, and if there is a situation where the young and the female alcoholic are not greatly to increase in number, then families should be more involved than they have been in the past and the employers perhaps, also. Al-Anon with its emphasis on the family needing support, and perhaps in turn providing it, should surely be primarily considered for a co-ordinating role. The alcoholic who is diagnosed and under any kind of treatment at a fairly early stage of his disease will sink or swim to some degree through the quality of support afforded by his near family. If the problem of alcoholism has gone on for many years, if the family unit has collapsed, if the individual is already homeless, then indeed the hostel, the social care home, the foster parent, will replace the family and the quality of support given by these family-equivalents whether with local social workers or psychiatric community nurses, or volunteers, should have the same central role in co-ordinating the care for these people as the family, because they are indeed the family.

The author presents a plea for a simplification of the structure of care. In trying to substitute for the hard-pressed family, with its looser ties and more transient relationships, the social worker has tried to pick up the burdens of a whole society under stress. In attempting this so gallantly but with inadequate resources, they have become too often not the family but the scapegoat. Let there be a multiplicity of enthusiasms and skills and attitudes, but let the machinery be streamlined so that the family, whether the natural family or the hostel or foster-family, are involved in day-to-day care, since alcoholism, as much as any other disease process, is genuinely a day-to-day problem.

Section V

Prevention and Education

THE ROLE OF LEGISLATION IN DIMINISHING THE MISUSE OF ALCOHOL

CHRISTOPHER CLAYSON

CHAIRMAN, CLAYSON COMMITTEE

SCOTLAND

The three main factors which are supposed to limit the misuse of alcohol are social, fiscal and legal. It is imagined that this proposition will apply to any country and the three factors should operate in harmony with each other. In proportion as one controlling factor fails the repercussions of the other two increase until the point can be reached at which all three fail and it is suggested that it is virtually the situation which now confronts the United Kingdom.

Social control largely for instance by means of education whilst attempting to accept and even to encourage widespread moderate drinking also attempts to reject immoderate drinking or drinking in inappropriate circumstances. The law which in its modern form was introduced in England in 1902 and in Scotland in 1903 was passed by a well intentioned Parliament which believed that it could prevent people doing what people wanted to do by putting difficulties in their way and would, therefore, help to discourage immoderate drinking. More specifically, the principles of the law were that :

1. if the number of places where alcohol could be purchased, or purchased and consumed, were limited;

2. if the number of hours in which alcohol can be so obtained were limited;

3. if the sale of alcohol to drunken persons is prohibited;

4. if the licence holder has to ensure that these restrictions are observed on pain of loss of his licence,

then the result would be a diminution in the misuse of alcohol. This assumption cannot be justified.

However, the astonishing thing is that the law so passed appeared to work for some years after its inception. It worked to even better effect in World War I when added restrictions were widely if temporarily in use and most of all it worked during the great depression in the 1930's when poverty was extreme. But the depression passed and consumption and misuse of alcohol started to increase again. This tendency has accelerated especially since about 1960 on account of the fact that we are spending an increasing proportion of our augmented incomes on alcohol.

This feature of recent years can be studied in the annual review of statistics in the Central Statistical Office for 1975. It shows among many other things, how the proportion of our average weekly household expenditure on commodities and services has changed between 1964 and 1974. Although we are of course spending more on everything, the actual proportions spent on different items have changed as follows, during the ten year period :

The proportion spent on food dropped by 15 per cent.
The proportion spent on heat, light and fuel dropped by 17 per cent.
The proportion spent on clothing dropped 10 per cent.
But the proportion spent on alcohol increased by 17 per cent.

It would seem that we are tending to restrain expenditure on essentials whilst allowing ourselves to spend more on alcohol.

One might think that this was where the third factor, namely fiscal control, should help but this is not so. For instance, in this country whatever is felt about the price of whisky, the cost of whisky expressed as a percentage of our disposable income has been diminishing steadily for the last twenty years. To be effective, fiscal control would have to be based on alcohol content and to increase steadily each year to a degree which all Chancellors have the wisdom to calculate but none as yet have had the courage to impose.

So all three methods of controlling the misuse of alcohol have been unsuccessful. The concern of British governments of recent years has been shown in various ways but in two particular respects. Firstly, in 1968 it established the Health Education Council for England, Wales and Northern Ireland, and the Scottish Health Education Unit. Secondly, in 1971 it created two Departmental Committees, one under Lord Erroll for England and Wales and the other, with which the author was associated, for Scotland. The task was to ascertain whether and in what way the licensing law could be changed in the public interest. The problem in Scotland is particularly difficult. The frequent and futile admonitions of the law and the equally frequent and equally futile admonitions of the medical profession, have identified five measurable indices for the misuse of alcohol and these are: drunkenness offences, drunk driving offences, admissions to hospital for alcoholism, deaths from alcoholism and deaths from cirrhosis of the liver. According to which index is chosen to study the misuse of alcohol in Scotland, it is two to seven times as bad as it is in England and Wales. This is not due to the fact that more alcohol is drunk in Scotland, but rather to the way in which it is drunk. The actual amounts drunk per adult per week are represented by an expenditure of 79 pence in the United Kingdom and 78 pence in Scotland - an insignificant difference. In Scotland however, the population drink under greater pressure. There are fewer licensed premises and registered clubs in proportion to population: not a large difference (6%) but it is there. Then there are 17% fewer permitted hours in which these establishments can be patronised as compared with England, including none at all in public houses on Sundays. Saturday is of course the most popular day for drinking, but almost as much is drunk on Sunday as on Friday. But since public houses are not available on Sunday the disagreeable pressure on hotel bars can be imagined. The Scots have a higher proportion of spirit drinkers among their regular drinkers as compared with the English (29% as against 17%) and are more favourably disposed to mixing drinks with an effect which appears to be more than additive. There are other peculiar features in the Scottish scene but since the same demand for drink is met by means of restricted facilities the pressure to drink is greater than in England and Wales.

The committee had no doubt that changes in the law were urgently needed. Needless to say it was impressed though not astonished by the number and weight of discordant testimonies. Some argued that since the law had failed it must be tightened; that that same freedom which extends the vices cannot at the same time diffuse improvements in social life. Others said that restrictions of any kind were an affront to human liberty and that all should be completely free to drink what, where and when desired. The Committee took the middle view of a co-ordinated relaxation of the law, bearing in mind that beneficial changes could not be induced overnight and that probably planning was for at least 25 years.

Of the 103 recommendations, some were purely technical or procedural and would be of little interest to this international conference but the proposals which had important social implications were in four main groups.

Firstly, the Committee proposed a new local licensing authority to control the number and types of licence providing for the purchase and consumption of liquor. In each local authority district this should consist of elected representatives of the people appointed by and serviced by the district authority but operating independently of it. In granting or refusing licences it must take into account not any mathematical question of need but many other matters including location of premises, the potential for nuisance, the history of past nuisance and the effects on amenities, and moreover it must give reasons for its decisions if asked to do so. It should be noted that at the present time the licensing authority cannot be asked for any of the reasons involved in taking its decisions.

Secondly, if such reasons appeared wrong in law or an altogether unreasonable exercise of descretion having regard to the facts of the case there should be the possibility of an appeal to a judge, in Scotland to the Sheriff Principal. This was to ensure no corruption, of which be it said many vague suggestions were received but only one reference to a proved case.

Thirdly, the Committee proposed additions to the range of licences which presently include hotels (full and restricted), public houses, restaurants and off-sales. The additional licenses would be :

(a) An entertainment licence for the convenience of patrons, in places of public entertainment.
(b) A residential licence for guest houses for the convenience, as the name implies, of residents only. This could correspond to the arrangement which has been operative in England and Wales for years.
(c) A refreshment house licence. This new establishment has been referred to as a cafe-pub. This is quite wrong and is most unfortunate. It is not a pub - cafe or otherwise - and has no bar. It is a refreshment house with table service for snacks, tea, coffee, milk, soft drinks or alcoholic beverages. It is a place where the family could remain together and the parents could have their alcoholic refreshments if they wished and the children their soft drinks, until, of course, they were old enough for alcohol. This is one of the measures we took aiming to introduce young people to alcohol sensibly as a normal part of human activity and attempting to dispel the prevalent conception which equates drink with toughness.
(d) In this connection, the Committee recommended for the same purpose a 'children's certificate' for certain public houses. A whole series of conditions for approval of such public houses were laid down. At the present time only a minority could meet these conditions in Scotland but the planning is for 25 years, and much importance is attached to this development in the future.

Fourthly, the Committee proposed extensions to the permitted hours for the purchase and consuming of liquor. It recommended a straight 11.00 a.m. to 11.00 p.m. period on weekdays and a 12.30 pm to 11.00 p.m. on Sundays, all at the discretion of the licensee. This represents, if fully implemented a 60% increase in the permitted hours.

This is necessarily a very brief summary of a whole range of interdependent recommendations. From the social point of view it is firmly believed that given full implementation of our proposals the following sequence of events will occur :

First, a reduction in the pressure to drink.
Second, an improvement in the quality of leisure.
Thirdly, less tolerance of the public of drunken behaviour, and
Fourthly, a restraining effect on the increase in alcoholism.

The last of these will be a long term effect. It should be noted that no law will change the established alcoholic.

Human laws cannot of themselves affect human attitudes - or only to a very limited extent. But our proposals on licensing law, coupled with our proposals on advertising and on taxing of liquor according to alcohol content, will support the educational and social approach to this enormous problem.

Earlier in the paper Lord Erroll's Report for England and Wales was mentioned. England and Wales on the one hand and Scotland on the other for historical reasons have always had their separate laws for liquor licensing, and consequently separate sources of advice on these matters. Not surprisingly, therefore, there were certain matters of substance on which our two Reports differed: in fact fourteen. The Erroll Report, for instance, saw no need for change in the different licensing system which operates in England and Wales whereas the Clayson Committee wished to see quite fundamental changes in Scotland. The Erroll Committee was prepared to reduce the age for the purchase and consumption of liquor to 17, whereas the Clayson Committee wished to retain the present age of 18; and to reduce the age for employment in a bar to 16, whereas the author's Committee proposed to retain it at 18. Instead of the eight or nine different categories of licence proposed by the Clayson Committee, Lord Erroll proposed only one but attaching those conditions the licensing authority saw fit for each establishment according to its purpose.

But whatever differences there were between the two Committees both were agreed on the essential and fundamental principle that the current law should be relaxed, and Lord Erroll in fact goes further than Clayson; indeed in certain respects the relaxation proposed for Scotland hardly achieves the degree of freedom which England already enjoys. The author is convinced that this restrictive licensing which Scotland has long endured is directly related to the more serious problem which Scotland now suffers.

Needless to say the proposals of the Clayson Committee have been widely criticised both in Scotland, and in England and Wales, and all of the criticisms cannot be detailed here. The chief arguments can be summarised in the proposition that more opportunities to purchase and consume mean more over indulgence and more alcoholism. This does not follow. If more opportunities to purchase and consume were genuinely of major importance in causing alcoholism, then there would be far more alcoholism today in England and Wales than in Scotland, whereas in actual fact, it is very much the other way round. It is more money which leads to over indulgence rather than opportunities to purchase and consume. For the last eleven years the great British public has been increasing its annual expenditure on alcohol at an average rate of £253 million per annum reaching the latest known total of £3,927 million in 1974. As a result all the manifestations of the misuse of alcohol are getting worse, and will continue to do so as long as society choses to spend its money in this way.

If however, facilities for drinking, especially for family drinking are improved; if more time is allowed for drinking in licensed establishments; if the social aspects of drinking are improved; in short if the pressure is taken off drinking; then that extra £253 million per annum will be spent with less harm to the individual. But one thing is certain; if the present licensing restrictions are maintained, and spending continues in the same way, those very evils we seek to diminish will be aggravated. So what will Parliament do ?

The Government's Licensing (Scotland) Bill was published a fortnight ago, and it is now known what is intended - for Scotland at least.

Many of the purely technical and procedural matters need not be commented upon, but some of the four main groups of proposals stated which were required for their social impact have been watered down - a highly appropriate metaphor.

1. The new licensing authority - this it seems is likely to be accepted more or less in the way we proposed and in the long term the authority will have better powers to improve the quality of licensed premises and their location; and it appears that in the fullness of time a more civilised distribution of licensed premises than at present may be seen.

2. <u>The new appeals procedure</u> which were proposed will be agreed.

3. <u>The new system of licences</u> has two major omissions which will operate against the achievement of our objectives. There will apparently be no residential licence for guest houses and no acceptance will be given to the proposed 'children's certificate' approving certain public houses.

So far as the former is concerned the decision means that in order to supply liquor a guest house proprietor must apply for a restricted hotel licence. This would then encourage or even require him to serve alcohol with meals to non-residents, which is not the purpose intended. Secondly, the refusal to grant a 'children's certificate' to approved public houses is regarded as most unfortunate. The Secretary of State for Scotland stated that the proposal should be judged in relation to "present day Scottish drinking practices and attitudes to drink", and he did not favour the proposal. The Government's own Advisory Committee on Alcoholism and the National Council on Alcoholism opposed the 'children's certificate' since few public houses today would be suitable. All this is very true. The Clayson Committee however, was not looking to present day drinking practices but to the future. As far as is known, no one, in Parliament or outside has mentioned the detailed conditions proposed for the future approval of public houses in relation to the admission of children in charge of their parents. A gradual development in social attitudes over a period of years was foreseen and hope was expressed that licensing authorities would not frustrate intentions by adopting too restrictive a policy.

4. <u>Permitted hours</u> - recommendations for permitted hours in Scottish public houses would have meant an optional extension of 60 per cent which would certainly reduce the pressure to drink. The Government proposals would only increase the permitted hours by 12% (with none in public houses on Sunday). This will not reduce the pressure to drink in the slightest, but in fact as public expenditure on alcohol continues to increase the result will actually be to aggravate the demand and its consequences, and this will especially apply to the very day on which the pressure to drink should be reduced, namely Sunday.

SUMMARY

1. The present restrictive licensing law in the United Kingdom and especially in Scotland has failed in its purpose of reducing the misuse of alcohol.

2. The proposals of the Departmental Committees in Scotland and in England and Wales would go far to repair this defect by legislation which would support the social control of the

the misuse of alcohol, and reduce the pressure to drink.

3. The United Kingdom Government's selection of the Departmental Committee's proposals for Scotland whilst improving certain technical aspects of the law will not, unless suitably amended, give adequate support to the social control of the misuse of alcohol.

ADDENDUM

Since this paper was delivered the Licensing (Scotland) Bill has received its Third Reading in the House of Commons (27th July, 1976).

Certain amendments to the Bill have in fact been made, the most important of which provides for the optional opening of Scottish Public Houses on Sundays.

REFERENCES

ADVISORY COMMITTEE ON ALCOHOLISM, (1976). Department of Health and Social Security Report.

CENTRAL STATISTICAL OFFICE (1975). Annual Review of Statistics H.M.S.O.

THE CLAYSON REPORT (1973). Report of the Departmental Committee on Scottish Licensing Law. Command 5354 H.M.S.O.

THE ERROLL REPORT (1972). Report of the Departmental Committee on Liquor Licensing. Command 5154. H.M.S.O.

HOUSE OF COMMONS OFFICIAL REPORT : SCOTTISH GRAND COMMITTEE (1975). Clayson Report on Scottish Licensing Law.

HOUSE OF COMMONS OFFICIAL REPORT (1975). Scottish Licensing Law.

SEMPLE, H.M. and YARROW, A. (1974). "Health Education, Alcohol and Alcoholism in Scotland". Health Bulletin. XXXII, 1, 31.

THE NEED FOR AND SOME RESULTS OF EVALUATION OF ENGLISH DRUG EDUCATION

NICHOLAS DORN

INSTITUTE FOR THE STUDY OF DRUG DEPENDENCE

LONDON

THE DRUG EDUCATION PROBLEM - CAUSE FOR CONCERN?

It may be speculated that the process of growth of drug education shows remarkable similarity with the process of growth of drug use, and might attempt to pursue the analogy with the aid of analyses based on sociological or on individual-psychological perspectives. When use of a new drug initially "takes off" in a certain population, there is a period during which it is used most unwisely, in inappropriate settings and at inappropriate times. Kids come to school stoned, take too much, freak out and generally make a mess of things until the drug and ways of dealing with it are fully absorbed into the sub-culture. At this early stage, the user is very involved with the drug and what he can achieve with it, apt to be an active proselytiser, and is rather defensive, brooking no criticism of his behaviour. Drug education also tends to "take off" at a certain point, and during its expansion phase a large number of approaches are enthusiastically advocated. Analysis of samples of these drug education capsules shows that they are frequently not what the seller claims them to be: in a recent review of drug education materials by our staff, many of those sold as "facts" turned out to be something else (in both meanings of the phrase). Many of the samples contained adulterants likely to produce anxiety in the consumer, but it is not known whether these adulterants were introduced accidently or whether they were included deliberately to provide a bigger "kick". One of the most alarming features of taking drug education is that no one can predict the effects, which depend on the person who takes it, the set and setting as well as on the formula taken. Many hope that drug education will solve their problems, but often find that the problems are still there, and sometimes worse afterwards.

Why do people do it? Is it a disease? What sort of person is likely to turn to drug education? There are no studies of the drug education abuser other than by observer participation, but many researchers believe that there is no typical personality - social factors that lead to one being in a particular place at a particular time may be important. Some professions are particularly at risk. Perception of drug education abuse as a means to gaining desired status and gaining attention may also play a part, especially when other routes are blocked. Few experimenters will become dependent, but those who do come to centre their lives on the activity, and become dogmatic and inflexible in their demands for more. Even for those heavily involved, there is a good chance of "maturing out" to less intense involvement with a wider range of education.

Seriously, it does look as though people may sometimes become intensely involved with drug education, possibly to the long-term detriment of themselves and those they influence. The acceptance of the need to evaluate drug education before widespread dissemination, just as one evaluates a new clinical drug, is a long time coming in the U.K. and we are beginning to see the signs of impatience of some people who have their own mixture, and want to get other people to try it.

The points to be made at this juncture are that if there are improvements to be made in drug education materials, they are improvements in quality rather than in quantity, and any such need is overshadowed by the need for adequate teacher-training. Secondly, we may observe that the lack of central co-ordination of drug education in the U.K. whilst having the advantage of flexibility and innovation, has the corresponding disadvantage of encouraging competing enthusiasms. There is still little awareness of a need to evaluate drug education materials before deciding whether to disseminate them widely: even those few persons who have visited North America seem to pick up the latest trend in methods and materials, rather than the need to evaluate. And when evaluation is mentioned, it is too often in terms of gaining impressions of effectiveness from staff or selected students, and serious thought is seldom given to more reliable forms of knowledge of effects.

A COMPLETED EVALUATION STUDY

It is against this background that the following research studies should be seen. An exporatory study in 1971 used depth interviews with children and teachers in thirty state schools in England and Wales. The report highlighted the state of flux of teaching about drugs, and recommended research to discover the effects of teaching secondary school pupils about drugs, in order to provide objective evidence on the basis of which advice could be given to teachers.

From 1972 to 1974 an evaluation study employed confidential anonymous questionnaires administered to 1,300 fourth, fifth and sixth year secondary pupils before, after and again two months after teaching about drugs planned and given by the pupils' teachers. Altogether, over 5,000 schoolchildren in over ninety schools in thirty Local Educational Authorities throughout England and Wales participated in various parts of this study.

The key implications of this evaluation research for planning of teaching about drugs within the health/social/moral sphere, presented in a booklet for teachers, Teacher Training Colleges and Local Education Authorities, (Dorn & Thompson 1974) were :

i. Teachers and researchers cannot rely on immediate effect of teaching as a guide to longer-term effects.

ii. Lessons have some 'preventive' and some 'counter-preventive' effects: teaching is neither 'turn-off' nor 'turn-on' in an overall sense.

iii. 'Prevention' is too broad and abstract a concept to put into practice. Teachers must define the purpose or goal of teaching more clearly if they wish to develop effective methods.

iv. There is real potential for improvement in the quality of teaching.

Some elucidation of these results is necessary, and we shall first briefly sketch out the research designs. The key results stem from a sample of 14-18 year-old secondary school students in a number of schools spread throughout England and Wales. The sample is not representative, but was chosen to include a range of schools in terms of mixed or single-sex, social class, urban-rural, drug-involvement, etc. Local Education Authorities were asked to nominate schools where one of five types of lessons given by teachers occurred. The study was therefore one of field research, rather than of controlled, randomised or laboratory design. With full guarantees of anonymity and confidentiality, fieldworkers visited schools and gave questionnaires before, immediately after and two months following lessons, some with and some without film, given and selected by the pupils' regular teacher. Thus the study looked at effects of the amount and types of lessons already current. It should be emphasised thet the lessons studied rarely exceeded ninety minutes in length, and that follow-up discussion generally covered a lesser time period: thus the results cannot be directly compared with those of programmes covering ten or 20 hours.

A measure of self-reported drug use was also taken; this was intended to be used to segment the sample to see if those with different experience reacted differently to education, and to aid in interpretation of results. On the understanding that readers' will not attempt to generalise the finding to British secondary schools in general, the following ever-used self-reports (excluding alcohol, nicotine, prescribed use) are offered from a sample of four and a half thousand: fourth formers - 6% ever-used; fifth formers - 9% ever-used; sixth formers - 12% ever-used. These forms averaged ages 14-15, 15-16, 16-18 respectively; at the time of the survey, the lowest age at which one could leave the school system was 15 years. Most commonly reported drug behaviour was use of cannabis on one or two occasions. Other results showed that although pupils had a concept of "drugs", they also differentiate to a certain extent between drugs in terms of their intentions to accept if offered, dangerousness, perceptions of user of that substance as like-unlike self, etc. For the non-using majority, the order of increasing "drugyness" was sedative and stimulant pills, cannabis, LSD, heroin, whilst for those who had already experimented, cannabis was seen as least dangerous, most likely to be accepted if offered, etc. Generally speaking, those who did not report experimenting but who reported knowing somebody who took drugs (about 50%) fell in between those without such acquaintances and those who had experimented. Such findings may not apply to all schools nor to any particular school, nor to the present time.

Because existing levels of drug use were so low, it was not possible to measure change in them over the relatively short period for which the lessons may be expected to have an effect. Therefore measurement of change was made in terms of pupils intentions to accept drugs if offered, their attitudes, and their perception of the drug-taker (all on five-point scales). Analysis of the data by multivariate analysis and by t-test showed that immediate effects of the lessons, impressions of which are apparently often taken as evidence of effectiveness, were substantially different from effects two months later after passage of time and formal and informal discussion. In particular, it was noted that the "shock" method of informing about drugs did not maintain its initially strong effect on intentions. In general, two months after the initial lesson, there were no significant decreases in intention to accept any drug, after any drug educational method, some increases in intention to accept cannabis (after certain drug lessons using film), a mixed effect on attitudes probably leaning to the anti-drug, and a general decrease in the tendency to see the "drug-taker" as less unlike oneself (particularly in lessons employing film, and including the "shock" condition). Studying these changes, which are mostly slight, in detail, we interpret them as evidence that the sorts of drug education studied makes these pupils more pro-drug in some ways and more anti-drug in others. There was apparently no necessary link between intention change, attitude change to drugs, and attitude change to drug-takers. (A one-year follow-up was unable to discern any behavioural effects of the lesson: (Dorn & Thompson, 1976)). Our conclusion is that given such results, it is misleading to use the concept of "prevention", and it is necessary to speak of changes in behaviour, intention, <u>or</u> attitudes to drugs, <u>or</u> attitudes to users. The types of lesson with these pupils were neither "turn-off" nor "turn on" and use of such terms should cease in favour of clear choices of education goals. (Similar conclusions have recently been reached by a Dutch study: De Haes and Schuurman 1975)). In a booklet, written in the light of the research and distributed to government agencies, Local Education Authorities and schools (Dorn & Thompson 1974), a list of some possible goals was given :

to minimise use of particular drugs

to minimise consequences of existing experimentation

to promote particular attitudes towards particular drugs

to increase knowledge

to increase pupils decision-making skills

to increase teacher-pupil communication and mutual understanding

Educators cannot effectively pursue all of these goals simultaneously, and so must choose a small number of goals for a particular locality, school, syllabus or class. It is possible, however, that such a recommendation will be brushed aside by reassertions of the abstract idea of "prevention" on the one side, and by references to "educating the whole person" on the other. We have tried to suggest that clear definition of goals helps to make development of coherant methods easier, but we are aware that the tradition of education in this broad area is shy of the question of what constitutes a desirable outcome.

DEVELOPMENT AND EVALUATION OF NEW METHODS

The basic philosophy and rationale for working in this field is that the process of making decisions about drug education involves considerations of personal and social values (desirability) on the one hand and of practical constraints and limitations (realism) on the other. The author is neither for nor against drug education, believing that its quantitive growth is probably the result of factors not very amenable to influence by researchers, and that the contribution of educational research is to improve quality rather than quantity. This improvement in quality will not be achieved by simple presentation of research data, nor by declarations that particular manifestations of drug education are "successes" or "failures". It is hoped to make a contribution by clarifying issues that may not even have been recognised as decision-points, and to provide the sorts of information that decision-makers will need to resolve these issues. Thus it cannot be presupposed that the questions posed in drug education circles are the only possible or the most relevant questions: as described above, currently trying to shift the focus of attention from techniques back to the question of goals which are to be served effectively, uneffectively or counter-effectively, by those techniques. The question "does it work?" presupposes an answer to the question "what's it meant to do?" The aim is to discover the questions and to make them explicit, not just to supply "answers" to confused questions.

So much for theoretical orientation. As far as the state of play in the U.K. is concerned, only very recent interest in most Government agencies or academic institutions in the question of evaluation of drug education is observed. (Whilst it has been suggested in North America that disinterest in evaluation is a product of the fear of failure, it is also possible that the prerequisite of stating in operational manner the real goals of the programme has been too taxing and has threatened to arouse too much individual and social conflict). In the current U.K. situation many of the signs that preceded the American drug education bonanza that itself led to the recent moratorium on new materials are noticed. In particular, the author points to lack of clear location of responsibility in Government agencies, the usual involvement of small agencies and individuals motivated by serious concern sometimes mixed with desire for publicity, status or funding, an increasing interest by commercial interests, the obvious poor quality of much of the existing materials, and most significantly, the need to do <u>something</u> with the thousands of schoolchildren now held in schools for an additional year since the Raising of the School Leaving Age (ROSLA) from 15 to 16 years. It is no good just talking about evaluation, and so an attempt is being made to demonstrate the practical advantages.

The Drug Education Development and Evaluation project was begun in March 1974 and will be completed in January 1977. Its general purpose is to discover the extent to which clearly delineated goals can be attained, using educational methods and materials developed specifically for the chosen goals. The three goals chosen for this project are :- increased knowledge about legal and illegal drugs, enhanced decision-making skills vis-a-vis legal and illegal drugs, and improved teacher-pupil communication about this topic. These goals are not claimed to be "correct", and other goals may be preferred in many circumstances. Materials designed to attain the goals were developed, tested-out in classrooms, and then modified. The third and final versions of these prototype materials have been subjected to evaluation in schools in England (and also, in an extention of the project, in Denmark). The evaluation uses a mix of methodologies, including comparison of experimental and control groups in each school, interviews, group discussions, classroom observations, and teachers' diaries. Analysis of these data is now proceeding. It is hoped to make some statements about these particular prototype materials, about education for decision-making and social and health education generally, and about such education in the context of English and Danish schools.

CONCLUSION

Drug education should be made effective and needs to be 'normalised' and integrated into wider curricula. Teachers must be given the skills they need to teach in this area. These statements are easy to make, but they beg the question: what sort of drug education? what sort of skills? Making decisions about drug education requires us to consider its goals and also to evaluate it in terms of those goals. This is a process only recently begun in the U.K., but it is entirely practical.

In looking to the future, one can say that it is the conceptual framework into which drug education and its evaluation falls, as much as the techniques employed within this specific conceptual framework, that pre-empts many of the possibilities for improvement. "Drug education" that is premised upon currently popular ideas may function so as to perpetuate the "drug problem". For this reason, evaluators cannot afford to propose purely technical solutions exclusively within the structure of currently popular beliefs about the nature of the problem (held by the general public, by teenagers, by the professions), but must help to develop alternative descriptions of the situation. (One might make a start by denying that drug experimentation is a result of holding specific attitudes to drugs or drug takers, or a result of a lack of 'values'. (Dorn 1975). If evaluators content themselves with researching into variations on one theme, then they will give to drug educators the false impression that such variations (eg 'affective' versus 'cognitive' approaches) really do represent contrasting forms of education.

REFERENCES

DORN, N. and THOMPSON, A. (1974) "Planning Teaching about Drugs, Alcohol and Cigarettes" Institute for the Study of Drug Dependence, London.

DORN, N and THOMPSON, A. (1976) "Evaluation of Drug Education in the Longer-Term is not an Optional Extra". Community Health, 7, 154-161.

DE HAES, W. and SCHUURMAN, J. (1975) "Results of an Evaluation Study of Three Drug Education Methods. "International Journal of Health Education, 16.

DORN, N. (1975) "Notes on Prediction of Behavioural Change in Evaluation of Drug Education". Journal of Drug and Alcohol Dependence, 1 15-25.

DEVELOPING A CO-ORDINATED APPROACH TO INTERPROFESSIONAL EDUCATION

MARCUS GRANT

ALCOHOL EDUCATION CENTRE

LONDON

In this paper I shall be speaking exclusively about professional education. The premise from which I work is that, in order to improve the quality of a service, the most significant action you can take is to improve the quality of the people providing that service. There are other actions, such as re-defining the aims of the organisations providing the service, re-organising the practices of the organisations, so as to provide a series of different service/client interfaces, or creating a new semantic and administrative structure. In the end, however, the thing which is going to have most effect upon clients coming into contact with an agency is the quality of the staff who operate that agency. Thus, effective professional education is the basis of any coherent service for alcoholics and drug addicts.

It is perhaps important at this stage to clarify the way in which I intend to use the word _professional_. Obviously, though drawing all necessary distinctions between specialists and generalists, there exists a number of different professions with a defined responsibility to provide services for these groups of clients. I intend to extend this to include not merely members of the so-called helping professions, but also a range of other professions, which, though having no direct responsibility for the treatment outcomes of alcoholics and drug addicts, bring their members into significant contact with them. I also include a number of people who, though working in a voluntary rather than strictly professional capacity, occupy crucial positions with regard to these groups of clients. I am, therefore, speaking not only of doctors and social workers, but also of policemen, teachers, journalists, telephone counsellors and a host of others.

Mr. Cartwright has defined the roles of agents working in this field as recognition, referral and treatment. We can add to these a fourth - prevention - and I would ask you to bear these four roles in mind, since I wish to return to them later in this paper.

The issues which I am discussing here are based upon the experience of the Alcohol Education Centre, which provides a range of conferences, courses and seminars for those working with alcoholics. Although this work has brought us into contact with the wide range of professional groups which I have just described, it may be that some of the remarks I make will be relevant only to the context of the services in the United Kingdom. I hope, however, that our experience of curriculum development, although not replicable throughout the world, will prove sufficiently broadly based to be of interest to those of you from other countries.

The current state of alcohol education is certainly adversely influenced by two powerful factors, which I will typify as half-truths and heresies. The truncated and partial nature of much professional education has already been mentioned at this conference. Dr. Spratley has made it clear that the evidence from the study carried out by the Maudsley Alcohol Pilot Project suggests a very scanty knowledge and very uneven quality of professional education. Our experience in assessing participants at Summer Schools on Alcoholism would confirm those findings. Even people in key positions, people called upon to make decisions of enormous therapeutic or organisational importance and complexity, do so from the basis of very little knowledge indeed. The position is further exarcerbated by the fact that such knowledge as they do possess tends to be exclusively that which is thought to be appropriate to their own particular profession. Doctors, for example, may well be taught that there is a connection between excessive alcohol consumption and cirrhosis of the liver, even if, as still happens, that is all they are taught about alcoholism. Policemen, to take another example, may well be aware of rates of arrest for drunkenness offences and may even have some knowledge about the effects of acute intoxication as a contributory factor in motor traffic accidents. It is, however, unlikely that the doctor and the policeman will share much common knowledge. It is still more unlikely that they will communicate with each other or co-operate in the handling of an alcoholic.

Pre-qualification professional education could be likened to the provision for each profession of a single piece of a rather large jigsaw. If the community is regarded as a whole and a survey is conducted of all the facilities available, there emerges a remarkably encouraging picture of what the complete jigsaw could look like. Most communities, unfortunately, are by their nature fragmented and, although the whole jigsaw may well be there, it is scattered and jumbled haphazardly through all the different agencies. Down at police headquarters is the top left-hand corner, over at the addiction unit is another piece from the bottom right, somewhere in the Salvation Army hostel you can find another piece, but it is impossible to bring all these pieces together to make the completed picture. Fragmented services result from fragmented training.

Even among specialists, there is a tendency to view the whole subject of dependency as something which is itself intrinsically fragmented. It is, of course, a new subject, in the sense of having emerged as a discreet area of scientific study only comparatively recently. Because it cuts across a number of traditional disciplines, the view has been taken that it must itself be an extremely fragmented area of knowledge. That same evidence could equally well lead to the opposite conclusion. The very fact that it crosses so many boundaries could indicate its inherent comprehensiveness. I am not necessarily arguing for the creation of a new speciality. Rather, that by amalgamation of apparently diverse fragments, the extent of the hidden commonality will emerge. Inevitably, with the current fragmentation of the philosophical and scientific bases of dependency studies, any doctor and any policeman, however excellent their efforts, are working in isolation and, therefore, at least to some extent, in competition.

Fragmentation in service provision can thus be related to the way in which half-truths emerge from pre-qualification professional education. Even where attempts are made to transcend such partialities, however, a second factor can frequently be seen to influence adversely post-qualification practice and in-service training. At the beginning of the 19th century, the heresy of antinomianism enjoyed considerable vogue. This heresy posits as an absolute the existence of a number of elect who are predestined to salvation. Since salvation is related to being a member of this elect and is not connected to the accumulation of good works, sins committed by the elect in no way vitiate their ultimate salvation. There now exists, I often think, a specialised branch of this heresy, a kind of alcohologist's antinomianism. The alcoholism specialist, painfully aware of the gaps in his basic education, is equally aware of the pressing need for him to make decisions about individual alcoholics for whom he is responsible. As a result, he comes to believe that he must act and that, in order to justify particular actions, he must believe them to be right, knowing that the only real basis for the belief is the fact that he has acted. Like the original antinomian, his view is circular and self-justifying, the final refuge of the confused and guilty specialist.

It arises, this heresy, and is maintained, because of the lack of a body of recognised common basic knowledge. There is as yet no coherent expression of such a body of knowledge. This is a subject which has been touched upon at this conference, and it is a subject central to the somewhat dubious validity of conferences at all. Certainly, the lack of a common body of knowledge is something which we should all feel most pressingly today.

Although some of the foregoing paragraphs of this paper might seem to give grounds for dismay, there is no need to be entirely discouraged. On reviewing the alcoholism field in the U.K., it soon becomes apparent that there exist several places where work of enormous importance is being carried out. Significantly, it is not so much in single institutions as in places where a number of interdependent facilities have developed. It has been suggested by our Department of Health that such places be considered centres of excellence. The list of such centres of excellence is not long, comprising little more than half a dozen names, perhaps with Camberwell, Manchester, Liverpool and Edinburgh heading the list. There is, however, another way in which these places might be described. They might be called not so much centres of excellence as <u>coagulations of expertise</u>. The working model having established itself, the models just go on working. The services feed themselves, reinforce themselves, excellence attracting excellence, so that these centres survive, to some extent, at the expense of the rest of the country. Thus, the cumulating expertise coagulates, preventing the flow of excellence out from the centre towards the periphery. Of course, what we must not do is to destroy the excellence, for that is an essential component in the growth of service provision. What is necessary is, if you like, the creation of a suitable anti-coagulant, which will ensure that the wealth of information and expertise will begin to flow and bring life to the rest of the country. I feel sure that this problem is not confined to the United Kingdom. One of the most urgent needs is not to develop new models, new approaches, but rather to ensure that people know about the existing models and approaches. Surprisingly often, whole sections of the community are starved of what they could so easily have, simply because the expertise has coagulated somewhere twenty miles away along the railway track.

One well-established approach to this problem is in attempting to encourage an interdisciplinary approach to alcoholism and drug addiction. At the Alcohol Education Centre, it has always been a basic assumption behind our curriculum development that, through working together and through sharing knowledge and experience, professional workers are likely to make most significant advances. This is true not only for conferences and summer schools, but also in the area of pre-qualification training. Let me return to my policeman with his drunken driver and my doctor with his cirrhotic. It is, of course, essential that both should be fully conversant with their specialised area of responsibility. It is also clear that it will be of no particular advantage to the policeman to have a very detailed knowledge of liver cirrhosis or to the doctor to be fully conversant with the procedures involved in bringing a charge of drunk and disorderly. What would be of enormous advantage, however, would be for these professionals to be in possession of a body of common knowledge, which they were aware that they shared with each other and with other professionals. Interdisciplinary communication could thus become a reality instead of a laudable but unrealistic aim. What is required is that the area of common knowledge should be accurately defined and then presented, possibly in a number of quite different ways, so as to be relevant to the different professions. After all, the reason for such a proposal is an eminently practical one. The knowledge will be of use to them, not only in helping them to understand their own responsibilities, but also in assisting the development of an integrated community response which could replace the present fragmentation and coagulation.

There is always a danger in rather wide-ranging papers such as this that there comes a point, not quite at the end, but with the end in sight, when the author feels compelled to retreat into a vast and all-embracing truism. The same phenomenon is apparent in films about alcoholism where the link-man sits, looking sincerely into the camera, and says: "Well, I think what emerges from this film is that there is no single answer to this problem, because the problem itself is multi-factorial." Usually, such truisms are advanced apologetically, as if they inhibited action. I would like to suggest that the very fact that we must think of a range of solutions is a source of encouragement and is in its own right a potent weapon. It relieves us of the obligation, a most limiting and inhibiting obligation, to advance a single doctrinaire solution. We are under no obligation to say that everybody can be cured by hypnosis, or by apomorphine, or even by group therapy. The freedom, therefore, to adopt a range of different approaches which will be suggested by particular circumstances, allows us a rare and enviable flexibility. If we do not use this flexibility to our advantage in designing appropriate services and in designing training programmes for the staff of the services, then we have nothing to blame but our own lack of imagination.

Inevitably, the final question to be discussed must relate to the practicability of my suggestions. Given that the view I have advanced here is in any way accurate and my suggested solution attractive, how is it to be translated into reality, how are the flow and the flexibility to be achieved in the immediate future? The crucial role becomes one of co-ordination. If facilities and their educational trappins continue to emerge in an ad hoc and arbitrary fashion, no real improvement can be expected. The existence of a co-ordinating agency or even a group of inter-communicating co-ordinating agencies is the necessary precursor to what I have described. The advantages of a co-ordinating rather than a policy-forming agency is its very lack of a party line. Recall, if you will, the four roles within the community's response to alcohol and drug problems: recognition, referral, treatment and prevention. Despite the fact that these categories are in no sense mutually exclusive, they are still going to display very different education requirements. Once a coherent body of common knowledge has been established, it may be that some professions will require in addition only a relatively small input of specialised information whilst others will require more elaborate, detailed and experience-related programmes of great complexity. Only through the activities of an efficient co-ordinating agency could such diverse educational modules be expected to bear meaningful relation to each other.

One simple example of how this co-ordinating function might be expected to work relates to this conference, which finishes today. Although the conference is nominally about alcohol <u>and</u> drugs, I think we are all aware that most sessions have been about alcohol <u>or</u> drugs. I say this, not to criticise the conference, which simply mirrors a traditional distinction which is present at a service provision level, but rather to indicate how the activities of a co-ordinating agency could distinguish between divisions which are helpful and divisions which inhibit progress. A co-ordinator can open doors, which, without his intervention, are likely to remain not only closed, but firmly bolted, for years to come.

Not only should we be open to new approaches, but also equally open to old approaches. Before I entered this field, I was working in English Literature, which is a subject very open to old approaches, and where workers are continually looking back upon existing material and re-evaluating it in the light of new knowledge. In the field of alcohol and drug abuse, however, there is a tendency to assume that new knowledge eradicates the old. Listen to the conversation at the bar during any conference. "Have you heard what John is doing in Texas?" "Did you hear about the man in Bangkok who cures addicts by acupuncture?" "Did you listen to the paper on transcendental meditation?" And do on. There is a strong and pervasive sense of everybody grabbing after what is new. This is to the disservice of us all, because it prevents us establishing that body of common knowledge which I have been emphasising in this paper. Given such a body of knowledge, new approaches could easily be evaluated and, where appropriate, assimilated. We should now be prepared to look at what has been achieved already, for an enormous amount has been achieved.

There are few people prepared to undertake this crucial task, although I am happy to see some of them sitting amongst us today. We have to know what is good from the past as well as what looks exciting for the future. We want to increase people's skills through education and we want to increase people's understanding through education. If indeed my metaphor of these coagulations of expertise is true, if there are coagulations of expertise throughout this country, I do not think that legislation is likely to change those and I do not even think a massive injection of money is going to help particularly. The only effective anti-coagulant is education and that is where we have to put not merely our faith, but actually a little bit of our time and a little bit of our own free-flowing expertise.

ALCOHOL CONTROL POLICY AS A STRATEGY OF PREVENTION: A CRITICAL EXAMINATION OF THE EVIDENCE

JAN DE LINT

SOCIAL STUDIES DEPARTMENT

ALCOHOLISM & DRUG ADDICTION RESEARCH FOUNDATION, TORONTO

INTRODUCTION

Much has been written in recent years about the proliferation of alcohol use, its effects on public health and the need to implement or reactivate control policies which place high priority on restricting alcohol availability (Archibald 1973; Bruun, K. et al. 1975; de Lint, et al. 1975a, 1975b, 1975c, 1975f; Room 1974; Who 1974).

In part, the argument for such policies has been well documented. Thus, there can be little doubt that severe taxation limits both the overall level of consumption and the rate of chronic excessive use. In other part, the argument is not too convincing. Specifically the exact nature and prevalence of physical health damage attributable to alcohol-use per se has thus far not been determined.

It should be recognised, of course, that in many jurisdictions no-one can afford to wait until the case for implementing or reactivating restrictive controls has been established beyond any doubt. Most certainly, it cannot be argued that the continuous proliferation of alcohol use, and the rapid increases in excessive consumption, are benefiting public health.

But, on the other hand, there ought to be concern about the quality of the argument favouring restrictive controls. The issue of alcohol policies is politically highly sensitive. Not only because the production and distribution of alcoholic beverages is of considerable economic importance, but also because beverage alcohol is a very versatile and popular food consumed by many at quite different occasions. Changes in government policies therefore affect a large number of consumers, as well as a significant sector of the economy. Hence it is important that the evidence presented in support of restrictive controls is both relevant and valid.

NO SIMPLE SOLUTION FOR COMPLEX PROBLEM

To select good quality evidence is not an easy task. The author does not know of any substance other than alcohol where such a wide variety of control policies, use patterns and probably effects on health and behaviour have been documented or presumed.

Thus, control policies may include the total or partial prohibition of the manufacture and sale of alcoholic beverages, the control of outlet frequency, the regulation of type and location of outlet, the control of hours and day of sale, the limitation of drinking age, price control, differential taxation, the monopoly and licence system of distribution, the control of alcohol advertising. Use patterns may range from the consumption of quite moderate quantities to the consumption of near-lethal amounts of alcohol, at frequent or only at rare occasions, in the form of beer, wine, or as a distilled beverage, with or without other foods, and so forth. Within a society, but even more so cross-culturally, a very impressive array of use patterns has been described.

Some of these uses have been implicated in health and other problems. For instance, on the basis of many clinical, follow-up, and life history investigations, it would appear that excessive alcohol use may affect almost every part of the human body - the brain, muscles, skin, bones, heart, digestive and respiratory tracts, liver, pancreas, and prostate glands (Bruun et al. 1975; Lelbach 1974; de Lint 1975a; Schmidt 1972; Seixas 1975, 1974; Sundby 1967; U.S.Department of Health, Education and Welfare 1971; Wilkinson et al. 1971).

In addition, excessive use may lead to physical dependency on alcohol as indicated by increased tolerance to its effects and withdrawal symptoms, to severe emotional depressions, and to a variety of other problems in the area of human functioning and general well being (Cahalan et al. 1974, Mendelson 1970, Room 1974, Seixas 1974).

WHAT IS ALCOHOL-RELATED ?

The complexity of assessing the efficacy of control measures in preventing alcohol problems can further be illustrated if the many ways in which a preventive programme may affect a use pattern or the many ways in which a use pattern can be linked to a problem are considered. What specifically is meant in each instance of damage by the term "alcohol-related" ? Premature aging of bone tissue, suicide, pneumonia, cirrhosis of the liver, accident, cancer of the aesophagus, financial ruin, divorce, depressed state of mind, and skin disease, are among the numerous problems which at times are labelled "alcohol-related". But, in each case, the link between the specific use pattern and the problem behaviour or condition may be quite different.

For instance, the route from a consumption pattern to a problem can be rather short and direct, as in the case of liver damage following chronic excessive alcohol use. Or, it may happen that such use initially leads to family and job difficulties, then to increased exposure to hazardous conditions such as living in cheap boarding houses with no fire escapes, to a depressed state of mind, to negligence, and eventually, following an intoxicating drinking episode, to an accidental death by burning. Thus, in the case of some problems labelled "alcohol-related", very few factors other than alcohol consumption are implicated; in the case of others, many factors may have contributed.

And, to add to these difficulties, it is painfully evident from the literature that gross discrepancies exist in the volume and the quality of data substantiating the many linkages that have been proposed between control policies, alcohol use patterns and alcohol problems. Some have been well documented; others remain impressionistic.

In view of the wide variety of control policies, use patterns and problems, in view of the vast complexities inherent in the ways in which all of these may be connected, and in view of the chronic lack of good quality data in many areas of alcohol research, it will not be possible in the foreseeable future to determine with certainty the relevance of each control policy with respect to each type of alcohol-related damage. However, some of the strengths and weaknesses in the argument favouring restrictive controls on alcohol availability will be discussed.

ALCOHOL CONTROL POLICIES AND LEVELS OF CONSUMPTION

First, the connection between such a control policy and average consumption will be examined. Here there are numerous other factors influencing the current proliferation of alcohol use in society. For example, the increased acceptance of different drinking occasions, the lack of public knowledge about the hazards associated with excessive alcohol use, the continuous diffusion of different use patterns into societies which traditionally had low levels of consumption, the marketing efforts of the alcohol industry, have been mentioned in this context (Bruun et al. 1975, de Lint 1975f).

And, undoubtedly, there are many other relevant culture-historical observations to be made to help explain the current trends in our life style and more specifically in our drinking habits (Edwards 1974).

Much of the research attention, however, has specifically focussed on the probable effects of legal restraints on the rates of alcohol use and alcohol problems. On the basis of this research, it would appear that minor variation in the density, location and type of outlet, in the hours and days of sale, or in many other government regulations governing the context in which drinking takes place (e.g. the decor, seating arrangement, entertainment offered), have no measurable effect on the rates of alcohol consumption. On the other hand, the sudden expansion in the number of on- and off-premise outlets after Prohibition, the opening of stores in isolated dry areas in Norway and Finland, the rapid rise in the number of outlets in Finland following the release of medium strength beer for retail distribution, the lowering of the drinking age, the overall policy of relaxing many of the control laws in recent years, would seem to have facilitated significant increases in the rates of consumption.

In the case of taxation policies, the many econometric studies that have been conducted have shown fairly consistently that alcoholic beverages tend to behave on the market like many other commodities. Thus, where prices have fallen, the consumption of alcoholic beverages increased, and in the few instances where prices increased sharply, the demand decreased.

Again, as indicated earlier in the paper, there are a number of difficulties associated with the interpretation of these findings. Beverage alcohol, unlike many other commodities, has a wide variety of usages, e.g. dietary, medical, ritual and social. Each of these use patterns may have a quite different demand elasticity, and thus the proportionate occurrence of each usage may respond differently to taxation and other restrictive measures.

Also, it should be remembered that the enactment, repeal, relaxation or tightening of control laws are rarely isolated events. For example, in Canada during the last decade, many changes have occurred. There has been an increase in the number of outlets, new types of drinking places are now permitted, hours of sale have been extended, drinking age has been lowered, the real cost of alcohol has gone down, some restrictions on alcohol advertising have been removed. In the same period a wider variety of use patterns has been accepted by a more sophisticated and affluent public. In short, what usually seems to happen, is that many of the aetiologically relevant cultural, economic and legal factors tend to occur at the same time and therefore, to show their separate effect, if any, on the rate of consumption is not possible.

In this field of study it will always be a bit foolish to state categorically that a specific change in a control law or in taxation will result in a specific decrease or increase of consumption. But, it should also be emphasized, that over the last two decades or so, in many jurisdictions a steady and at times very rapid increase in alcohol consumption has occurred together with a relaxation of control laws and a decrease in the cost of alcohol.

CURRENT TRENDS IN CONSUMPTION

And the latest available statistics leave little doubt that trends towards higher rates of consumption continue unabated :

TABLE I : PERCENTAGE INCREASES OR DECREASES IN THE RATE OF ALCOHOL CONSUMPTION PER CAPITA 15 YEARS AND OLDER (1960 = 100)

COUNTRY	1970	1973
France	- 12	- 12
Italy	+ 9	+ 10
Spain	+ 42	+ 55
Luxembourg	+ 18	+ 34
West Germany	+ 58	+ 65
Portugal	+ 3	+ 17
C.S.S.R.	+ 40	+ 41
Switzerland	+ 15	+ 54
Austria	+ 22	+ 47
Belgium	+ 13	+ 24
Hungary	+ 42	+ 43
Australia	+ 24	+ 28
New Zealand	+ 16	+ 26
East Germany	+ 44	+ 58
Yugoslavia	+ 53	+ 31
U.S.A.	+ 24	+ 36
Denmark	+ 59	+ 80
Canada	+ 22	+ 42
Great Britain	+ 22	+ 47
Sweden	+ 35	+ 36
Netherlands	+104	+166
Poland	+ 22	+ 48
Eire	+ 48	+ 84
Finland	+ 64	+100
Norway	+ 23	+ 50

Note that in most of these countries per capita consumption increased by more than 40% during the last 13 years or so and that in quite a few instances increases in excess of 70 and 80% occurred. Indeed, compared to 1950 and 1960, very few developed countries remain where the consumption of alcohol can be said to be a relatively infrequent occurrence and where the overall level of consumption is still low :

TABLE II : DISTRIBUTION OF COUNTRIES ACCORDING TO THEIR ALCOHOL CONSUMPTION PER CAPITA 15 YEARS AND OLDER, 1950, 1960, 1970 and 1973

ALCOHOL CONSUMPTION IN LITRES OF ABSOLUTE ALCOHOL	NUMBER OF COUNTRIES			
	1950	1960	1970	1973
< 9	18	15	10	5
10 - 19	6	9	13	18
20 +	1	1	2	2
TOTAL:	25	25	25	25

RATES OF EXCESSIVE CONSUMPTION

Many investigations have shown that these higher levels of consumption invariably imply higher rates of excessive use (Bronetto 1963; Bruun et al. 1975; de Lint 1975b; de Lint et al. 1971; Makela 1971; Popham et al. 1975; Popham 1970; Seeley 1960; Skog 1973, 1971).

And, although it may be true as some critics have suggested, that the evidence to date is not sufficient to infer the exact nature of the relationship between average consumption and excessive use (de Lint 1975d), no known evidence exists which is inconsistent with the theory that overall levels of consumption rise and fall with rates of excessive use. In other words, the author does not know of any population, whether selected on the basis of occupation, geography, or some other way, where a high level of overall consumption prevails in the absence of a higher rate of heavy drinking.

EXCESSIVE CONSUMPTION AND ALCOHOL PROBLEMS

The argument favouring legal restraint on alcohol availability then proceeds to connect heavy drinking to a variety of alcohol problems. Thus surveys indicate that "in general, the onset of heavy drinking precedes the social consequences" (Cahalan et al. 1974). Experimental evidence leaves little doubt that tolerance to the effects of alcohol and the likelihood of withdrawal symptoms vary with the duration of drinking and the volume of consumption (Mendelson 1970). Clinical studies reveal that those who seek help at alcoholism clinics for their addiction and related problems consume at the average 25 cl of absolute alcohol daily (Lelbach 1974; Lundquist 1972; Schmidt et al. 1968; Wilkinson et al. 1969).

LONG-TERM HEALTH CONSEQUENCES

In short, many problem behaviours such as physical dependency, loss of job, family breakdown, clinical alcoholism, are closely related to excessive consumption. And, if rates of excessive consumption were to increase, a higher rate of occurrence of these problems can be expected as well.

There has been a tendency in recent years to ignore the more traditional justifications for implementing restrictive alcohol control policies, alcohol dependency, acute or chronic intoxication, clinical alcoholism, and to focus rather on the long term physical health damages associated with excessive use (Archibald 1973; Bruun et al. 1975; de Lint 1976). It is questionable at times, whether this change in emphasis and direction has been all that desirable, there can be little doubt that the numerous clinical, retrospective and follow-up investigations of recent years have described a very impressive array of alcohol-related health damages. For instance, among a sample of 825 male patients treated for alcoholism in a clinic in Australia, acute alcoholic liver disease, chronic bronchitis, peripheral neuropathy, hypertension, gastritis, traumatic injuries, cirrhosis of the liver, epilepsy, chronic peptic ulceration, chronic brain syndrome, pneumonia, were among the conditions frequently diagnosed.

Follow-up investigations of heavy drinkers have shown death to occur two to four times more often that is to be expected, usually from such causes as tuberculosis, cancers of the upper digestive and respiratory tract, cardiovascular diseases, cerebrovascular lesions, pneumonia, peptic ulcer, cirrhosis of the liver, suicide and accidents (de Lint et al 1975a; Schmidt et al. 1975, 1976, 1972).

TABLE III : OCCURRENCE OF PHYSICAL HEALTH PROBLEMS IN A CLINICAL SAMPLE OF MALE ALCOHOLICS

Diagnosis	Number of cases	Per cent of total patients
Acute alcoholic liver disease	210	25.5
Chronic bronchitis	151	18.3
Peripheral neuropathy	146	17.7
Hypertension (labile or fixed)	139	16.8
Alcoholic gastritis	116	14.6
Traumatic injuries*	101	12.2
Cirrhosis of the liver	69	8.4
Epilepsy	68	8.2
Chronic peptic ulceration*	65	7.9
Chronic brain syndrome	63	7.6
Acute confusional states	52	6.3
Pneumonia	49	5.9
Gout	20	2.4
Pancreatitis	18	2.2
Cardiomyopathy	17	2.1
Pulmonary tuberculosis	16	1.9
Cardiac beri-beri	6	0.7
Patients with one or more diagnosis	576	69.9
Patients with no evidence of disability	214	25.9
Patients not examined	35	4.2
Total patients	825	100.0

* At admission or prior diagnosis.
Source: Wilkinson et al. 1971

However, a careful examination of the cause-specific mortality would indicate that a substantial proportion of all excess deaths in samples of excessive users is in whole or in part attributable to acute intoxication and to conditions and behaviours other than chronic excessive alcohol use, such as cigarette smoking, depressed state of mind, neglect of proper nutrition, exposure to environmental hazards. Indeed, one may well ask whether there would have been a much elevated rate of death among samples of chronic excessive alcohol users if these other behaviours and conditions had not also occurred ?

CAUSE SPECIFIC MORTALITY

For example, as noted earlier in the paper, several follow-up studies reported a considerable number of deaths from cancers of the buccal cavity, pharynx, larynx, lung and oesophagus (Gabriel 1935; Nicholls et al. in press; Pell et al. 1973; Schmidt et al. 1972; Sundby 1967). These findings have been substantiated by :

1. retrospective studies of patients with such cancers which revealed a high proportion of alcoholics among them (Cappellano 1960; Gsell 1962; Josserand et al. 1951; Morice 1954), and;

2. correlational studies which showed spatial covariance between deaths from these cancers and indices of alcoholism (Lasserre 1967; Schwartz 1966).

But to what extent can the apparently high rate of death from these cancers to excessive alcohol use be attributed ? In the case of lung cancer, there would appear to be little doubt that it is almost entirely attributable to their heavy smoking and not their heavy drinking. Doll and Hill (1964) in a study of British doctors, concluded that "there is evidence of so close a relation (between the number of cigarettes smoked and mortality) that it becomes increasingly difficult to envisage any other feature correlated with smoking as being the real and underlying cause", and thus this conclusion has since been confirmed in many other studies (U.S. Surgeon General 1971). However, in deaths from cancer of the buccal cavity, pharynx, larynx and oesophagus, it would appear from a number of retrospective investigations that both the heavy use of alcohol and of tobacco are aetiologically relevant (Doll & Hill 1964; Keller et al. 1965; Nicholls et al. in press; Schmidt et al. 1972; Schwartz 1966; Weir et al. 1970; Wynder et al. 1961).

One persistent difficulty in the study of mortality among heavy drinkers is the significant variation from case to case in the diagnosis and recording of causes of death. For instance, deaths due to acute and chronic alcoholism are often attributable to a variety of other conditions. For example, in one follow-up study (Schmidt et al. 1972) 32 deaths from "acute" or "chronic" alcoholism were reported. In ten of these pneumonia, in six coronary thrombosis, in three acute or chronic myocarditis and in two cerebral degeneration were listed on the death certificate as contributory causes. In another study (Sundby 1967) many of the deaths certified as due to alcoholism were in fact due to cardiomyopathy with heart failure.

A large percentage of excess deaths in samples of alcoholics (Brenner 1967; Nicholls et al. in press; Pell et al. 1973; Schmidt et al. 1972) has been attributed to cardiovascular diseases. In these deaths the role of heavy drinking is also quite uncertain. Undoubtedly some are the result of cigarette smoking so common among alcoholics (Bates 1965; Best et al. 1967; Cartwright et al. 1959; Doll and Hill 1964; Dorn 1959; Doyle et al. 1964; Dreher et al. 1968; Friedman 1967; Hammond et al. 1958; Sundby 1967; Walton 1972; Zeiner-Henriksen 1971). A few may be attributable to alcoholic cardiomyopathy, a clinically well described but relatively rare condition among heavy drinkers. Also of relevance here is the significantly high rate of occurrence of hypertension among alcoholics, a well established risk factor in coronary heart disease (D'Alonzo et al. 1968; Leclainche et al. 1970). For example, in one recently conducted prospective study it was estimated that about 25 percent of excess coronary deaths in the alcoholic sample were the result of hypertension (Pell et al. 1973). Whether there are other factors explaining the higher mortality from cardiovascular diseases in alcoholic samples cannot be answered at present.

The high rates of death from pneumonia frequently reported for alcoholic populations (Gabriel 1935; Giffen et al. 1971, Gorwitz et al. 1970; Lipscomb 1959; Nicholls et al. in press; Schmidt et al. 1972; Sundby 1967) are again difficult to interpret. In many instances a variety of other morbid conditions, e.g. heart disease, liver cirrhosis, delirium tremens, apoplexy, are present (Sundby 1967). Indeed, the ascription of a death to pneumonia rather than to any of the other conditions appears to be somewhat arbitrary. And, in addition, several of the attributes of alcoholism such as gross intoxication, inadequate clothing, extensive exposure to cold, poor nutrition, are of known significance in the development of this disease (Alassandri et al. 1944; Bogue 1963; Chomet et al. 1967; Jellinek 1942).

Probably no major cause of death has been so consistently linked to alcoholism for so many years as cirrhosis of the liver. Several recent reviews of the relevant literature makes it quite unnecessary to document the very significant contribution of alcoholism to mortality from this cause. It may suffice to point out that in follow-up studies the ratios of observed/expected deaths from liver cirrhosis range from eight to 23 (Brenner 1967; Nicholls et al. in press; Salum 1972; Schmidt et al. 1968; Sundby 1967) and also that liver cirrhosis is frequently mentioned on the death certificate of alcoholics as a contributing cause of death (de Lint et al. 1970; Nicholls et al. in press; Tashiro et al. 1963).

The many follow-up and retrospective studies in this field have clearly indicated that volume and duration of alcohol consumption are by far the most important factors in alcoholic cirrhosis (Lelbach 1974). The type of beverage alcohol consumed (e.g. beer, wine or distilled spirits) and the pattern of drinking (e.g. periodic bouts vs daily inveterate) are apparently of little importance (Lelbach 1974).

Accidental poisonings, motor vehicle accidents, accidental falls, accidents caused by fire and homicides also contribute very significantly to the mortality of alcoholics (Brenner 1967; Dahlgren 1951; Gillis 1969; Gorwitz et al. 1970; Storby 1953; Sundby 1967). These deaths have been attributed to the acute effects of alcohol, to the style and conditions of life of alcoholics and to personality traits associated with alcoholism (Goodwin 1973; Graves 1960; MacFarland et al. 1962; Payne et al. 1962; Schmidt et al. 1954). Again, it is difficult to determine the relative importance of each of these factors in accidental deaths. In one sample drunkenness appeared on the death certificate in 28% of cases as a contributary cause and it is conceivable that in many more instances a state of intoxication also existed (Schmidt et al. 1972). For example, of deaths caused by fire many are attributed to careless smoking but intoxication may also have occurred. Likewise in deaths from freezing and in traffic fatalities, alcohol use is often not mentioned on the death certificate although intoxication must have played an important role in at least some cases. Certainly, the results of numerous retrospective investigations of accidental deaths would support the impression that intoxication is an important factor (LeRoux et al. 1964; Potondi et al. 1967; Spain et al. 1951; Waller 1972). Studies of blood alcohol levels in persons admitted to the emergency services of general hospitals after involvement in home, transportation or occupation accidents have shown that many were drinking heavily prior to the accident (Birrell 1965; Borkenstein et al. 1964; Demone et al. 1966; Gjone 1963; Haddon 1963; Hindemarsh et al. 1934; Im Obersteg et al. 1967; Johnsen et al. 1966; Kirkpatrick et al. 1967; Rydberg et al. 1973; Tonge 1968; Weschsler et al. 1969). Specifically, in the case of traffic accidents, high blood alcohol levels have frequently been found (Bonnischen et al. 1968; Haddon et al. 1961; Hossack 1972; McBay 1972; Tonge 1968; Waller 1972; Waller et al. 1966, 1969).

The literature on alcoholism and suicide is quite extensive. Firstly, clinicians have frequently referred to the self-destructive aspects of chronic intoxication as suicidal behaviour (Menninger

1938; Palola et al. 1962). Secondly, significant similarities in the life histories of suicidal persons and alcoholics have been noted (Gadourek 1963). Thirdly, follow-up studies have reported high rates of suicide among alcoholics (Ciompi et al. 1969; Dahlgren 1951; Gabriel 1935; Gillis 1969; Kendell et al. 1966; Kessel et al. 1961; Lipscomb 1959; Nicholls et al. in press; Nørvig et al. 1956; Schmidt et al. 1972; Sundby 1967), and fourthly, retrospective studies of suicides and attempted suicides have found alcoholism to be often indicated in the life history of the deceased or of the patient, (Batchelor 1954; Dahlgren 1951; Dorpak et al. 1960; Harenko 1967; James 1966; James et al. 1963; Mayfield et al. 1972; Murphy et al. 1967; Palola et al. 1962; Robins et al. 1959; Robins et al. 1957; Sainsbury 1955; Schmidt et al. 1954; Stengel et al. 1958; Zmuc 1968). Suicides among alcoholics are usually attributed to predisposing social and personality factors (Clinard 1963; Faris 1961; Lundquist 1970; Menninger 1938; Mowrer 1942; Rushing 1969a; Wallinga 1949), and to social isolation resulting from the common interpersonal disturbances in the alcoholic's way of life (Gadourek 1963; Palola et al. 1962; Rushing 1969b). Relatively little attention has been given to alcohol-induced depressive states as a direct cause of suicides. Indeed, it would be extremely difficult to determine to what extent a depressed state is the psychological consequence of chronic excessive drinking or is brought about by a process of social deterioration.

In short, what many investigations have demonstrated is a rather non-specific association between, on the one hand, the conglomerate of habits and conditions of life usually labelled "alcoholism" and, on the other, deaths from a wide variety of causes.

It would seem that the characteristics of alcoholics, their drinking behaviour, smoking habits, emotional state, life style, are uniquely expressed in their mode of dying. Thus, in death from poisonings, falls and fire, the acute effects of alcohol use are indicated, whereas in deaths from liver cirrhosis the chronic effects are evident. In suicides the depressed state of mind of the alcoholic appears to be the more important factor, whereas the high rate of death from cancer of the lung is attributable to heavy smoking.

For these and other reasons, it has been a rather difficult task to try to assess the total impact of alcohol use per se on current mortality and morbidity rates (de Lint 1975d, Schmidt et al. 1973) or to estimate how much more physical health damage is likely to result from further increases in consumption. Only, in the case of deaths and diseases which are largely attributable to the use of alcohol can higher rates of occurrence be predicted if consumption continues to increase. For example, within the last ten years, rates of death from liver cirrhosis among males 15 - 74 rose by more than 20% in almost all of the 25 developed countries referred to earlier in this paper (Appendix I). In 11 countries, rates of death from liver cirrhosis increased by more than 50% during these years.

THE ARGUMENT FOR RESTRICTING ALCOHOL AVAILABILITY

Since the relaxation of alcohol control policies has undoubtedly facilitated the proliferation of alcohol use in society and therefore higher rates of excessive use and related problems, it would seem reasonable in the context of public health to try to stabilize these trends and to control alcohol availability.

However, there are a number of conditions which may well impede the implementation of such a programme.

First, no matter how much evidence can be marshalled in support of legal restraints on alcohol availability, it should be recognised that the many studies which have yielded these data all have some methodological shortcomings. In the field of alcoholism research this is simply inavoidable. But because government control of drinking is a controversial issue, these methodological weaknesses tend to be exploited by the adversaries of controls while the degree of consistency in the vast amount of evidence linking alcohol availability, the overall level of consumption and the rates of alcohol problems tends to be ignored. As mentioned earlier, the quality of the argument for control policies can be much improved. It is wrong to virtually ignore the problems of acute and chronic intoxication, of clinical alcoholism, of physical dependency, as valid justifications for implementing legal restraints on alcohol availability and to place too much emphasis on the long term consequences to physical health. Many deaths and diseases in samples of excessive drinkers are largely attributable to their deviant life style, their smoking habits, their neglect of proper nutrition, their unhappy state of mind, and not to their excessive alcohol use. Therefore, it is difficult to predict whether the incidence of these deaths and diseases will at all be affected by restrictive control policies. On the other hand, if the trends in consumption are not stabilized, rates of death from cirrhosis of the liver and from several other alcohol-sensitive causes will most certainly continue to climb.

A second impediment to the implementing or reactivating of controls on alcohol availability would be the existence of some rather popular notions about alcoholism which deny the aetiological relevance of alcohol availability.

For example, there is the suggestion that heavy alcohol consumption by alcoholics is symptomatic of some disorder peculiar to them. This point of view is particularly favoured by spokesmen for the alcoholic beverage industry but it is also endorsed by many members of Alcoholics Anonymous. There are indeed significant individual differences in the likelihood of becoming an alcoholic but it is also well known that in exceedingly wet environments, where beverage alcohol is readily available and frequently consumed, excessive drinking is a much more prevalent behaviour. Restricting alcohol availability cannot prevent a very vulnerable person from becoming a heavy drinker but it can prevent less vulnerable persons from indulging (Jellinek 1954; de Lint et al. 1971; Popham 1959).

Another example is the suggestion that alcohol problems are rooted in the mysticism associated with alcohol use, in the ambivalent attitudes towards drinking. Thus, it is argued, young people should be introduced to alcoholic beverages at an early age so that they may learn to drink moderately and come to regard the activity as of no greater significance than eating. Restrictive control measures are seen as reinforcers of an unhealthy ambivalence towards drinking and as impediments to the adoption of so-called healthy drinking styles. In fact, the Co-operative Commission on the Study of Alcoholism in the U.S.A. has recommended several years ago that "drinking should become more 'civilized' e.g. the convivial use of beverage alcohol and drinking with meals should be encouraged, the so-called 'beverage of moderation' (beer) should be stressed, and drinking should become an incidental part of routine activities" (Plaut 1967).

The rather naive idea behind this recommendation was that these so-called desirable drinking patterns would eventually replace so-called undesirable drinking patterns. Besides the logical flaws in the argument, why would vastly different consumption behaviours serving quite different purposes überhaupt replace one another, there exists no evidence in the alcohol literature to substantiate this theory. The Finnish social experiment is particularly illuminating. For a number of years, so-called desirable drinking practices involving beer and wine were actively promoted by its State Alcohol Monopoly on the advice of its scientists believing that the integration of these practices would gradually replace the traditional weekend benders and thus reduce the rate of alcohol problems. Needless to say, these expectations were not borne out. Within a relatively short time this change in policy led to rather dramatic increases in both alcohol consumption and alcohol problems (Bruun et al. 1975; W.H.O. 1974).

A third impediment to the implementing or reactivating of controls on the availability of alcohol is that such programmes are perceived by many as a somewhat radical departure from the existing preventive strategies.

Although one may argue that the relaxation of control policies can, with equal justification, be looked at as a somewhat radical departure from the traditional public health responsibility of government, that sort of reply will not dissuade the critics. After all, the relaxation of control policies seem to have been in tune with many other economic and cultural developments of the post-war period. Among these, as already mentioned, the rapid diffusion and acceptance of a wide variety of alcohol use patterns, a trend undoubtedly facilitated by the very efficient marketing efforts of large multinational industries, by trade agreements between countries seeking closer economic alliance, by increased tourism,by a growing sophistication and affluence on the side of the consumer and by a decline in the political and moral influence of the Temperance movement. In this context, it is pointed out that prices of alcoholic beverages do not respond well to inflationary pressures as they are largely determined by taxation. Only continuous efforts on the part of government to adjust taxes on alcohol from year to year if not from month to month can prevent significant decreases in the real price of alcoholic beverages. However, since the issue of alcohol controls is politically highly sensitive, there is an understandable reluctance on the side of government to act in this area.

It may well be that the trend towards a relaxation of control policies has reached rock bottom. The results of several recently conducted surveys in North America seem to indicate that a surprisingly large segment of the general public now supports legal restraints on alcohol availability and no longer favours a continuation of a policy of relaxation (de Lint 1975f).

If there is indeed such a change in public sentiment, there is a need to be all the more concerned with the quality of the argument favouring legal restraints on alcohol availability.

APPENDIX I

RATES OF DEATH FROM LIVER CIRRHOSIS PER
100,000 MEN AND WOMEN AGED 15 - 74, 1962 and 1972
IN A NUMBER OF COUNTRIES AND THE PERCENTAGE INCREASES AND DECREASES

COUNTRY	MEN			WOMEN		
	1962	1972	%	1962	1972	%
France	57.03	61.23	7	24.78	24.88	0
Italy	36.30	55.90	54	12.89	18.90	47
Spain	24.75	37.72	52	12.35	15.00	21
Luxembourg	N/A	57.81		N/A	22.90	
West Germany	29.46	40.92	39	11.71	16.26	39
Portugal	N/A	58.81		N/A	22.57	
Switzerland	28.00	26.46	-6	5.48	7.74	41
C.S.S.R.	14.03	28.94	106	7.34	10.09	37
Austria	39.73	56.01	41	11.66	16.30	40
Belgium	14.18	17.65	24	8.22	9.44	15
Hungary	13.98	20.44	46	6.52	8.69	33
Australia	9.35	N/A		3.83	N/A	
New Zealand	4.38	7.79	78	2.28	4.22	85
Yugoslavia	10.57	25.75	144	5.17	9.41	82
U.S.A.	22.27	28.42	28	10.88	14.07	29
Denmark	8.13	13.53	66	7.79	8.22	6
Canada	10.18	18.32	80	5.12	8.59	68
England & Wales	3.57	4.34	22	2.81	3.57	27
Sweden	8.63	16.74	94	4.05	5.60	38
Netherlands	4.77	5.87	23	2.50	3.12	25
Poland	7.38	14.10	91	5.00	7.33	47
Eire	N/A	5.81		N/A	4.01	
Finland	5.24	7.55	44	3.14	3.21	2
Norway	5.79	6.63	15	3.02	4.07	35

REFERENCES

ALCOHOLIC BEVERAGE STUDY COMMITTEE (1973). "Beer, Wine and Spirits: Beverage Differences and Public Policy in Canada". Brewers Association of Canada, Ottawa, 164.

ALESSANDRI, H. et al (1944). "El alcoholismo como factor de enfermedad en un servicio de medicina interna". Revista Medica de Chile, 72. 199.

ARCHIBALD, H.D. (1973). "Changing Drinking Patterns in Ontario - Some Implications". Addiction, Fall.

BATCHELOR, I.R.C. (1954). "Alcoholism and Attempted Suicide". Journal of Mental Science, 100. 451.

BATES, R.C. (1965). "The Diagnosis of Alcoholism". Applied Therapeutics, 7. 466.

BEST, E.W.R. et al. (1967). "Summary of a Canadian Study of Smoking and Health". Canadian Medical Association Journal, 96. 1104.

BIRRELL, J.H.W. (1965). "Blood Alcohol Levels in Drunk Drivers, Drunk and Disorderly Subjects and Moderate Social Drinkers". Medical Journal of Australia, 2. 949.

BOGUE, D.J. (1963). "Skid Row in American Cities". University of Chicago Press, Chicago, Illinois.

BONNICHSEN, R.B. et al (1968). "Alkoholens roll vid svenska trafikolyckor". Alkoholfrågan, 62. 202.

BORKENSTEIN, R.F. et al (1964). "The Role of the Drinking Driver in Traffic Accidents". Department of Police Administration, Indiana University.

BRENNER, B. (1967). "Alcoholism and Fatal Accidents". Quarterly Journal of Studies on Alcohol, 28. 517.

BRONETTO, J. (1963). "Alcohol Price, Alcohol Consumption and Death by Liver Cirrhosis, Canada, U.S.A. and several European Countries". Alcohol Addiction Research Foundation Mimeo

BRUUN, K. et al (1975). "Alcohol Control Policies in Public Health Perspective". A Collaborative Project of The Finnish Foundation for Alcohol Studies, World Health Organisation (EURO) and The Addiction Research Foundation of Ontario, The Finnish Foundation for Alcohol Studies, 25.

CABOT, R.C. (1904). "The Relation of Alcohol to Arteriosclerosis". Journal of the American Medical Association, 43. 774.

CAHALAN, D. et al (1974). "Problem Drinking Among American Men". Rutgers Centre of Alcohol Studies, Monograph No.7, College and University Press, New Haven.

CAPPELLANO, R. (1960). "Incidencia da siflis, etilimso e tabagismo nos pacientes portadores de câncer das vias aero-digestivas superiores". Revista Brasileira de Cirurgia, 39. 518.

CARTWRIGHT, A. et al (1959). "Distribution and Development of Smoking Habits". Lancet, 2. 725.

CHOMET, B. et al (1967). "Lobar Pneumonia and Alcoholism: An Analysis of 37 Cases". American Journal of Medical Science, 253, 300.

CIOMPI, L. et al (1969). "Mortalité et causes de décès chez les alcooliques". Sozial Psychiatre, 4. 159.

CLINARD, M.B. (1963). Sociology of Deviant Behaviour. Holt, Rinehart and Winston, New York.

DAHLGREN, K.G. (1951). "On Death Rates and Causes of Death in Alcohol Addicts". Acta Psychiatrica et Neurologica Scandinavica, 26. 297.

D'ALONZO, C.A. et al (1968). "Cardiovascular Disease Among Problem Drinkers". Journal of Occupational Medicine, 10. 344.

DEMONE, H.W. Jr et al (1966). "Alcohol and Non-Motor Vehicle Injuries". Public Health Report, 81. 585.

DOLL, R. and HILL (1964). "Mortality in Relation to Smoking: Ten Years' Observations of British Doctors". British Medical Journal, 1. 1399, 1460.

DORN, H.F. (1959). "Tobacco Consumption and Mortality from Cancer and Other Diseases". Public Health Report, 74, 581.

DORPAT, T.L. et al (1960). "A Study of Suicide in the Seattle Area". Journal of Comparative Psychiatry, 6. 349.

DOYLE, J.T. et al (1964). "The Relationship of Cigarette Smoking to Coronary Heart Disease". The Second Report of the Combined Experience of the Albany, New York and Framingham, Massachusetts Studies. Journal of the American Medical Association, 190. 886.

DREHER, K.F. et al (1968). "Smoking Habits of Alcoholic Out-Patients I, II". International Journal of Addiction, 2. 259, 3. 65.

EDWARDS, G. (1974). "Alternative Strategies for Minimizing Alcohol Problems. Coming Out of the Doldrums". Expert Conference on the Prevention of Alcohol Problems, Berkeley, California.

FARIS, R.E.L. (1961). Social Disorganisation. Harper and Company, New York.

FRIEDMAN, G.D. (1967). "Cigarette Smoking and Geographic Variation in Coronary Heart Disease Mortality in the United States". Journal of Chronic Diseases, 20. 769.

GABRIEL, E. (1935). "Uber die Todesursachen bei Alkoholikern". Neurologische und Psychiatrische Abhandlungen, 153. 385.

GADOUREK, I. (1963). Riskante gewoonten en zorg voor eigen welzijn. Wolters, Groningen.

GIFFEN, P.J. et al (1971). "The Chronic Drunkenness Offender: Ages and Causes of Death of the Chronic Drunkenness Offender Population". Addiction Research Foundation, mimeo, Toronto, 13.

GILLIS, L.S. (1969). "The Mortality Rate and Causes of Death of Treated Alcoholics". South African Medical Journal, 43. 230.

GJONE, R. (1963). "Kraniecerebrale trafikkskader". Tidsskrift for Den Norske, Laegesorening, 83. 424.

GOODWIN, D.W. (1973). "Alcohol in Suicides and Homicides". Quarterly Journal of Studies on Alcohol, 34. 144.

GORWITZ, K. et al (1970). "Some Epidemiological Data on Alcoholism in Maryland". Quarterly Journal of Studies on Alcohol, 31. 423.

GRAVES, J.H. (1960). "Suicide, Homicide and Psychosis". 26th International Congress on Alcohol and Alcoholism, Stockholm.

GSELL, V.O. et al (1962). "Etiological Factors in Carcinoma of the Esophagus". Deutsche Medizinische Wochenschrift, 87. 2173.

HADDON, W. Jr (1963). "Alcohol and Highway Accidents". 3rd International Conference on Alcohol and Road Traffic, BMA House, London.

HADDON, W. Jr et al (1961). "A Controlled Investigation of the Characteristics of Adult Pedestrians Fatally Injured by Motor Vehicles in Manhattan". Journal of Chronic Diseases, 14. 655.

HALL, E.M. et al (1953). "Portal Cirrhosis; Clinical and Pathological Review of 782 Cases from 16,600 Necropsies". American Journal of Pathology, 29. 993.

HAMMOND, E.C. et al (1958). "Smoking and Death Rates. I. Total Mortality. II. Death Rates by Causes". Journal of the American Medical Association, 166. 1159, 1294.

HARENKO, A. (1967). "Alkoholin osallisuus myrkytysitsemurhayrityksissä". Helsingissä v. 1962-64, Suom. Lääkärilehti, 22. 109.

HINDEMARSH, J. et al (1934). "Alkoholuntersuchungen bei Unfallverletzten". Acta Chirurgica Scandinavica, 75. 198.

HIRST, A.E. et al (1965). "The Effect of Chronic Alcoholism and Cirrhosis of the Liver on Atherosclerosis". American Journal of Medical Science, 45. 143.

HOSSACK, D.W. (1972). "Investigation of 400 people killed in Road Accidents with Special Reference to Blood Alcohol Levels". Medical Journal of Australia, 2. 255.

IM OBERSTEG, J. et al (1967). "Unfall unter der Einwirkung von Arzneimitteln und Alkohol". Schweizerische Medizinische Wochenschrift, 97. 1039.

JAMES, J.P. (1966). "Blood Alcohol Levels following Successful Suicide". Quarterly Journal of Studies on Alcohol, 27. 23.

JAMES, J.P. et al (1963). "Blood Alcohol Levels following Attempted Suicide". Quarterly Journal of Studies on Alcohol, 24. 14.

JELLINEK, E.M. (1954). "International Experience with the Problem of Alcoholism". Alcoholism Research Symposium, Fifth International Congress on Mental Health, Toronto.

JELLINEK, E.M. (1942). "Death from Alcoholism in the United States in 1940". A Statistical Analysis". Quarterly Journal of Studies on Alcohol, 3. 465.

JOHNSEN, C. et al (1966). "Förekomst av alkoholpåverkade i ett kirurgiskt akutklientel". Socialmedicinsk Tidsskrift, 43. 108.

JOSSERAND, A. et al (1951). "De l'influence aggravante de l'ethylisme sur l'evolution des epitheliomas buccopharyngés". Lyon Médical, 184. 165.

KELLER, A.Z. et al (1965). "The Association of Alcohol and Tobacco with Cancer of the Mouth and Pharynx". American Journal of Public Health, 55. 1578.

KENDELL, R.E. et al (1966). "The Fate of Untreated Alcoholics". Quarterly Journal of Studies on Alcohol, 27. 30.

KESSEL, N. et al (1961). "Suicide in Alcoholics". British Medical Journal, 2. 1671.

KIRKPATRICK, J.R. et al (1967). "Blood Alcohol Levels of Home Accident Patients". Quarterly Journal of Studies on Alcohol 28, 734.

LASSERRE, O. et al (1967). "Alcool et cancer; étude de pathologie géographique portant sur les departements francais". Bulletin de l'Institut National de la Sante et de la Recherche Medical, 22. 53.

LECLAINCHE, X. et al (1970). "Alcoolisme et maladies associées". Bulletin de l'Academie Nationale de Medecine, 153. 373.

LELBACH, W. (1974) "Organic Pathology Related to Volume and Pattern of Alcohol Use". Research Advances in Alcohol and Drug Problems 1. 93. R.J.Gibbins, et al. (eds.) John Wiley & Sons, New York.

LELBACH, W.K. (1966). "Leberschäden bei Chronischem Alkoholismus". Acta Hepatosplen, 13. 321.

LeROUX, R.C. et al (1964). "Violent Deaths and Alcoholic Intoxication". Journal of Forensic Medicine, 11. 131.

de LINT, J. (1976). "Epidemiological Aspects of Alcoholism". International Journal of Mental Health, 5. 29. (In Press)

de LINT, J. et al (1975a). "Alcoholism and Mortality". The Biology of Alcoholism, Vol. IV. B.Kissen and H.Begleiter (eds.) Plenum Publishing Corp. New York.

de LINT, J. (1975b). "The Epidemiology of Alcoholism: The Elusive Nature of the Problem, Estimating the Prevalence of Excessive Alcohol Use and Alcohol-related Mortality, Current Trends and the Issue of Prevention". Alcoholism: A Medical Profile. N.Kessel, A.Hawker & H. Chalke, (eds.) First International Medical Conference on Alcoholism. Edsall & Co. London.

de LINT, J. (1975c). "Control Laws and Taxation as Preventive Strategies". Triennial Refresher Course for Alumni of the Summer School of Alcohol Studies, Rutgers University, New Brunswick, New Jersey.

de LINT, J. (1975d). "Estimating Rates of Excessive Alcohol Use and Alcohol-Related Health Damage". 21st International Institute on the Prevention and Treatment of Alcoholism, Helsinki.

de LINT, J. et al (1975e). "Mortality Among Patients Treated for Alcoholism: A 5-Year Follow-Up". Canadian Medical Association Journal, 113. 385.

de LINT, J. et al (1975f). "Control Laws and Price Manipulation as Preventive Strategies". Expert Conference on the Prevention of Alcohol Problems, Berkeley, California. Addiction Research Foundation, mimeo.

de LINT, J. et al (1971). "Consumption Averages and Alcoholism Prevalence: A Brief Review of Epidemiological Investigations". British Journal of Addiction, 66. 97.

de LINT, J. et al (1970). "Mortality from Liver Cirrhosis and Other Causes in Alcoholics. A follow-up Study of Patients with and without a History of Enlarged Fatty Liver". Quarterly Journal of Studies on Alcohol, 31. 705.

LIPSCOMB, W.R. (1959). "Mortality Among Treated Alcoholics: A Three Year Follow-Up Study". Quarterly Journal of Studies on Alcohol, 20. 596.

LUNDQUIST, G.A.R. (1972). "Alkoholberoendets yttringer och förlopp". Alkoholfrågan, 66. 3.

LUNDQUIST, G.A.R. (1970). "Alcohol Dependence and Depressive States". 16th International Institute on the Prevention and Treatment of Alcoholism, Lausanne.

MacFARLAND, R.A. et al (1962). "The Epidemiology of Motor Vehicle Accidents", Journal of the American Medical Association, 180. 289.

MÄKELÄ, K. (1971). "Alkoholinkulutuksen jakautama". Alkoholikysymys, 39. 3.

MAYFIELD, D.G. et al (1972). "Alcoholism, Alcohol Intoxication and Suicide". Archives of General Psychiatry, 27. 349.

McBAY, A.J. (1972). "Alcohol and Highway Fatalities". A Study of 961 Fatalities in North Carolina during the past six months of 1970". North Carolina Medical Journal, 33. 769.

MENDELSON, J.H. (1970). "Biological Concomitant of Alcoholism" I. II. New England Journal of Medicine, 283. 24. 71.

MENNINGER, K.A. (1938). Man Against Himself Harcourt, Brace, New York.

MORICE, A. (1954). "De l'action d'alcool sur le développement du cancer de l'oesophage". Bulletin de l'Academie Nationale de Medecine.

MOWRER, E. (1942). Disorganization: Personal and Social. Lippincott, Philadelphia.

MURPHY, G.C. et al (1967). "Social Factors in Suicide". Journal of the American Medical Association, 199. 303.

NICHOLLS, P. et al (in press). "A Study of Alcoholics Admitted to Four Hospitals. II. General and Cause-Specific Mortality During Follow-Up." Quarterly Journal of Studies on Alcohol.

NORVIG, J. et al (1956). "A Follow-Up Study of 221 Alcohol Addicts in Denmark". Quarterly Journal of Studies on Alcohol, 17. 633

PALOLA, E.G. et al (1962). "Alcoholism and Suicidal Behaviour". Society, Culture and Drinking Patterns, 511. C.Snyder and D.J.Pittman (eds.) John Wiley & Sons, New York.

PAYNE, C.E. et al (1962). "Traffic Accidents, Personality and Alcoholism". Journal of Abdominal Surgery, 4. 21.

PELL, S. et al (1973). "Five Year Mortality Study of Alcoholics". Journal of Occupational Medicine, 15. 120.

PLAUT, T.F.A. (1967). "Alcohol Problems: A Report to the Nation". The Co-operative Commission on the Study of Alcoholism, Oxford University Press, New York.

POPHAM, R.E. et al (1975). "The Effects of Legal Restraint in Drinking". The Biology of Alcoholism, IV. B.Kissin and H.Begleiter, (eds.) Plenum Publishing Corp, New York.

POPHAM, R.E. (1970). "Indirect Methods of Alcoholism Prevalence Estimation: A Critical Evaluation". Alcohol and Alcoholism 678, R.E.Popham (ed.) University of Toronto Press, Toronto.

POPHAM, R.E. (1959). "Some Social and Cultural Aspects of Alcoholism". Canadian Psychiatric Association Journal, 4. 222.

POTONDI, A. et al (1967). "Uber die spontanverletztungen betrunkener" Acta Chirurgica Academial Scientarium Hungarical, 8. 337.

ROBINS, E. et al (1959). "The Communication of Suicidal Intent: A Study of 134 Consecutive Cases of Successful (completed) Suicides". American Journal of Psychiatry, 115. 724.

ROBINS, E. et al (1957). "Some Interrelations of Social Factors and Clinical Diagnosis of Attempted Suicide: A study of 109 Patients". American Journal of Psychiatry, 114. 222

ROOM, R. (1974). "Minimizing Alcohol Problems". Expert Conference on the Prevention of Alcohol Problems, Berkeley, California.

RUSHING, W.A. (1969a). "Suicide as a Possible Consequence of Alcoholism". Deviant Behaviour and Social Progress, 323. W.A. Rushing. (ed.) Rand McNally, Chicago.

RUSHING, W.A. (1969b). "Deviance, Interpersonal Relations and Suicide". Human Relations. 22. 61.

RYDBERG, U .et al (1973). "The Alcohol Factor in a Surgical Emergency Unit". Acta Medicinae Legalis et Socialis, 22. 71.

SAINSBURY, P. (1955). Suicide in London: An Ecological Study. Chapman and Hall, London.

SALUM, I. (ed.) (1972). "Delirium Tremens and Certain Other Acute Sequels of Alcohol Abuse. VIII. Mortality". Acta Psychiatrica Scandinavica, 235. 86.

SCHMIDT, H. et al (1954). "Evaluation of Suicide Attempts as a Guide to Therapy". Journal of the American Medical Association, 155. 552.

SCHMIDT, W. et al (1975/76). "Heavy Alcohol Consumption and Physical Health Problems: A Review of the Epidemiological Evidence". Drug and Alcohol Dependence, 1. 27.

SCHMIDT, W. et al (1973). "Mortality of Alcoholics". Alcohol Health and Research World, Summer 16.

SCHMIDT, W. et al (1972). "Causes of Death of Alcoholics". Quarterly Journal of Studies on Alcohol, 33. 171.

SCHMIDT, W. et al (1968). "Alcohol Consumption of Alcoholics". Addiction Research Foundation, Toronto, mimeo.

SCHWARTZ, D. (1966). "Alcool et cancer; étude de pathologie géographique". Cancro, 19. 200.

SEELEY, J.R. (1960). "Death by Liver Cirrhosis and The Price of Beverage Alcohol". Canadian Medical Association Journal, 83. 1361.

SEIXAS, F.A. et al (eds.) (1975). "Medical Consequences of Alcoholism". Annals of the New York Academy of Sciences, 252.

SEIXAS, F.A. (1974). "Criteria for the Diagnosis of Alcoholism". Criteria Committee, National Council on Alcoholism.

SKOG, O.J. (1973). "Less Alcohol - Few Alcoholics ?" The Surveyor, 7. 7.

SKOG, O.J. (1971). "Alkoholkonsumets fordeling i befolkningen". Statens Institutt for Alkoholforskning, Oslo.

SPAIN, D.M. et al (1951). "Alcohol and Violent Death. A One-Year Study of Consecutive Cases in a Representative Community." Journal of the American Medical Association, 146. 334.

STARE, F. (1961). "Myocardial Infarction in Patients with Portal Cirrhosis". Nutrition Reviews, 19. 37.

STENGEL, E. et al (1958). Attempted Suicide, Chapman and Hall, London.

STORBY, Å. (1953). "Olycksfallsfrekvensen i ett alkoholistmaterial". Svenska Läkartidningen, 55. 2100.

SUNDBY, P. (1967). "Alcoholism and Mortality". National Institute for Alcohol Research, 6. Universitetsforlaget, Oslo.

TASHIRO, M. et al (1963). "Mortality Experience of Alcoholics". Quarterly Journal of Studies on Alcohol, 24. 203.

TONGE, J. (1962). "Blood Alcohol Levels in Patients Attending Hospital after Involvement in Traffic Accidents". Journal of Forensic Medicine, 15. 152.

TONGE, J. et al (1964). "Fatal Traffic Accidents in Brisbane from 1935 to 1964". Medical Journal of Australia, 2. 811.

U.S. DEPARTMENT OF HEALTH, EDUCATION AND WELFARE (1971). "First Special Report to the U.S. Congress on Alcohol and Health". DHEW Public No. (HSM) 72-9099. Washington, D.C. U.S.Government Printing Office, 45.

U.S. SURGEON GENERAL, REPORT TO (1971). "The Health Consequences of Smoking". U.S.Department of Health, Education and Welfare, U.S. Government Printing Office, Washington D.C.

WALLER, J.A. (1972). "Non-Highway Injury Fatalities. 1. The Role of Alcohol and Problem Drinking, Drugs and Medical Impairment". Journal of Chronic Diseases, 25. 33.

WALLER, J.A. et al (1966). "Alcoholism and Traffic Deaths". New England Journal of Medicine, 275. 532.

WALLER, J.A. et al (1969). "Alcohol and Other Factors in California Highway Fatalities". Journal of Forensic Science, 14. 429.

WALLINGA, J.V. (1949). "Attempted Suicide: A Ten Year Survey". Diseases of the Nervous System, 10. 15.

WALTON, R.G. (1972). "Smoking and Alcoholism: A Brief Report". American Journal of Psychiatry, 128. 1455.

WESCHSLER, H. et al (1969). "Alcohol Level and Home Accidents". A Study of Emergency Service Patients". Public Health Report, 84, 1043.

WEIR, J.M. et al (1970). "Smoking and Mortality: A Prospective Study" Cancer, New York, 25. 105.

WILKINSON, P. et al (1971). "Physical Disease in Alcoholism: Initial Survey of 1,000 Patients". Medical Journal of Australia, L. 1217.

WILKINSON, P. et al (1969). "Epidemiology of Alcoholism: Social Data and Drinking Patterns of a Sample of Australian Alcoholics". Medical Journal of Australia, 1. 1020.

WORLD HEALTH ORGANISATION (1974). "Twentieth Report of the WHO Expert Committee on Drug Dependence". Technical Report Series, 551.

WYNDER, E.L. et al (1961). "A Study of the Etiological Factors in Cancer of the Oesophagus". Cancer, 14. 389.

ZEINER-HENRIKSEN, T. (1971). "Cardiovascular Disease Symptoms in Norway". A Study of Prevalence and a Mortality Follow-Up". Journal of Chronic Diseases, 24. 553.

ZMUC, M. (1968). "Alcohol and Suicide". Alcoholism, 4. 38.

PARENTS, CHILDREN AND LEARNING TO DRINK

R. J. McKECHNIE

THE ALCOHOL RESEARCH & TREATMENT GROUP

CRICHTON ROYAL, DUMFRIES

Whenever statistics relating to childhood or teenage drinking are reported there is a general stirring of concern and calls for tighter controls. These cries arise from three assumptions which should be considered.

1. Children are innocent creatures who don't know or care about alcohol.
2. Children and teenagers shouldn't know or care about alcohol, and
3. Early introduction to alcohol consumption leads to heavy drinking and alcoholism.

In respect of the first assumption, Jahoda and Cramond (1972) have shown that by the age of six two-fifths of children recognise the smell of beer or whisky as alcohol and can arrange drinks into alcoholic and non-alcoholic groups even before they can use the word 'alcohol'. Six-sevenths recognise drunken behaviour on film as due to drink, and 57% have tasted alcohol. Davies and Stacey (1972) report that by the age of 14, 85% of girls and 92% of boys have tasted alcohol and McKechnie et al (1975) that by the age of 15, 93% of girls and 95% of boys have tasted alcohol. Before creating another hue and cry, the author points out that these figures represent positive responses to the question "Have you ever tasted alcohol?". A positive response may refer to a single instance on a special occasion and it does not mean that all these children are regular drinkers. If you ask yourself when you had your first taste of alcohol you will realise that the figures are not so incredible as some would make out. Thus the first assumption that children do not know about alcohol is simply untrue.

The second and third assumptions have to be considered together since they are linked together in many peoples reasoning. Although the second assumption that children should not know about alcohol may arise from the original assumption of innocence, the belief that early drinking predisposes one to alcoholism leads to the notion that children should be kept ignorant with regard to alcohol. It is this third assumption that early introduction leads to early consumption and therefore alcoholism which gives rise to greatest concern. However, it is an assumption and it seems as though there is little or no evidence to support the case and there is evidence which may suggest otherwise.

Ullman (1962) reports that "addictive drinkers had their first drink at a later age than non-addictive drinkers" and Davies and Stacey (1972) report that : "the abstinent and the heavy drinking categories (in their sample) are unique in having certain things in common. It appears that both abstaining and heavy drinking teenagers have parents with strong negative attitudes towards adolescent drinking and that these parents are generally more disciplinarian than those of any other drinking group. In addition it appears that the heavy drinking teenager is often introduced to alcohol later in life than those in other drinking groups and is thus more likely to have learned to use alcohol for himself rather than with the help of his parents".

So, rather than supporting the case that early introduction to drinking leads to heavy consumption there is some evidence which suggests the contrary. So our third assumption begins to look somewhat shaky. It is inadequate for another reason and that is that it looks at age alone and does not consider the environment in which that introduction to drinking takes place. Notice in the quote from Davies and Stacey an implicit assumption that one learns how to drink. They suggest that those who do not learn with the help of parents have to learn for themselves despite presenting a great deal of evidence that those who have failed to learn from their parents before adolescence learn at that stage from their peers and that may be fraught with problems, because of the nature of adolescent drinking.

The relationship between parental attitude and drinking behaviour and some related data from a study by the author which reports a normative pattern of learning of drinking behaviour should be considered.

It is generally held that children from homes of alcoholics are over-represented in alcoholic population. Alonso-Fernandez (1970) suggests that between 35 and 55% of alcoholics in Spain come from alcoholic homes and Bosma (1972) suggests the figure of 52% in America. While some writers claim that these figures support the notion of genetic transmission of alcoholism, others argue that the high incidence is due to faulty learning of drinking behaviour. The idea that parental attitude and behaviour might be important as aetiological factors in problem drinking is not new.

Jackson and Conner (1953) report that adult alcoholics regarded their parents as having inconsistent attitudes about drinking. McCord and McCord (1960) report a higher incidence of alcoholism in those whose parents held conflicting views about alcohol and/or differed in their drinking behaviour. Globetti and Chamblin (1966) found a relationship between parental disapproval of drinking and a high percentage of problem drinkers among Mississippi high school students. Sadoun et al (1965) show that although drinking practices in France and Italy are remarkably similar they have very different alcoholism rates. Even within France there are regional variations; in fact there is a negative correlation between consumption and alcoholism (as measured by deaths from cirrhosis) taken region by region. So we have to look elsewhere for aetiological factors. Sadoun suggests that unlike Italian parents who regard the moderate use of wine by children as normal and unimportant, most French parents are reported to have firm and often quite rigorous attitudes with two-thirds of them being opposed, 16% of them very strongly, to their children drinking in any form. Strictest control over childhood drinking in France was reported by respondents from the North East and North West which are the two regions with the highest incidence of deaths from cirrhosis and lowest per capita consumption.

In the study by McKechnie et al (1975) the relationship between parental attitude and frequency of drinking was examined. When comparing those whose parents are "against" young people drinking with those "not against" it was found that parental disapproval is associated with a low frequency of drinking in the home relative to parental tolerance. Outside the house the relationship is reversed, i.e. those whose parents disapprove drink more frequently than those whose parents are tolerant.

One of the contradictions of the data is that although more than half of the parents are reported to disapprove of teenage drinking the place where most children will have had a drink is in their own home and supplied by their parents (Davies and Stacey, 1972; Jahoda and Cramond, 1972; McKechnie et al, 1975).

The places where teenagers reported having had a drink were looked at :

Own Home	Home of Adult Relative	Home of own friend	Open air	Dance	Public House and Hotel
167	140	101	81	73	67

This data was examined to see if there was a pattern or sequence which might suggest a developmental scale followed by most of the children. The above data might suggest that most children have had a drink in their own home, almost all who have had a drink in the home of an adult relative have had a drink in their own home, and those who have had a drink in the home of one of their own friends, have had a drink in the two previous situations and so on. If the sequence were fixed and the situations formed a "perfect scale" then each of the smaller numbers would be drawn from the immediately above larger number. Scalogram analysis (Green 1965) allows one to calculate how close the actual pattern of scores matches the ideal pattern. The statistic is called an Index of Reproducability and ranges from 0 (no match) to 1 (exact match). All values above .5 are accepted as showing significant match. For the data an Index of .82 is obtained showing a close resemblance to the ideal pattern.

The sequence may vary from country to country or even from region to region but might imply the existence of an implicit training programme for young people. There is probably nothing new about the fact that children are introduced to alcohol in their own homes by their parents, by other adults and then their drinking develops into a peer group activity engaged in outwith parental control.

It is known from Davies and Stacey (1972) that adolescents view drinking as associated with toughness, sociability and anti-authoritarian attitudes and it may be unhealthy to be introduced to alcohol for the first time in such context. Therefore, it is argued that pre-adolescent introduction to alcoholic beverage associated with other values is extremely important and that this training rightly should take place within the family.

This kind of training has probably existed for generations and has served most people well in that they handle alcohol in appropriate ways and manage to avoid the serious pitfalls encountered by a small proportion of the population normally labelled "alcoholic".

When this sequence was first reported in public, it was suggested that it was spurious as a similar finding would be made if one considered any other beverage say tea or coffee. In fact the similarity between this sequence and what we suppose might apply to tea or coffee suggests that alcohol like tea or coffee is socially prescribed and children learn about it in a similar kind of way, if perhaps a little later.

One way in which this sequence might be used is to identify those who are learning about alcohol in a "socially deviant" manner and longitudinal studies might confirm the hypothesis that early deviant learning leads to later difficulties in handling alcohol.

In the sample there are six subjects who show markedly different patterns from the norm. Their parents are either "tend to be against" or "strongly against" drinking by young people. Two of these six subjects have only had a little (less than 1 pint) alcohol in a place other than someone's home and are unusual for not having had drinks in their own home. The other four drink one or more times per week and in fairly large amounts in "friends' homes" and "in places other than someone's home". This latter group of four accounts for 21% of boys who drink once or more per week in places other than someone's home. Another finding that supports the importance of this sequence is that in Ullman's (1962) study :

(a) "Most of the addictive drinkers had their first drink in a place other than a private home or a place where liquor is usually sold". and
(b) "More addictive than non-addictive drinkers had their first drink in the company of persons outside the family".

In conclusion it has to be accepted :

(a) that children and teenagers are not innocent little creatures who neither know nor care about alcohol,
(b) the assumption that early drinking leads to heavy consumption is not proven and there is some evidence to suggest the contrary.
(c) that drinking by teenagers is a normal phenomenon. A conclusion arrived at also by Maddox (1962) in the States although our children appear to be more precocious.
(d) When a clearer picture of normal drinking development is available for any culture then deviant learning and therefore risk may become identifiable.

REFERENCES

ALONSO-FERNANDEZ, F. (1970). "La Psychotherapie Comme Traitment de Base des Alcoholiques." Revista Alasme, 16, 217-223.

BOSMA, W.G.H. (1972). "Children of Alcoholics - A Hidden Tragedy". Maryland State Medical Journal, 21, 34-36.

DAVIES, J. and STACEY, B.G. (1972). Teenagers and Alcohol. H.M.S.O. LONDON, 463, II.

GLOBETTI, G. and CHAMBLIN, F. (1966). Sociology-Anthropology Series No. 3, Mississippi State University, New Haven.

GREEN, B.F. (1965). "A Method of Scalogram Analysis using Summary Statitstics". Psychometrika, 21, 79-88.

JACKSON, J. and CONNER, R. (1953). "Attitudes of Parents of Alcoholics, Moderate Drinkers and Non Drinkers Towards Drinking". Quarterly Journal Studies on Alcohol, 14, 590-613.

JAHODA, G. and CRAMOND, J. (1972). Children and Alcohol, H.M.S.O. London, 463, 1.

MADDOX, G.L. (1962) "Teenage Drinking in the United States". Pittman and Snyder. Society Culture and Drinking Patterns, Wiley, New York.

McCORD, W. and McCORD, J. (1960). Origins of Alcoholism Tavistock Publications, London.

McKECHNIE, R.J., CAMERON, D., CAMERON, I.A. and DREWERY, J. (1975) "Teenage Drinking in South-West Scotland". British Journal of Addiction.

SADOUN, R., LOLLI, G., and SILVERMAN, M. (1965). Drinking in French Culture. Monograph 5. Rutgers Center of Alcohol Studies.

ULLMAN, A.D. (1962). "First Drinking Experience as Related to Age and Sex". Pittman and Snyder, "Society Culture and Drinking Patterns. Wiley, New York.

ALCOHOL AND EDUCATION

BRUCE RITSON

THE ANDREW DUNCAN CLINIC

THE ROYAL EDINBURGH HOSPITAL, EDINBURGH

In London of the last century devastating outbreaks of cholera were commonplace. Dr John Snow, a physician of that time, noted in one outbreak that all the cases in one area of Soho drew their water from a single pump. Using this observed association between two events, he removed the handle from the Broad Street pump and the epidemic abated. This well known early essay demonstrated the value of epidemiological method in preventive medicine.

With alcoholism, in common with many other diseases, the chain of association between cause and effect is much more complex, and there is no one gesture which can eradicate the problem. Epidemiologists in recent years, notably de Lint and Schmidt, have strengthened the view that, without a high overall level of alcohol consumption in a population, there will be little alcoholism. The aim of this paper is to illustrate some inconsistencies in our attitude towards prevention and to suggest some priorities in our preventive strategy.

The author is indebted for some of the ideas put forward here, to a series of discussions at the Advanced School of the Alcohol Education Centre last year where the author had the privilege of being tutor to the section on prevention and education.

There are those who view alcoholism as a disease which has its origins almost exclusively in the drinker rather than alcohol itself and accordingly eschew all attempts to influence the availability of alcohol and its use in the general population. This view seems too narrow, first because it ignores the evidence that availability and ease of access to drink are significantly associated with alcoholism - not of course with the clarity of the Broad Street pump but with sufficient strength to convince most epidemiologists, and second because we cannot confine ourselves singly with alcoholism but need to consider alcohol related problems such as road traffic accidents, public drunkenness and domestic violence where alcohol abuse plays a significant part. Not only do these seem valid concerns in their own right but they are frequently the prodromal features of alcoholism itself.

Any debate on primary prevention readily turns to the issue of educating children about drinking. This often provokes ruminations about the role of the family as the primary purveyor of social learning, the decline of parental responsibility and so on. But the family unit, unlike the child, is not universally accessible to educational influences. Everybody has to go to school and for these few years each succeeding generation is a captive audience. For practical and apparently logical reasons, the temptation to invest most of our resources in educating the young is enormous.

It is not suggested that this venture be withdrawn but doubtful factors are worth consideration. The first concerns the value of the exercise and the second its cost effectiveness.

Educational programmes are notoriously difficult to evaluate and most attempts to do so have produced disappointing, equivocal or even contrary results to those expected,(Globetti 1974). Dorn and Thomson (1976) in the field of drug education recently expressed similar doubts about the value of education, particularly if this implies the mere passing on of facts. Children are aware of the facts about tobacco but this may not influence attitudes on behaviour: and even more disturbing it is clear that changes in attitude may not correlate particularly closely with behavioural change. Jahoda and Crammond (1972), in their study of Glasgow schoolchildren, showed that most eight year olds appeared to accept conventional morality about the dangers of drink; but this did not seem to influence behaviour as they grew older and came to reject these values which they then associated with the authority that they wished to overthrow.

No one is so naive as to imagine that a single film and discussion about alcohol and its effects with a group of children aged 12 - 15 will seriously influence their behaviour at a party two months or even two years later when they will daily be confronted with the role model of alcohol use presented by their parents, their peers and the media.

It is not the intention to discuss the value of education. It seems only reasonable that children should have a chance of discussing alcohol which will form a significant part of the adult world they are entering. Equally, it is to be hoped that they would have a chance of developing the capacity to criticise and evaluate information whether it comes in a drinks advertisement or a health education leaflet. They might also consider the general use of drugs such as alcohol as a means of dealing with stress. Such tasks seem a fundamental challenge to all education and are beyond the competence of the author to discuss. They form the substance of all time education for living and it is only reasonable that knowledge of alcohol and drinking practices should be available to those responsible for education.

Convenient as children may be as an object of education, there is some question whether they are the most useful focus of limited funds available specifically for alcohol education. Hawkes recently questioned the assumption that the individual is the appropriate object of education. He urges differentiation between "education that has as its objective the persuasion of the individual, and education whose goal is the rearrangement of the environmental contingencies which affect individuals".

Rather than focusing our attention on the individual and changing him so that he can learn to cope with the environment, it is equally valid to direct health education towards the environment itself as the object. Historically, many of the most successful public health ventures have been effected by altering the conditions under which people exist rather than altering the individual himself (for instance atmospheric pollution and bronchitis).

It might be argued that the environment itself cannot be educated but those who have most control of the environment could well become the focus of education. At present, most of the dice appear to be loaded in favour of consuming ever increasing quantities of alcohol and a few ways in which the environment might usefully be influenced to give an individual a more real choice should be considered - some of these changes are modest and require no major alteration of law or custom.

A mundane example was offered to the author in a public bar at lunchtime recently, when two people wished to have a sandwich and a drink. One wished to have beer and the other coffee. They were told that coffee, although available, could only be taken with a full three course lunch and not with a snack. It is rare enough for any Scottish pub to offer coffee but why is the balance so much tilted in favour of the alcoholic drink ?

The author recently had the opportunity of visiting a cafe pub in Brighton and observed a group of young people drinking together, some taking coffee, others soft drinks and still others beer - there was no pressure to take alcohol. The proprietor, who seemed to be making an adequate living, commented that in this setting it was much easier for the young girl with her boyfriend to opt for coffee when offered a drink than it is in most pubs. True freedom of choice is often not present, as a tour of most Scottish towns will readily reveal - it is hard to find there any focus of social entertainment which is not intimately concerned with alcohol. If the leisure time activity of the young is inevitably linked with alcohol through lack of alternatives, it is tempting to hypothesise that alcohol and happy occasions will become linked so that the one comes to stand for the other, and the individual becomes conditioned to associate alcohol and fun or a good time to such an extent that one cannot exist without the other.

Availability is a potent factor in promoting alcoholism, (Murray 1975) - and that company directors are conspicuously more at risk than other occupational groups. Yet access to free or expense account alcohol is frequently a "perk" of the executive ranks. There is surely some modest inconsistency in a society which wants to teach children about alcohol and its hazards and provides the same drug almost as a reward to our captains of industry. It is not intended that alcohol should be banned in staff canteens and executive dining rooms but the public health implications of what is being done merit consideration.

Two modest changes in the drinking environment have been suggested which could be possible without major social upheavals. In fact, the second might even enhance the efficiency of British industry.

More major and significant endeavours in social engineering bring the health educationist into the arena as a political lobbyist. Fiscal and licensing measures are amongst the most potent means of influencing drinking practices. There is even some evidence of effectiveness here; a rare finding in the field of prevention. The introduction of the Breath Test was probably one of the most effective public health measures of the past decade and immediately saved more than one thousand lives in the first year of operation.

There is some evidence that increasing the tax on alcohol can bring about a reduction in consumption (Nyberg 1967). Such rises require to be very large to produce any lasting change and may bring other unforseen consequences such as a shift to home brewing, the consumption of alcohol substitutes, illicit distillation, smuggling and (perhaps easiest) restricting other parts of the family budget so that drinking can continue. Changes in beverage choice are more vulnerable to price alterations. For instance, British sales of whisky fell by 12½% in the eight months after the 1975 Budget. It is interesting in this respect that increased expenditure on drink in the past decade has far outstripped the increases on other commodities such as clothing and food. A further anxiety about fiscal controls is that they hit hardest those families which have the smallest incomes.

Allowing for these reservations, use of increasing taxation as a control strategy may be worth consideration. Scottish data show that between 1950 and 1970 the price of a bottle of whisky as a percentage of mean weekly per capita disposable income fell steadily while the whisky consumption and admissions to hospital for alcoholism rose at the same rate, (Semple and Yarrow 1974). We therefore appear justified in regarding health as one of the factors to be taken into account when the Treasury considers the tax on alcohol.

In saying this, it is not suggested that it should be the only or even the most important factor but weighing the health aspect of alcohol consumption should be a recognised part of the decision making process. This calls for education with the Government and the Treasury as objects.

Some have suggested that alcohol could at least be "index linked" so that the price keeps in step with inflation - this alone would help to ensure that the relative cost of alcohol did not lag behind other commodities. It would also depoliticise the rise in alcohol taxation which can so readily become an annual preoccupation at budget time.

Dr. Clayson, in another conference paper, discussed licensing hours and legislation. It is clearly a further area of control and influence. The aim of his Committee's Report was to relax the pressure to drink in Scotland by influencing the nature of the public house and the conditions under which drinking occurred. The Committee also stepped away briefly from concern with licensing to mention taxation on fortified British wines. The price of these is paradoxically low and makes them a source of disproportionately cheap alcohol. No heed has been taken of the committee's recommendation that the tax on these should be related to their alcohol content as it is for most other drinks. Where and by whom can this idea be debated further in society ?

The control of alcohol advertising inevitably suggests itself as a further preventive measure. Some restrictions of course already exist and it is probable that amongst established drinkers advertising mostly influences brand choice. Nonetheless, advertising in recent years appears consciously to aim at enlisting new drinkers, notably women and young men, by linking social and sexual sophistication with particular drinks. It was revealing to read last week of a £250,000 advertising campaign for whisky which will lay special emphasis on "the female and younger markets" (Observer 4th April 1976).

So far, the ways in which we might usefully identify new objects for education have been concentrated upon. Finally, consideration should be given to education as an aid to early identification.

The problem drinker finds that his situation becomes increasingly chaotic as alochol damages his domestic life, his capacity for work and his health. He realises that he has problems and may attribute them to a difficult home life, authoritarian employer, mindless bureaucracy or whatever; but he may not see the common link with alcohol. The individual himself and those around him can all portentially facilitate identification of the alcohol problem that constitutes the unifying theme in the turmoil that surrounds him. Viewed in this way, early identification becomes everybody's business and a proper focus for public education. The Scottish Health Education Unit's recent campaigns have tried to establish an awareness of alcohol problems in the community so that identification and perhaps in some cases, management, may occur within the family or even the individual himself. We do not know how many spontaneous "cures" occur as a result of self recognition, facilitated by educational campaigns. Certainly there are many case reports of individuals who came to realise they had a drink problem and took themselves in hand. Wilkins in his study of problem drinkers in general practice noted many men who developed alcohol dependence at some stage in their lives but had subsequently overcome this.

Some individuals are particularly well placed to facilitate early recognition - barmen, personnel officers, teachers, police, casualty staff and social workers provide examples of groups that merit particular educational focus.

If a broader view of alcohol problems is allowed, the range of potential helpers extends well beyond the socio-medical professions. If those who are likely to meet alcoholics have some understanding of alcohol problems then there is a chance of turning the crisis of an arrest, accident, or domestic breakdown from a disaster into an opportunity for change.

REFERENCES

DORN, N. (1976) Drug and Alcohol Dependence, I, 15.

DORN, N. and THOMSON, A. (1976) Community Health, 7, 154.

GLOBETTI, G. (1974) "Research on methods and programmes of Drug Education" - Addiction Research Foundation, Ontario, 97, 112.

HAWK, D. (1974) "Research on methods and programmes of Drug Education" - Addiction Research Foundation, Ontario, 161-168.

JAHODA, G. and CRAMMOND, J. (1972) Children and Alcohol, H.M.S.O.

de LINT, J. and SCHMIDT, W. (1971) British Journal of Addiction, 66, 97.

MURRAY, R. (1975) Journal of Alcoholism, 16, 23.

NYBERG, A. (1967) "Consumption and price of alcoholic beverages" Finnish Foundation of Alcohol Studies, 15.

SEMPLE, B.M. and YARROW, A. (1974) Health Bulletin, 32, 114.

WILKINS, R. (1974) The Hidden Alcoholic in General Practice, Elek.

CHAIRMEN

Surgeon Vice-Admiral Sir Dick Caldwell,
Executive Director, Medical Council on Alcoholism, London.

Professor Sir Cyril A.Clarke,
President, Royal College of Physicians, London.

The Rt.Hon.The Lord Cohen of Birkenhead.

Professor J.R.M.Copeland,
Department of Psychiatry, University of Liverpool.

Miss P.Farrell,
Director of Social Services, Metropolitan Borough of Sefton, Liverpool.

Mr.J.Gordon,
Chairman, Alcohol Education Centre, London.

Mr.M.Grant,
Educational Director, Alcohol Education Centre, London.

Mr.R.Grant,
Assistant Chief Constable, Merseyside Police.

Major P.D.V.Gwinner,
Consultant Psychiatrist, Alcoholism Treatment Unit, Royal Victoria Hospital, Southampton.

Dr.I.Hindmarch,
Lecturer in Psychology, University of Leeds.

Dr.W.J.E.McKee,
Regional Medical Officer, Mersey Regional Health Authority.

Dr.J.S.Madden,
Consultant Psychiatrist, Regional Addiction Unit, Moston Hospital, Chester.

Dr.T.Malcolm,
Consultant Psychiatrist, West Cheshire Hospital, Chester.

Dr.R.Rathod,
Consultant Psychiatrist, St.Christophers Day Hospital, Horsham.

Professor W.Linford Rees,
President, The Royal College of Psychiatrists, London.

Dr.D.Robinson,
Senior Research Sociologist, Institute of Psychiatry, Addiction Research Unit, London.

Mr.D.Rutherford,
Director, National Council on Alcoholism, London.

Dr.A.Balfour Sclare,
Consultant Psychiatrist, Duke Street Hospital, Glasgow.

Miss M.Storey,
Regional Nursing Officer, Mersey Regional Health Authority.

Mr.A.Tongue,
Director, International Council on Alcohol & Addictions, Lausanne.

Dr.R.Walker,
Consultant Physician, Walton Hospital, Liverpool.

CONTRIBUTORS

Dr. M.J. Akhtar,
Consultant Psychiatrist, The General Hospital Harton Lane, South Shields NE3 4OPL.

Dr. D. Cameron,
Drury House, Narborough, Leicester.

Mr.A.K.J.Cartwright,
Research Director, Maudsley Alcohol Pilot Project, The Bethlem Royal and Maudsley Hospital, London.

Dr.C.Clayson,
Chairman, Clayson Committee, Scotland.

Professor R.M.Costello,
Assistant Director, Alcoholism Treatment Unit, San Antonio, Texas.

Mr.B.Coyle,
Senior Social Worker, Douglas Inch Clinic for Forensic Psychiatry, Glasgow.

Mr.T.Crolley,
Senior Probation Officer, Probation and After-Care Service, Liverpool.

Mrs.S.Delahaye,
Researcher, Department of Psychiatry, University of Manchester.

Mr.N.Dorn,
Assistant Director, Institute for the Study of Drug Dependence, London.

Dr.C.Fazey,
Senior Lecturer, Department of Social Studies, Preston Polytechnic

Mr.J.Fischer,
Social Worker, Charing Cross Clinic for Drug and Alcohol Problems, Glasgow.

Dr.J.Flanzer,
Assistant Professor, School of Social Welfare, The University of Wisconsin, Milwaukee.

Mr.M.Grant,
Educational Director, Alcohol Education Centre, London.

Major P.D.V.Gwinner,
Consultant Psychiatrist, Alcoholism Treatment Unit, Royal Victoria Hospital, Netley, Southampton.

Dr.J.R.Hamilton,
Lecturer in Forensic Psychiatry, University Department of Psychiatry, Royal Edinburgh Hospital, Edinburgh.

Mrs.A.Hawker,
Researcher, Medical Council on Alcoholism, London.

Dr.E.Heilmann,
The Medizinische Poliklinic, Der Westfalischen Wilhelms-Universitat Munster, West Germany.

Dr.I.Hindmarch,
Lecturer in Psychology, Department of Psychology, The University of Leeds.

Dr.B.D.Hore,
Consultant Psychiatrist, Regional Alcoholism Unit, Manchester.

Mr.W.H.Kenyon,
Executive Director, Merseyside Lancashire and Cheshire Council on Alcoholism, Liverpool.

Dr.M.A.Lavenhar,
Associate Professor and Director, Division of Biostatistics, New Jersey Medical School, New Jersey.

Dr.T.Levinson,
Psychologist, The Clarke Institute of Psychiatry, Toronto.

Mr.J.de Lint,
Senior Scientist, Social Studies Department, Alcoholism and Drug Addiction Research Foundation, Toronto.

Dr.G.Lowe,
Psychologist, Department of Psychology, University of Hull.

Mr.R.J.McKechnie,
Senior Psychologist, Alcohol Research and Treatment Group, Crichton Royal, Dumfries.

Dr.J.S.Madden,
Consultant Psychiatrist, Regional Addiction Unit, Chester.

Mr.M.Meacher,
Parliamentary Under Secretary of State, Department of Health and Social Security, London.

Dr.G.M.St.L.O'Brien,
Professor and Dean, School of Social Welfare, University of Wisconsin, Milwaukee.

Dr.D.Owens,
Consultant Physician, Walton Hospital, Liverpool.

Dr.M.A.Plant,
Sociologist, Medical Research Council Unit for Epidemiological Studies in Psychiatry, University Department of Psychiatry, Edinburgh.

Mr.C.E.Reeves,
Principal Investigator, Social Sciences Research Council, California.

Dr.K.H.Reuband
Zentralarchiv Fur Empirische Sozialforschung, University of Koln, Germany.

Dr.E.B.Ritson,
Consultant Psychiatrist, The Andrew Duncan Clinic, The Royal Edinburgh Hospital, Edinburgh.

Dr.D.Robinson,
Senior Research Sociologist, Addiction Research Unit, Institute of Psychiatry, University of London.

Dr.E.S.M.Saad,
Consultant Psychiatrist, Department of Psychiatry, Bridgewater Hospital, Manchester.

Dr.A.B.Sclare,
Consultant Psychiatrist, Department of Psychiatry, Duke Street Hospital, Glasgow.

Dr.F.A.Seixas,
Medical Director, National Council on Alcoholism Inc., New York.

Mr.S.J.Shaw,
Medical Sociologist, Maudsley Alcohol Pilot Project, The Bethlem Royal and Maudsley Hospital, London.

Dr.R.G.Smart,
Associate Research Director, Addiction Research Foundation, Toronto.

Dr.T.A.Spratley,
Consultant Psychiatrist, Maudsley Alcohol Pilot Project, The Bethlem Royal and Maudsley Hospital, London.

Dr.R.Walker,
Consultant Physician, Walton Hospital, Liverpool.

Mr.D.Warren-Holland,
Director, Phoenix House, London.

Professor E.Wilkes,
Professor of Community Care and General Practice, Department of Community Medicine, University of Sheffield.

Dr.R.H.Wilkins,
Lecturer, Department of General Practice, University of Manchester.

INDEX

Absorption of ethanol, 23
Abstainers
  hospital admissions and, 132
  proportion in population, 79
Abstinence
  as goal of treatment, 293, 297, 366
  attitudes towards, 311
  neuroticism and, 294
Aetiology of alcoholism, 255, 269
  factors involved, 106
  financial factors, 116
  hereditary factors, 116
  loss of control concept, 294, 297
  role of alcohol in, 123
  treatment and, 107
Aetiology of drug abuse
  research and, 88
Age
  alcoholism and, 172, 182, 362
  drug abuse and, 144, 145, 146, 147
  frequency of admission to units, 5
Age limit for drinking, 103
Agencies for alcoholism, 335
  characteristics of clients, 345
  choice of by patient, 341
  community, 382
  comparisons of, 340, 345, 347, 383
  co-ordination and co-operation, 351
  description of, 335
Agencies for alcoholism (cont'd)
  fear of alcoholics, 384
  hopelessness among, 382
  in South London District, 381
  interprofessional education, 417
  lack of knowledge and skill, 383
  multiple use of, 344, 348
  quality of care, 384
  quality of staff, 417
  responsibility and role, 381, 383
  source of referral to, 341, 347
Aggression, 102, 134
Alcohol
  cultural function of, 48
  habituation to, 193, 194
  importance in alcoholism, 123
  ingestion of, 191
Alcohol dehydrogenase, 23
Alcohol denial, 64
Alcoholic at Risk Register, 63
Alcoholic personality, 106
Alcoholics Anonymous, 98, 189, 214, 267, 293, 336, 351, 364
  comparison with other agencies, 348
  results of, 344
Alkaline phosphatase, 26
Anaemia, 16, 30
Antabuse, 107, 109, 215, 218, 219

Anxiety, 106, 110
Arrhythmias, 30
Ascites, 28
At risk patients, 63
Attitudinal barriers, 286
Aversion therapy, 108

Barbiturates, 117
Behaviour
  alcohol and, 135
  change in, 388
  modification, 108, 110
  patterns, 87
  subcultures and, 155
Beri-Beri heart disease, 24, 30
Blood alcohol levels, 192
  discrimination of, 111
Bone marrow, alcohol toxicity, 30

Cannabis, 35, 38, 41
Cardiomyopathy, 30, 434
Cardivascular system, effects of alcohol on, 30, 434
Care, co-ordination of, 395, 417
Children
  drinking patterns in, 97, 98, 111
  effect of alcoholism on, 244
  learning to drink, 451, 458
  of alcoholic mothers, 185
  of alcoholics, 244
Church Army hostels, 335
Cirrhosis, 25, 26, 58
  as index of alcoholism, 125
  incidence of, 60
  mechanisms of, 59
  mortality rates, 128, 130, 133, 434, 440
  occupation and, 133
  symptoms and signs of, 27
  Vitamin B_{12} and folic acid in, 17
Cobalt toxicity, 31
Community Alcohol Teams, 385
Community involvement, 355
  improvement in, 384

Community involvement (cont'd)
  services offered, 364
  Maudsley Pilot Project, 379
Consumption of alcohol, 356
  age of introduction and, 452
  alcoholism and, 431
  among females, 183
  as guide to problem, 380
  control policies and, 427
  current trends in, 429
  excessive, 430
  increase in, 381
  Ledermann model, 123
  long-term consequences, 431
  measurement of, 77
  mortality rates and, 433
  national rates of, 71, 72, 74, 131, 429
  per capita, 71, 74, 78, 112
  restriction of, 437
Contagion theory, 86
Controlled drinking, 199, 294
  acceptance of, 295
  attendance at meetings, 204
  Donwood Institute study, 299
  follow-up, 301
  procedure, 201
    changes in, 202
  results of, 302
    factors of success, 302
  search for common features, 297
  self reporting, 204, 206
Convictions for drunkenness, 73, 74, 115, 131, 271
  among women, 183
  in England and Wales, 357
  national differences, 126
  number of court appearances, 273
  types of offenders, 271, 272
Co-ordination for care, 395, 417
Councils on alcoholism, 352
Counselling of alcoholics, 366
  See also Group Counselling
Covert sensitisation, 109
Crime, alcoholism and, 250, 331, 332, 333

Cultural function of alcohol, 48
Cultural influences in teenage drinking, 98
Cultural orientation, drug use and, 164

Definitions of alcoholism, 64, 118, 337, 363
Definitions of drug users and takers, 142
Delinquency, 141
Delirium tremens in women, 182
Dependency, 107, 363
  aetiology of, 189
  definitions, 118
  hereditary factors, 116
  illness concept of, 117
  personality and, 117
  psychiatric view of, 115
  treatment of, 277
    administrative approach, 279
    interdisciplinary approach, 287
    legislative approach, 282
    medical approach, 278
    psychological approach, 285
    sociological approach, 286
    sympathetic approach, 284
    thirteen steps, 284
'Dependency-conflict,' 87
Depression, 436
  in women, 183, 184
Deprivation in childhood, 269
Detection of alcoholics by general practitioners, 67
Detoxification, 271, 278, 300
  Ontario system, 321
  place of, 362
Detoxification centres, 271, 273
  aims of, 322
  current position of, 327
  effect on arrest rates, 323
  effect on population, 324
  evaluation of, 323
  future research on, 326
  halfway houses, 322, 326
  limitations of, 327
  results of, 274, 275, 325
Diarrhoea, 25
Disease, alcoholism and, 51
Disease concept of alcoholism, 105, 239, 297, 298, 457
Donwood Institute, 300
Drinking, reasons for, 134
Drinking patterns
  alteration in, 111
  desirable, 438
  in children, 97, 98, 111
  parental attitudes to, 452
  in students, 95
  instruction on, 199
  of young people, 95
  post-alcoholism, 111
Drug education
  aims of, 413
  evaluation of, 374, 375, 409
    research study, 410
  improvements in, 410
  new methods, 413
  problem of, 409
Drug use and abuse, 36
  aetiologies of, 85
  alcohol and, 39
  among young people, 26, 27
  'British system,' 43
  contagion theory, 86
  convictions for,
    age and, 144
    social circumstances, 139
  criminal activities and, 140
  cultural orientation, 164
  definitions, 49, 142
  dependence, 35, 43, 50, 86, 107
  determinators of causality, 153
  diversification of, 158
  education and,
    See Drug education
  environmental factors, 89
  familial factors, 160
  frustration and, 162
  hospital admissions, 117
  incidence of, 115
  increase in, 131
  interaction partners, 164
  interdisciplinary approach to, 420

Drug Use and Abuse (cont'd)
  intervention programmes, 369
  pathological model of, 151, 152
    basis of, 157
    test of, 158
  patterns of, 41
  people involved, 39, 48
  personal relationships and, 153, 158, 160, 161
  prevalence of, 38
  problems, 42, 48, 50
  progression in, 158
  reasons for, 40, 151
  rehabilitation,
    Featherstone Lodge Project, 387
    re-entry stage, 391
    See also Treatment
  situational mood and, 163
  social class and, 143
  social disturbance and, 161
  sociology of, 47, 152, 154
  subcultural model of, 151, 154
    test of, 163
  treatment of, 277
    administrative approach, 279
    aims and objectives, 370
    encounter groups, 390
    evaluation of, 369
    experimental designs, 372
    follow-up, 372
    future directions in evaluation, 373
    interdisciplinary approach, 287
    legislative approach, 282
    medical approach, 278
    New Jersey evaluation model, 372
    obtaining data on, 377
    outcome measures, 371
    Phoenix House, 287
    psychological appraoch, 285
    sociological approach, 286
    sympathetic approach, 284
    thirteen steps, 284
  trends in, 38
  underlying problems, 392
  unhappiness and, 161

Dutch courage, 134

Education
  about smoking, 99
  alcoholism and, 98, 418, 457
Ejaculatory disorders, 7, 8, 10, 11
Electric shocks in aversion therapy, 109
Employment, 133, 361
  See also Unemployment
Encephalopathy
  portal, 28
  Wernicke's, 30
Encounter groups, 390
Environmental factors, 60, 89, 106
Ethanol, 23, 24
Expenditure on alcohol, 128, 129, 402
Expenditure on alcoholism, 361

Familial factors in drug use, 160
Family
  of alcoholics
    involvement in therapy, 239
    needs of, 247
    problems for, 242
Family focused treatment, 239
  food and shelter, 251
  intervention steps, 248
  physical safety, 249
  self-actualisation, 255
  social interaction, 254
  socio-economic environment, 253
  training of personnel, 256
Featherstone Lodge Project, 387
Fiscal policies, 190, 402, 425, 428, 439, 460
Foetal alcohol syndrome, 119, 185
Folic acid, 15-18
Frustation, drug use and, 162

Gastritis, 24
Gastro-intestinal bleeding, 27
General practice, hidden alcohol alcoholic in, 63

General practitioners, treatment of alcoholism and, 382
Genetic factors, 60
Group counselling for alcoholics, 309
  defaulting, 317
  details of members, 313
  follow-up, 313
  methods, 310
  results of, 314
  sources of referral to, 310, 312
  topics discussed, 311
Group therapy
  in alcoholism, 108, 199, 206, 215, 220, 221, 229, 267
  in drug dependency treatment, 286

Habituation, 193, 194
Haemopoietic system, 15
  effects of excess alcohol on, 30
Hallucinogenic drugs, 220
Hashish, 164, 165
Heart, arrhythmias, 30
Heart disease, beri-beri, 24, 30
Helping Hand Hostel, 335
Hemosiderosis, 20
Hepatitis, 25, 26, 58
  iron metabolism in, 20
Hepatoma, 27
Hereditary factors in alcoholism, 116, 195
Heroin, 35, 39, 86, 279
Hidden alcoholic, in general practice, 63
Histology of liver, 26
Hospitals
  admissions to, 73, 74, 78, 129, 362
    female, 183
    for drug abuse, 117
    incidence of, 358
    national differences, 127
  treatment in, 111, 218
Hostels, 335, 352
  following detoxification, 275
Hyperlipidaemia, 24
Hypertension, portal, 28
Hyperuricaemia, 24
Hypoglycaemia, 24
Hypogonadism, 9

Illness concept of dependency, 117
Impotence, incidence of, 3, 8, 9, 10
Incidence of alcoholism, 64, 67, 71, 75, 118, 240
  increase in, 359
  indicators of, 356, 403
  per capita consumption and, 112
  size and nature of, 380
  validity of per capita consumption, 71
Individualized behavior therapy, 298
Intestine
  effects of alcohol excess on, 25
  malabsorption, 25
Iron binding capacity, 20
Iron metabolism in porphyria, 20, 21

Jealousy, 3

Korsakov psychosis, 30

LSD, 38, 41
  in treatment of alcoholism, **220**, 221
Law, alcoholism and, 331, 401
Ledermann model of alcoholism, 123
  criticisms of, 125
Legislation, role in preventing alcoholism, 331, 401
Libido, diminished, 7, 8, 9, 12
Licensing laws, 401, 403, 428, 461
  changes in, 360, 406
  national differences in, 405
Liver
  alcohol metabolism in, 23

Liver (cont'd)
  damage, 58, 427
    hypogonadism in, 9
    mechanism of, 25
    in young alcoholics, 266
    porphyria and, 20
    prognosis, 29
    vitamin B_{12} and, 16
  effects of excess alcohol on, 25
  failure, 26
  fatty, 58
    hepatitis complicating, 26
    mechanism, 25
  fibrosis, 26
  histology, 26
    in cirrhosis, 27
    in hepatitis, 59
  toxicity of alcohol to, 59
Liver biopsy, 59
Liver function tests, 26
Loss of control concept, 87, 294, 297
Lung cancer, 433

Macrocytosis, 15
Malabsorption, 25
Males, sexual disorders in, 3
Mallory-Weiss syndrome, 24
Malnutrition, 24, 25, 59
Marriage, alcoholism and, 362
Maturity, 331
Maudsley Alcohol Pilot Project, 75
Mental illness hospitals, admissions to, 359
Merseyside, Lancashire and Cheshire Council on Alcoholism, 355
Metabolism of alcohol, 23
Mortality rates of alcoholics, 431
Motorists, proceedings against, 358

Narcotics anonymous, 284
'Need for personalized power,' 87
Nervous system effects of alcohol on, 30
Neuropathy, peripheral, 30
Neurosis, alcoholism and, 294
New Jersey Medical School Treatment Evaluation Model, 372
Nicotinamide adenine dinucleotide (NAD), 23
Nutrition, alcoholism and, 251

Occupation
  alcoholism and, 133, 361
  cirrhosis and, 133
Oesophagus
  carcinoma of, 25
  effect of alcohol on, 25
  varices, 27
Omega House, 391
Ontario detoxification system, 321
Operant conditioning, 110
Othello syndrome, 4

Pancreatitis, 29, 60
Parents, and children learning to drink, 451
Parental attitudes to drinking, 453
Pedestrians, drunkenness among, 357
Peptic ulceration, 25
Peripheral neuropathy, 30
Personal intake of alcohol, 200
Personality
  and alcoholism, 106, 117, 133
  dependence and, 117
Personal relationships, drug use and, 153, 158, 160, 161
Phoenix House, rehabilitation of drug offenders at, 387
Physicians, alcohol and, 57
Physiological factors in alcoholism, 106
Physiology of alcohol ingestion, 191
Pneumonia, deaths from, 434

Police, drug abuse and, 147, 148
Population surveys, 75
Porphyria cutanea tarda, 19
Portal encephalopathy, 28
Portal hypertension, 28
Prevention of alcoholism
  alcohol control policy, 425
  control of advertising, 461
  control policies, 426
    consumption and, 427
  education and, 457
  fiscal measures, 402, 425, 428, 439, 460
  legal measures, 401, 428, 461
  restriction of availability, 437
Probation service, in alcoholism, 331
Psychiatrists, alcoholism and, 57
Psychology, alcoholism and, 105
Psychotherapy, 108, 112, 214, 219
Public attitudes
  to alcoholism, 37
  to drinking, 132, 181
  to drug dependence, 37

Questionnaires, design of, 99

Reactive alcoholics, 305
Regional treatment units for alcoholism, 352
Rehabilitation of drug offenders, 334
  Featherstone Lodge Project, 387
  re-entry stage, 391
Religion and alcoholism, 132
Replication studies, 89
Research into alcoholism, 119
  control groups, 88
  design of questionnaires, 99
  experimental designs, 372
  on agencies 337
  on cultural influences, 98
  problems, 88
  replication studies, 89
  sampling, 88
Research into drug abuse, 85
'Revolving door' system, 322

Scoline, 109
Self respect, loss of, 332
Sex behavior, deviant, 7, 8
Sex hormone-binding-globulin, 10
Sex incidence of alcoholism, 362
Sexual disorders, 3
  duration of alcoholism and, 5, 6, 7, 11
  investigation of, 4
Sheffield, Council for Alcoholism in, 396
Sickness absenteeism in alcoholics, 171
Smoking, education in, 99
Social class of alcoholics, 6, 173
Social drinking in therapy
  See Controlled drinking
Social factors in drug use, 152, 154
Social reinforcement in treatment of alcoholism, 200
Social stability, 340, 346, 351
Sociologist view of alcoholism and drug dependence, 47
Spare Time Activity Questionnaire, 63
Stages of alcoholism, 241
Stomach, effect of alcohol excess on, 24
Students, drinking patterns, 95
Subcultures, drug use and, 154, 163
Suicide, 435
Susceptibility to alcohol, 86, 106

Teenage drinking, 95, 451
  attitudes towards, 102
  cultural influences in, 98
  education and, 459
  first experience of, 100
  frequency of, 100, 101
  hangovers following, 101
  incidence of, 97

Teenage drinking (cont'd)
  parental attitudes to, 454
  reasons for, 97
  results of excess, 102
Tension, reduction of, 107, 110
Testosterone levels, effect of alcohol on, 9, 10
Tobacco dependency, 117
Tobacco smoking, 86
Treatment of alcoholism
  abstinence as goal of, 293, 297, 366
  agents and agencies for
    See under Agencies
  aims of, 293
  antabuse, 215, 218, 219
  aversion therapy, 108, 293
  behavior modification, 100, 108
  benefits of hospital admission, 218
  causation and, 107
  community alcohol teams, 385
  community based services, 355, 379
    improvement in, 384
  components, 212, 213
  comprehensive, 112
  controlled drinking, 199, 294, 295
    acceptance of, 295
    attendance at meetings, 204
    Donwood Institute Study, 299
    factors in success, 302
    follow-up, 301
    procedure, 201
    results of, 302
    search for common features, 297
    self reporting, 204, 206
  coordination of care, 395
  counselling, 366
  covert sensitization, 109
  deterrent, 107
  detoxification, 278
    effect on arrest rates, 323
  detoxification centres, 271, 321
    aims of, 322
    current position, 327
Treatment (cont'd)
  detoxification centres (cont'd)
    effect on population, 324
    evaluation of, 323
    future research on, 326
    limitations of, 327
    results, 325
  evaluation of, 119
  exclusion criteria, 211, 212
  expected success rates, 213
  failures, 217
  family-focused, 239, 244
    food and shelter, 251
    intervention steps, 248
    management, 245
    physical safety, 249
    self-actualisation, 255
    social interaction, 254
    socio-economic environment in, 253
    training in, 256
  group counselling, 309
    defaulting, 317
    details of patients, 313
    follow-up, 313
    methods, 310
    results of, 314
    sources of referral to, 310, 312
    topics discussed, 311
  group therapy, 108, 199, 206, 215, 220, 229
    with young alcoholics, 267
  halfway houses, 322, 326
  historical survey, 209
  individualized behaviour therapy, 298
  interprofessional education in, 418
  LSD in, 220, 221
  numbers involved, 412
  of young alcoholic, 263
  outside hospital, 111
  personnel involved, 118, 365
    training of, 239, 256
  physiological, 107
  programming, 209
  psychotherapy, 108, 112, 214, 219

Treatment (cont'd)
  regional units, 352
  results of, 211
  'revolving door,' 322
  role of probation service, 331
  social reinforcement in, 200
  social stability and, 217
  success in, 219
    factors, 219, 221, 222
  type of alcoholic and, 305
  withdrawal, 278
  young alcoholics, 227
    compared with others, 267
    follow-up, 232, 234
    goals, 230, 234
    group structure, 229
    group therapy, 267
    interviews, 231
    modifications of therapy, 230
    results, 232
    selection of patients, 228
Treatment of drug abuse
  administrative approach, 279
  aims and objectives, 277, 370
  by substitution, 279
  encounter groups, 390
  evaluation of, 369
    future directions in, 373
    guidelines for, 376
  experimental designs, 372
  Featherstone Lodge Project, 387
  follow-up, 372
  interdisciplinary approach, 287
  intervention programmes, 369
  legislative approach, 282
  medical approach, 278
  New Jersey evaluation model, 372
  obtaining data on, 371
  outcome measures, 371
  psychological approach, 285
  sociological approach, 286
  sympathetic approach, 284
  thirteen steps, 284
Tremor, 28
Types of alcoholism, 241, 305

Unemployment in alcoholics, 171
Urine, porphyrins in, 20

Vagrant alcoholics, 272
Vitamin B_{12}, 15-18

Wernicke's encephalopathy, 24, 30
Withdrawal, 30
  approaches to, 278
Women
  alcohol problems in, 181, 240
    infants of, 185
    partners of, 184

Young alcoholic
  See also Teenage drinking
  characteristics of, 265
  clinical experiences with, 267
  compared with others, 265
  definition of, 263
  drinking programme, 227
    follow-up, 232, 234
    goals, 230, 234
    results, 232
    selection of patients, 228
  liver damage in, 266
  treatment of, 263
    group structure, 229
    group therapy, 267
    interviews, 231
    modifications, 230
Young people, drinking patterns of, 95